The Politics of Disease Control

NEW AFRICAN HISTORIES

SERIES EDITORS: JEAN ALLMAN, ALLEN ISAACMAN, AND DEREK R. PETERSON

David William Cohen and E. S. Atieno Odhiambo, *The Risks of Knowledge*

Belinda Bozzoli, *Theatres of Struggle and the End of Apartheid*

Gary Kynoch, *We Are Fighting the World*

Stephanie Newell, *The Forger's Tale*

Jacob A. Tropp, *Natures of Colonial Change*

Jan Bender Shetler, *Imagining Serengeti*

Cheikh Anta Babou, *Fighting the Greater*

Jihad Marc Epprecht, *Heterosexual Africa?*

Marissa J. Moorman, *Intonations*

Karen E. Flint, *Healing Traditions*

Derek R. Peterson and Giacomo Macola, editors, *Recasting the Past*

Moses E. Ochonu, *Colonial Meltdown*

Emily S. Burrill, Richard L. Roberts, and Elizabeth Thornberry, editors, *Domestic Violence and the Law in Colonial and Postcolonial Africa*

Daniel R. Magaziner, *The Law and the Prophets*

Emily Lynn Osborn, *Our New Husbands Are Here*

Robert Trent Vinson, *The Americans Are Coming!*

James R. Brennan, *Taifa*

Benjamin N. Lawrance and Richard L. Roberts, editors, *Trafficking in Slavery's Wake*

David M. Gordon, *Invisible Agents*

Allen F. Isaacman and Barbara S. Isaacman, *Dams, Displacement, and the Delusion of Development*

Stephanie Newell, *The Power to Name*

Gibril R. Cole, *The Krio of West Africa*

Matthew M. Heaton, *Black Skin, White Coats*

Meredith Terretta, *Nation of Outlaws, State of Violence*

Paolo Israel, *In Step with the Times*

Michelle R. Moyd, *Violent Intermediaries*

Abosede A. George, *Making Modern Girls*

Alicia C. Decker, *In Idi Amin's Shadow*

Rachel Jean-Baptiste, *Conjugal Rights*

Shobana Shankar, *Who Shall Enter Paradise?*

Emily S. Burrill, *States of Marriage*

Todd Cleveland, *Diamonds in the Rough*

Carina E. Ray, *Crossing the Color Line*

Sarah Van Beurden, *Authentically African*

Giacomo Macola, *The Gun in Central Africa*

Lynn Schler, *Nation on Board*

Julie MacArthur, *Cartography and the Political Imagination*

Abou B. Bamba, *African Miracle, African Mirage*

Daniel Magaziner, *The Art of Life in South Africa*

Paul Ocobock, *An Uncertain Age*

Keren Weitzberg, *We Do Not Have Borders*

Nuno Domingos, *Football and Colonialism*

Jeffrey S. Ahlman, *Living with Nkrumahism*

Bianca Murillo, *Market Encounters*

Laura Fair, *Reel Pleasures*

Thomas F. McDow, *Buying Time*

Jon Soske, *Internal Frontiers*

Elizabeth W. Giorgis, *Modernist Art in Ethiopia*

Matthew V. Bender, *Water Brings No Harm*

David Morton, *Age of Concrete*

Marissa J. Moorman, *Powerful Frequencies*

Ndubueze L. Mbah, *Emergent Masculinities*

Judith A. Byfield, *The Great Upheaval*

Patricia Hayes and Gary Minkley, editors, *Ambivalent*

Mari K. Webel, *The Politics of Disease Control*

Kara Moskowitz, *Seeing Like a Citizen*

The Politics of Disease Control

Sleeping Sickness in Eastern Africa, 1890–1920

Mari K. Webel

OHIO UNIVERSITY PRESS · ATHENS, OHIO

Ohio University Press, Athens, Ohio 45701
ohioswallow.com

Printed in the United States of America
Ohio University Press books are printed on acid-free paper ∞ ™

29 28 27 26 25 24 23 22 21 20 19 5 4 3 2 1

Library of Congress Cataloging-in-Publication Data

Names: Webel, Mari K., author.
Title: The politics of disease control : sleeping sickness in eastern Africa, 1890-1920 / Mari K. Webel.
Other titles: New African histories series.
Description: Athens : Ohio University Press, 2019. | Series: New African histories | Includes bibliographical references and index.
Identifiers: LCCN 2019031424 | ISBN 9780821423998 (hardcover) | ISBN 9780821424001 (paperback) | ISBN 9780821446911 (pdf)
Subjects: LCSH: African trypanosomiasis–Africa, Eastern–Epidemiology–History–19th century. | African trypanosomiasis–Africa, Eastern–Epidemiology–History–20th century. | Public health–Political aspects–Africa, Eastern–History–19th century. | Public health–Political aspects–Africa, Eastern–History–20th century. | Imperialism. | Epidemics–Africa, Eastern–History.
Classification: LCC RA644.T69 W43 2019 | DDC 616.9363096875–dc23
LC record available at https://lccn.loc.gov/2019031424

For Josh

And for my parents, Max and Kathryn Webel

Contents

Illustrations

MAPS

FIGURES

Acknowledgments

My research has had the generous support of the German Academic Exchange Service (DAAD) Summer Language Study Grant and Postdoctoral Research Grant, the Council on Library and Information Resources (CLIR) Mellon Fellowship for Dissertation Research in Original Sources, the Social Science Research Council—International Dissertation Research Fellowship, the Berlin Program for Advanced German and European Studies, the Mellon Interdisciplinary Graduate Fellows Program of the Institute for Social and Economic Research and Policy (now the Interdisciplinary Center for Innovative Theory and Empirics) at Columbia University, the American Council of Learned Societies—Mellon Dissertation Completion Fellowship, and the American Historical Association Bernadotte Schmitt Grant. At the University of Pittsburgh, the completion of this project has been supported by the Dietrich School of Arts and Sciences, the Richard D. and Mary Jane Edwards Endowed Publication Fund, and the University Center for International Studies Hewlett International Grant.

I am grateful to my editors in the New African Histories series at Ohio University Press—Jean Allman, Allen Isaacman, and Derek Peterson—for their thoughtful guidance; the manuscript's anonymous readers also provided insightful and constructive comments. My sincere thanks to Gillian Berchowitz, Rick Huard, Nancy Basmajian, and the Ohio University Press staff for shepherding this book so expertly through development and completion, and to Brian Edward Balsley for his thoughtful and diligent cartographic expertise.

Earlier versions of chapter four were published, in part, as "Medical Auxiliaries and the Negotiation of Public Health in Colonial North Western Tanzania" in the *Journal of African History* 54, no. 3 (2013): 393–416 and as "Ziba Politics and the German Sleeping Sickness Camp at Kigarama, Tanzania, 1907–14" in the *International Journal of African Historical Studies* 47, no. 3 (2014): 399–423.

My deepest appreciation to Gregory Mann, Volker Berghahn, David Rosner, and Deborah Coen of Columbia University for their support of a transnational, intercolonial history of health, research, and everyday life with sleeping sickness at its center. Julie Livingston's probing questions and intellectual creativity were also central to how this book took shape. Marcia Wright and Nancy Leys Stepan were the bedrock of my doctoral studies, serving as mentors and models for a life of teaching and research. My thanks also to faculty whose support was instrumental during my years at Columbia: Betsy Blackmar, Matt Connelly, Victoria de Grazia, Barbara Fields, Carol Gluck, Matt Jones, Adam McKeown, Susan Pedersen, Sam Roberts, Pamela Smith, and Lisa Tiersten. The enduring collegiality and friendship of Bill McAllister, now of the Interdisciplinary Center for Innovative Theory and Empirics (INCITE), has been both a great benefit to my work and a great joy. I am also grateful for the critical and constructive engagement of the 2009–11 Mellon Interdisciplinary Graduate Fellows at INCITE (formerly ISERP) at Columbia University.

A postdoctoral fellowship in African Studies and Global Health, Culture, and Society at Emory University gave me a stimulating and supportive intellectual home for crucial years in this project's development. Clifton Crais and Peter J. Brown were ever conscientious and supportive mentors. Many other Emory faculty welcomed me in a model of collegiality and fellowship. Among them, I thank Kristin D. Phillips, Mary Frederickson and Clint Joiner, Uriel Kitron, Peter Little, Kristin Mann, Elizabeth McBride, Amy Patterson, Tom Rogers, Sita Ranchod-Nilsson, Pamela Scully, Sydney Spangler, Nathan Suhr-Sytsma, and Subha Xavier, as well as Aubrey Graham, Kara Moskowitz, and Jill Rosenthal. At Georgia Tech, Anne Pollack and John Krige offered me another set of engaging interdisciplinary interlocutors. My thanks also to Jeffrey Koplan and the staff at the Emory Global Health Institute and to Paul Emerson, Moses Katabarwa, and Frank O. Richards, Jr., of the Carter Center.

Since coming to the University of Pittsburgh, I've enjoyed a wonderful group of colleagues who work every day with energy, creativity, and dedication. My particular thanks to Raja Adal, Laura Lovett, James Pickett, Lara Putnam, Marcus Rediker, and Amir Syed for their critical attention to specific pieces of the book. My gratitude, too, to colleagues past and present: Reid Andrews, Elizabeth Archibald, Keisha Blain, Bill Chase, Sy Drescher, Urmi Engineer, Niklas Frykman, Larry Glasco, Michel Gobat, Laura Gotkowitz, Janelle Greenberg, Maurine Greenwald, Bernie Hagerty, Lannie Hammond, Diego Holstein, Holger Hoock, Vincent Leung, Irina Livezeanu, David Luesink, Pat Manning, Elspeth Martini,

Jamie Miller, Ruth Mostern, Carla Nappi, Tony Novosel, Patryk Reid, Jessica Jordan Ricketts, Paul Ricketts, Pernille Røge, Rob Ruck, Jomo Smith, Scott Smith, John Stoner, Gregor Thum, Liann Tsoukas, Bruce Venarde, Molly Warsh, Katja Wezel, and Emily Winerock as well as graduate students Jack Bouchard, Marcy Ladson, Jake Pomerantz, and Kelly Urban in the Department of History. Outside of my own department, fantastic colleagues also abound: Michael Dietrich, Veronica Dristas, Felix Germain, Michael Goodhart, Macrina Lelei, anupama jain, Jessica Pickett, Michele Reid-Vazquez, Philipp Stetzel, Emily Wanderer, Jacques Bromberg, and Benno Weiner. I am especially grateful to Yolanda Covington-Ward for her mentorship and guidance and to Donald Burke at the Graduate School of Public Health and Thuy Bui in the School of Medicine for their collaborative spirit. Chris Lemery, Arif Jamal, and the Interlibrary Loan Office staff in the University of Pittsburgh Library System have always been an invaluable resource.

This project has benefited significantly from the expertise of scholars near and far who have been exceptionally generous with their time and energy. Particular thanks to Simon Ditchfield, Lukas Engelmann, Paul Finkelman, Jennifer Foray, Jeremy Green, Nancy Rose Hunt, Mark S. R. Jenner, Jennifer Lee Johnson, Neil Kodesh, Guillaume Lachenal, Stacey Langwick, Thomas F. McDow, Michelle Moyd, Deborah Neill, Rhiannon Stephens, Binyavanga Wainaina, and Jim Webb. Randall Packard offered his critical acumen on the work in progress at several key moments, including a seminar at Johns Hopkins and a manuscript colloquium at Pitt (and many conference chats besides), for which I am deeply grateful. I am also fortunate to be in a field where constructive engagement is a hallmark of seminars and conferences, and thank the countless participants, known and unfamiliar, who have workshopped and responded to portions of this book over the years.

A dynamic, far-flung, and supportive circle of colleagues has been one of the greatest pleasures of this itinerant, strange career. For their warmth and brilliance, I thank Melissa Creary, Julie Weiskopf, Jennifer Tappan, Melissa Graboyes, Aimee Genell, Sarah Cook Runcie, Claire Edington, Marian Moser Jones, Alex Cummings, Christine Evans, Alvan Azinna Ikoku, Dominique Kirchner Reill, Daniel Fridman, Uri Shwed, Rozlyn Redd, Anderson Blanton, Alison Bateman-House, Courtney Fullilove, Maura Finkelstein, Rich McKay, Michael Brown, Marissa Mika, Julia Cummiskey, Heidi Morefield, Kirsten Moore-Sheeley, Dinah Hannaford, and Adam Rosenthal. This book would not be what it is without Brandon County, whose keen eye and intellectual generosity have pulled me out of the weeds many times.

The transnational nature of this project has generated wonderful opportunities to connect with colleagues and scholars in Tanzania, Germany, the United Kingdom, Italy, and Belgium, and to accumulate a deep debt to scores of archivists and archives staff members throughout. I am grateful to the Institute for the History of Medicine—Charité in Berlin and to Volker Hess for providing me with a home base in Berlin on multiple occasions over the years. My thanks to Franz Göttlicher at the Bundesarchiv in Lichterfelde and to the tireless and cheerful staff who support researchers there every day. At the Robert Koch Institute in Berlin, Ulrike Folkens, Heike Tröllmich, and Henriette Senst were invaluable. I am also grateful for the helpful staff of the Auswärtiges Amt, the Geheimes Staatsarchiv-Preußisches Kulturbesitz, and the Berlin Staatsbibliothek, and for Wolfgang Apelt in the Archives of the Bethel Mission in Wuppertal and for Martina Koschwitz in the Archives of the Bernhard Nocht Institute in Hamburg. My gratitude to Heinz-Peter Brogiato and Bruno Schelhaas at Leibniz-Insitut für Länderkünde and Giselher Blesse at the Grassi Museum in Leipzig and the staff of the Staatsarchiv in Hamburg. Many other colleagues in Berlin also enriched my research and time there: Andreas Eckert, Silke Strickrodt, and Manuela Bauche; Annette Hinz-Wessels, Runar Jordaen, Marion Hulverscheidt, and Sascha Topp; and Christoph Gradmann and Wolfgang Eckart when they passed through at critical moments in the project's development. In Berlin, the 2007–09 Berlin Program of the Freie Universität and Karin Goihl gave me an intellectual home, as did Freyja Hartzell, Stephen Gross, Chase Gummer, Melissa Kravetz, Aeleah Soine, and Jeffrey Saletnik. My thanks to Helen Bömelburg and Damien Butaeye, Darren De Ronde, Elmar Ostermann, and Katharina Bolze and the Familie Bolze, for their warm hospitality over years of work in Hamburg and Berlin.

I am grateful to the Tanzanian Commission for Science and Technology for its support of my research. In Tanzania, I was fortunate to be welcomed by faculty and graduate students at the University of Dar es Salaam: Frederick Kaijage, Yusufu Lawi, Bertrand Mapunda, Henry Muzale, Musa Sadock, and Zubeida Tumbo-Masabo. My thanks also to Conso Musale at UDSM. John Rajabu led me through the East Africana Collection at the UDSM library and its wonderful maps. At the Tanzania National Archives, I benefited from the daily assistance and guidance of Ally Y. Ally, Laurent Mwombeki, Grayson Nyanga, Mamsanga Mbarouk, and Sospeter Mkapa, among many other behind-the-scenes staff. Fr. Donald Anderson opened the door to Atiman House and the White Fathers Provincial Archive to me, generous with both his knowledge of the White Fathers' history and his own time. In Bukoba, Bishop Method Kilaini graciously shared his family history and his own scholarship. Fr. Elpidius Rwegoshora's help and dedication have forever convinced me to trust in serendipitous meetings. I am thankful for the warmth extended to me by the Catholic community in the Bukoba Diocese, with the support of Bishop Nestor Timanywa, and to Fr. Deogracias Mwikira especially. Thanks also to the Rev. Lawrence Nshombo of the Lutheran congregation in Bukoba. Over the years, Melissa Graboyes and Alfredo Burlando, Charlotte Miller and Mattar Ali, Amy Jamison, and Beate Kasonta offered me ready friendship and hospitality in Tanzania.

In London, my thanks to Richard Meunier at the Archives of the London School of Hygiene and Tropical Medicine, Joanna Corden at the Archives of the Royal Society, and the gracious and efficient staff at the Wellcome Library. Maureen Watry and Adrian Allen in Liverpool guided me through the archives of the School of Tropical Medicine. My thanks to Anne Clark, Ivana Frlan, Jenny Childs, Anne George, Marc Eccleston, and others at the Cadbury Research Library at the University of Birmingham; the staff of the Royal Commonwealth Society Archives at Cambridge University; and Dan Gilfoyle and countless staff members at the National Archives in Kew. In Italy, Fr. Stefaan Minaert at the White Fathers Generalate made my research in Rome incredibly productive, as did Fr. Juan Rios and Fr. Julien Corbier. In Belgium, Pierre Dandoy and Alain Gérard, as well as Rafaël Storme at the Ministère des Affaires Etrangères et de Commerce Extérieur–Archives Africaines in Brussels were exceptionally generous with their time. My gratitude and affection to Gill and Jon Epstein and Dani Serlin, my London family, for making every landing there a smooth one.

All translations from German are my own. Katja Wezel and Gregor Thum helpfully consulted on occasion. Brandon County and Alissa Martin Webel each collaborated with me on several translations from French, with Brandon lending his particular expertise on material in White Fathers diaries and journals. I benefited greatly from the expertise of several translators of Kiswahili and Oluhaya in Tanzania and in the United States, and I acknowledge with deep appreciation Arnold Kisiraga, Irene Rwegalulira, and particularly Elpidius Rwegoshora and Nyambura Mpesha for their work in transcribing and translating both manuscript and oral historical sources. A project of such scope and duration relies on the generosity and assistance of many people; I remain fully responsible for any and all errors or omissions in this work.

This book came together over many years, in New York, Berlin, Dar es Salaam, Detroit, London, Atlanta, Pittsburgh, and Pike County, Illinois, with the enduring support of my family and friends. My love and gratitude to Rachel Allison, Lauren Oster, and Naila-Jean Meyers for caring about these epidemics and parasites, and giving me an escape from them, in New York and beyond. Deepest appreciation also to Heidi Reiner and Alex Yacoub, David and Jenny Yeend, Koren McCaffrey and Jacob Waldman, Stephen Yuhan, Joe Soldevere, Clara Burke, Amira Wolfson, Sara and Micah Myers, Kristin and Evan Ray, Lauren Herckis and Rory McCarthy, Katharina Bolze, Alexia Huffman D'Arco, Mike Bocchini, Helen Bömelburg, Eleanor Gregory Miles, and Valerie and Grant Shirk and Jen Gadda and Ben Wilhelm and their families. My immediate and extended family have been steadfast supports: Baird and Alissa Webel, Chris Hume, Jay and Karen Hume, and Nicole Pelly; Ann Williamson; Steve and Janet Webel; Craig Williamson and Renu Tipirneni; Marian and Larry Kobrin; and Rachel Kobrin. Max B. Webel, Alexandra Webel, Sophie Hume, Noa Kobrin-Brody, Adin Kobrin-Brody, Jake Hume, Tanner Hume, and Hudson Pelly deserve special mention for being wonderful companions over the years. I thank Asher Simon William Kobrin for his recent enthusiasm about how my book is going and his excellent high-fives.

My parents, Max and Kathy Webel, have always been with me. Each and every day, I am grateful for their abiding and tenacious love, their sharp minds and good sense, and the sanctuary they have always provided me. Finally, and with deepest affection, I hold a full heart's worth of love and gratitude for Josh Kobrin, who has kept my chin up and my eyes clear. Josh has been the bedrock of the best years of life (yet) and ever my greatest champion. My thanks to him for all the lightning bolts, the early mornings, and the uncountable ways he has supported me.

Introduction

AROUND 1900, many people living on the northern shores of the great Nyanza (Lake Victoria) began to die after wasting into thinness and falling into a nodding, impenetrable sleep. Their strength had been diminished and their ability to care for themselves was gone. Similarly, around the vast and deep Lake Tanganyika, wasting sickness and a deadly sleepiness began to affect people on the lake's western shore, driving their flight from villages and migration to areas not yet touched by illness. The first people afflicted were primarily those who traveled to trade and work around the region's growing commercial hubs on the lakes, those who farmed on the fertile edges of the Lake Victoria basin and the Lake Tanganyika valley, and those whose lives took them to the shores of the lake to fish, to draw water, or to row across the vast inland seas. In these areas, they were bitten by various insects as they went about their daily routines. They were already contending with the irregular rains and droughts that in recent years had brought widespread hunger and insecurity and coping with outbreaks of illnesses that struck people down swiftly and without respite.[1] They had survived the disruption and violence of European colonial incursions that had divided the region into Belgian, British, and German spheres of influence after 1880. But this wasting sleepiness that led to the deaths of increasing numbers of people on the lakes' shores was something different.

In the first years of the twentieth century, the process of making sense of this illness had just begun for people living on the Ssese Islands of Lake Victoria, in the kingdoms of the Haya people on the lake's western shores, and in the coastal lowlands of Lake Tanganyika. Around Lake Victoria, people named this new form of illness and death *kaumpuli*, *botongo*, *isimagira*, *mongota*, *tulo*, or *ugonjwa wa malale*; on the shores of Lake

Tanganyika, people called the sickness *malali, ugonjwa wa usingizi*, or *ugonjwa wa malale*. European observers in the region identified a disease, naming it *maladie du sommeil, Schlafkrankheit*, or *sleeping sickness*. These diverse names reflect differing experiences rather than a unified and uniform understanding. As illness increased, African elites, affected individuals and their communities, colonial officials, missionaries, researchers, and a few scattered ethnographers began to document the arrival of this sleeping sickness, which seemed to be new to the area and unprecedented in its scale and severity.[2]

While evidence exists that sleepy, wasting illnesses were known and recognized as serious by some populations around Africa's Great Lakes (the interlacustrine region), their greater extent in the early twentieth century was novel and alarming. Tens of thousands of people died around Lake Victoria alone in the first few years of the 1900s; other epidemics peppered the continent simultaneously. As historical phenomena, these epidemics of sleeping sickness loom large in studies of African life. Scholars have argued that the expansion of sleeping sickness and its staggering mortality rates related to colonial incursion and subsequent colonial economic imperatives.[3] Equally compelling are studies that demonstrate how colonial disease prevention efforts attempted to completely reconfigure African lives and livelihoods.[4] But such emphasis on the causes of these epidemics and on extensive prevention efforts that followed has effectively concentrated our attention on the actions of European colonial regimes at the expense of understanding African intellectual worlds and existing systems of managing illness and disaster. Scholars have paid scant attention to how people responded to widespread illness at the time—what intellectual resources they drew upon, how they acted in response.[5] In the interlacustrine region, many populations linked new illnesses directly to past experiences of sickness and death. Their strategic responses drew on the intimate histories, experiences, and memories that loomed large as family members or neighbors began to sicken and die in new ways. Affected people also engaged with European colonial officials and European missionaries, relatively recent arrivals in the region. While German, British, and Belgian empires were expanding in the Great Lakes region, the area's social, political, economic, and ecological dynamics also shifted. Between 1902 and 1914, the overlap between the habitat of a particular biting fly and the spaces and lands used daily by people in the region would ultimately catalyze some of the most ambitious, extensive, and disruptive colonial public health campaigns of the twentieth century.

This book is a history of public health and politics in Africa's Great Lakes region in the early twentieth century. It focuses on epidemic sleeping sickness and colonial and African efforts to prevent it, drawing on case studies from colonial Uganda, Tanzania, and Burundi. It fits sleeping sickness into local people's pasts and presents in order to highlight the experiences and intellectual worlds of the vast majority of the people who sickened and died at the time. It argues that African systems of managing land, labor, politics, and healing were central in shaping the trajectory, strategies, and tactics of colonial public health campaigns around Lake Victoria and Lake Tanganyika. African engagement with, evasion of, or negotiation within anti–sleeping sickness measures shaped the very nature of the campaigns, as people sought to make colonial interventions work within their own frameworks and colonial officials were forced to respond to (if not accommodate) this engagement in order to maintain their programs. Possibilities for negotiation opened up through the mutability and uncertainty of biomedical knowledge and practice as well as through the evolving nature of new political and economic relationships. In these changing circumstances, multiple players—such as the German scientists, British officials, Ziba royalty, Rundi or Bwari commoners, Belgian doctors, or Ssese islanders in my case studies—interacted to shape anti–sleeping sickness measures.

Following Frederick Cooper's conceptualization of colonial power as "arterial . . . concentrated spatially and socially . . . and in need of a pump to push it from moment to moment and place to place," I argue that sleeping sickness provided just such a "pump" for the movement of new energy and resources into rural communities in the Great Lakes region, but that unpredictable points of friction and openness within African life shaped its ultimate direction and impacts.[6] The individual and communal goals and ethics of diverse stakeholders sometimes aligned to produce the programs that European policymakers envisioned, but sometimes tilted so drastically in another direction as to require a fundamental reconceptualization of colonial public health practice. In this early era of colonial civilian administration, amid processes of engagement, negotiation, contestation, and accommodation, populations living around Lake Victoria and Lake Tanganyika asserted their own moral politics and therapeutic judgements to shape sleeping sickness control. The situated, spatial dynamics of interlacustrine intellectual worlds—their place-centered politics, therapies, mobilities, and social relations—fundamentally defined the field within which colonial interventions took place.[7]

At the center of this study is sleeping sickness. From a biomedical standpoint, sleeping sickness, known today as human African trypanosomiasis, is an infection caused by two different trypanosome parasites (*Trypanosoma brucei rhodesiense* and *T. b. gambiense*). It is transmitted exclusively by several species of a biting fly (*Glossina spp.*) known widely as *tsetse.* Human African trypanosomiasis caused by either subspecies of parasite is generally fatal when untreated. It is, importantly, a disease of two stages; a person may not know that they have been infected for weeks, if not months, after being bitten by a fly. The first stage of illness, following transmission of the parasite by an infected fly, involves fever, malaise, local swelling of the eyelids and face, headache, and gland inflammation as the parasite becomes established in the blood, lymph, and other tissues. Inflammation of the cervical lymph glands on the back of the neck, known as Winterbottom's sign, has been considered a telltale sign of the disease for centuries. As the parasite moves into the central nervous system and causes inflammation, "progressive neurological disturbances" appear, manifesting in changes in behavior and mood, tremors in the fingers and tongue, difficulty walking, wasting and weakness, and deeply disrupted sleep patterns. Disrupted nighttime sleep and excessive daytime sleepiness, culminating in a coma-like inability to be awakened, characterize late stages of infection and give the disease its colloquial name.[8] The parasites causing human disease, *T. b. gambiense* and *T. b. rhodesiense*, cannot be differentiated by appearance during microscopic examination, but cause radically different clinical manifestations of disease.[9] Clinicians distinguish them by the speed of their progress to second-stage illness and death. *T. b. rhodesiense* causes the acute form of disease, moving swiftly, with outward signs of advanced disease appearing as early as two months after infection, and an average duration absent treatment of around six months until death. *T. b. gambiense* presents, by contrast, as a chronic illness, with a slow progress and an average of around two years absent treatment before coma and death.[10] The two parasites have different and distinctive geographic distribution on the African continent. Historically limited in their spread to the north by the Sahara Desert, *T. b. rhodesiense* has predominated across southern and eastern Africa, while *T. b. gambiense* has predominated in western and central Africa, with possible convergence points at Lake Victoria. Species of flies that transmit the disease prefer two common ecologies in eastern Africa—either the damp environments and thick vegetation found near many bodies of water or in forests (riverine tsetse or forest-dwelling tsetse) or the dense grasses and brush of open grasslands (savannah tsetse). Cattle and wild ruminants are important reservoirs for *T. b. rhodesiense* and implicated in outbreaks of human illness, but no nonhuman reservoir exists for *T. b. gambiense.*[11]

This consensus about the etiology and transmission of sleeping sickness has evolved over the course of the twentieth century. During the period discussed in this book, however, neither Africans nor Europeans understood the illness consistently on these biomedical terms.

RECONSIDERING SLEEPING SICKNESS CONTROL AND COLONIAL PUBLIC HEALTH

We now understand that epidemic sleeping sickness exploded in communities around Lake Victoria and Lake Tanganyika at the turn of the twentieth century, concomitant with apparently unprecedented mortality—an estimated 250,000 people purportedly died around Lake Victoria alone—before 1920. Parallel epidemics in the Congo River basin killed hundreds of thousands of people.[12] The epidemic followed several difficult decades for the region's populations, during which internal political conflict, drought, famine, cattle disease, sand fleas (*Tunga penetrans*) and other epidemics struck in succession, preceding and alongside European colonial incursion.[13] The wide extent of sleeping sickness across regions of eastern and central Africa in the late 1890s connected to new, extractive colonial economies and the widespread disruption of ecological and agricultural circumstances brought by the imposition of European colonial rule. Across a wide territory, African political authorities acted to cope with this seemingly new form of misfortune and severe illness. In 1902, British scientists at work in Uganda identified the causative parasite and fly carrier. Thereafter, with rising fears of the impact of sleeping sickness on colonial economies, European colonial administrations kicked prevention and control campaigns into high gear.

Between 1902 and 1914, German, British, and Belgian colonial authorities in the Great Lakes region imposed myriad measures to try to control the disease's spread. Anti–sleeping sickness measures were European authorities' first attempts to focus specifically on African health as part of wider colonial health concerns, in contrast to attending primarily to European survival in the tropics in the prior decades.[14] These measures ranged widely, from the forced depopulation of the lakeshores to the local eradication of crocodiles to experimental chemotherapies to the deforestation of fly habitats to the internment of the sick in isolation camps. Colonial authorities sought to alter how African communities fished, farmed, hunted, traveled, and sought healing, often under coercion and sometimes by force. Anti–sleeping sickness measures took place concurrently with increasingly strong assertions of colonial influence in royal politics, pressure to cultivate cash crops, and efforts to enumerate and locate populations to facilitate taxation and control mobility. Likewise, they occurred amid increasingly frequent efforts on the part of targeted populations to evade the brunt of such political and economic impositions. Sleeping sickness prevention and control measures differed across colonial regimes, but all involved strategies aimed at breaking the cycle of transmission by limiting contact between humans and flies.[15] Prior to World War I, there was no durable pharmaceutical cure for sleeping sickness and the drugs being tested had serious and sometimes deadly side effects. Drug treatments that were later developed were often toxic and difficult for patients to endure.[16] The majority of people infected with trypanosome parasites ultimately died. After the 1920s, mortality rates seemed to drop off precipitously across Africa for several decades, before the disease roared back to life among the rural African poor in the 1970s and 1980s.[17]

Epidemic sleeping sickness is often understood as a great rupture in turn-of-the-century Africa. Both the disease and colonial responses to it had significant and enduring impacts on African lives and livelihoods. While I, too, share an interest in understanding the nature and extent of the disruption that the epidemics in the Great Lakes region caused, diverse evidence indicates that these epidemics also had strong continuities with past experiences and illnesses. Widespread illness and death in new forms may have shaken communities deeply, but people did not meet either at a standstill. In this book, I seek to disrupt and expand our histories of sleeping sickness by orienting around affected communities and how they responded to and made sense of illness amid colonial control measures. I center key local contexts of colonial public health—place, politics, and mobility—in examining how sleeping sickness prevention measures functioned. Each requires attention to a deeper past. People living on the shorelines of the Great Lakes drew on intellectual and practical resources based on past experiences and utilized established strategies to address widespread illness. Interlacustrine societies' ideas, practices, and strategies, in turn, shaped the horizons of possibility for a particular colonial intervention that is a core concern of this book: the sleeping sickness isolation camp. In the camps established by German authorities at Lake Victoria and Lake Tanganyika, colonial medical officers concentrated on identifying and diagnosing cases, isolating the sick, and experimentally treating people with a variety of drugs; camps also served as a base for work to destroy fly vector habitats, all within a wide catchment area.[18] But these sleeping sickness camps had contingent, unpredictable stories, rife with negotiation, conflict, hope, misunderstanding, and shrewd calculation. Their history offers new insight on the continued importance of African intellectual worlds and of established systems of healing in how new colonial public health programs functioned.

This book argues that reorienting explorations of sleeping sickness around interlacustrine African concerns can generate productive new insights for an admittedly well-studied phenomenon in African history. Such a reorientation requires viewing sleeping sickness prevention and control from a different perspective, subordinating biomedical priorities and scientific detail to focus instead on the social, environmental, and political contexts of public health. To illustrate this shift and its consequences, consider two German colonial maps (figures I.1 and I.2) produced during the sleeping sickness epidemic. Figure I.1 is a 1907 map depicting Lake Victoria and its immediate environs and figure I.2 is a map of the northeastern littoral of Lake Tanganyika and its environs, circa 1913. Each map resulted from the combined efforts of colonial cartographers, medical researchers, and countless auxiliaries and assistants in the early twentieth century.[19] The Lake Victoria map emphasizes three spaces, each roughly equidistant on the three sides of the lake in German colonial territory, and highlights known outbreaks of human illness around the northern arc of the lakeshore. Colonial borders are important on the Lake Victoria map, which draws the eye to where British Uganda and German East Africa meet as bright red hotspots, concentrations of human cases in German territory; important, too, are sketches of green along the lakeshore, depicting the range of the tsetse fly vector and suggesting the epidemic's potential spread. A map-reader anticipates a problem—what would happen if the green and red zones should overlap?—and thus also considers the potential location of some checkpoint or intervention in those areas of impending overlap of fly vectors and human disease, to keep the disease from spreading. The Lake Tanganyika map shows a series of stations, evenly spaced along the lake, where eight sleeping sickness camps (*Lager*) in colonial Burundi were located. Shaded areas along the lakeshore and adjacent rivers indicate that colonial geographies prioritized particular ecologies, denoting areas where fly habitats had been "*saniert*"—cleared away.

These two maps encourage an aerial imagining of a colonial public health problem and the campaign that solved it: tactically precise, strategically balanced, rationally comprehensive, and covering all bases. The mapped campaign seems proportional: sensible for the management of both manpower and resources and fitting with contemporary epidemiological practice. These maps and their makers' perspectives capture colonial public health as it emerged in the early twentieth century to begin considering epidemic diseases among colonized populations: a top-down, hierarchical apparatus of the state, targeting specific problems in geographically focused campaigns, and prioritizing the implications of illness for the imperial economic bottom line.[20]

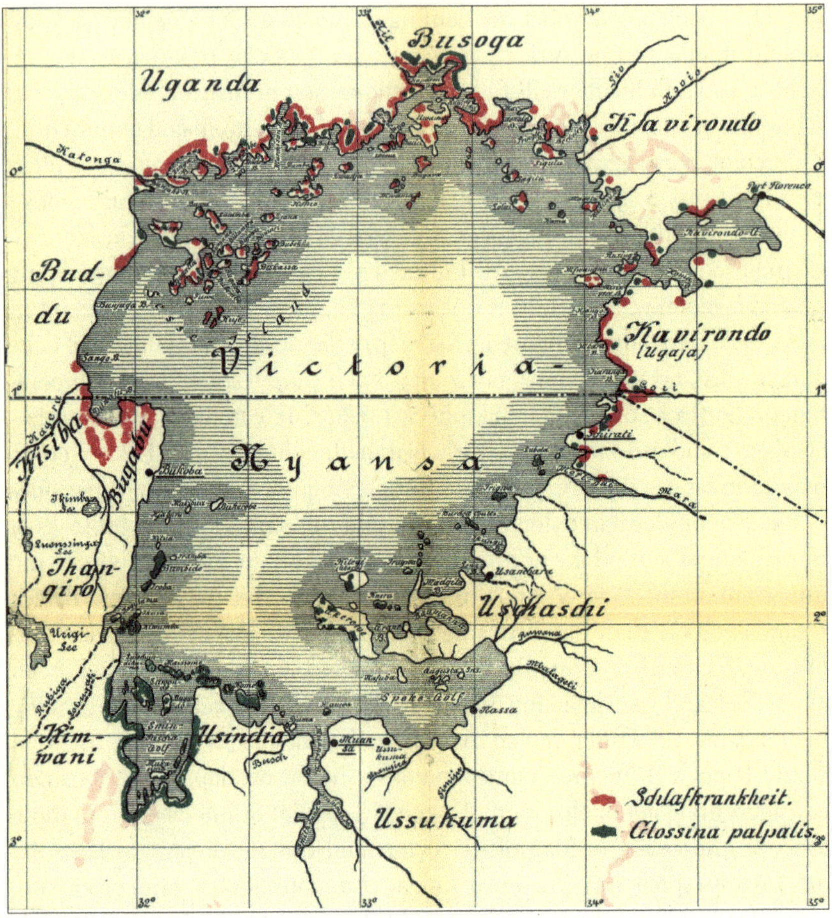

FIGURE I.1. **Overview Map of the Extent of Sleeping Sickness in East Africa, 1907.** *Courtesy of the Geheimes Staatsarchiv-Preussisches Kulturbesitz, Berlin-Dahlem.* This map of the known extent of sleeping sickness in German East Africa accompanied materials submitted to a meeting of the Imperial Health Council's Committee for Maritime and Tropical Medicine in November 1907, after Robert Koch's expedition to eastern Africa. Areas with confirmed cases of sleeping sickness are shaded red; areas with the *Glossina palpalis* (tsetse) fly vector are shaded green. The colonial border between British and German territories bisects Lake Victoria. Source: Geheimes Staatsarchiv-Preussisches Kulturbesitz, 1 HA. rep. 8, no. 4118, "Aufzeichnung über die Sitzung des Reichsgesundheits-rats (Ausschuss für Schiffs- und Tropenhygiene und Unterausschuss für Cholera)," 18 Nov. 1907.

FIGURE I.2. (*opposite*) **Detail of "Plan—Tanganyika," c. 1913.** *Courtesy of the Bundesarchiv, Berlin-Lichterfelde.* This map shows the German sleeping sickness campaign's field of work on the northwestern littoral of Lake Tanganyika in colonial Burundi. Seven camps dot the shoreline between Kigoma and Usumbura (modern Bujumbura). "Cleared" areas where tsetse fly habitats had been destroyed are shaded in along much of the lake shore and river courses descending toward the lake; areas where clearing is not planned are noted with cross-hatching. The Ubwari peninsula opposite Rumonge is not depicted. Source: Bundesarchiv, Berlin-Lichterfelde, R86/2632, Report, 12 May 1914.

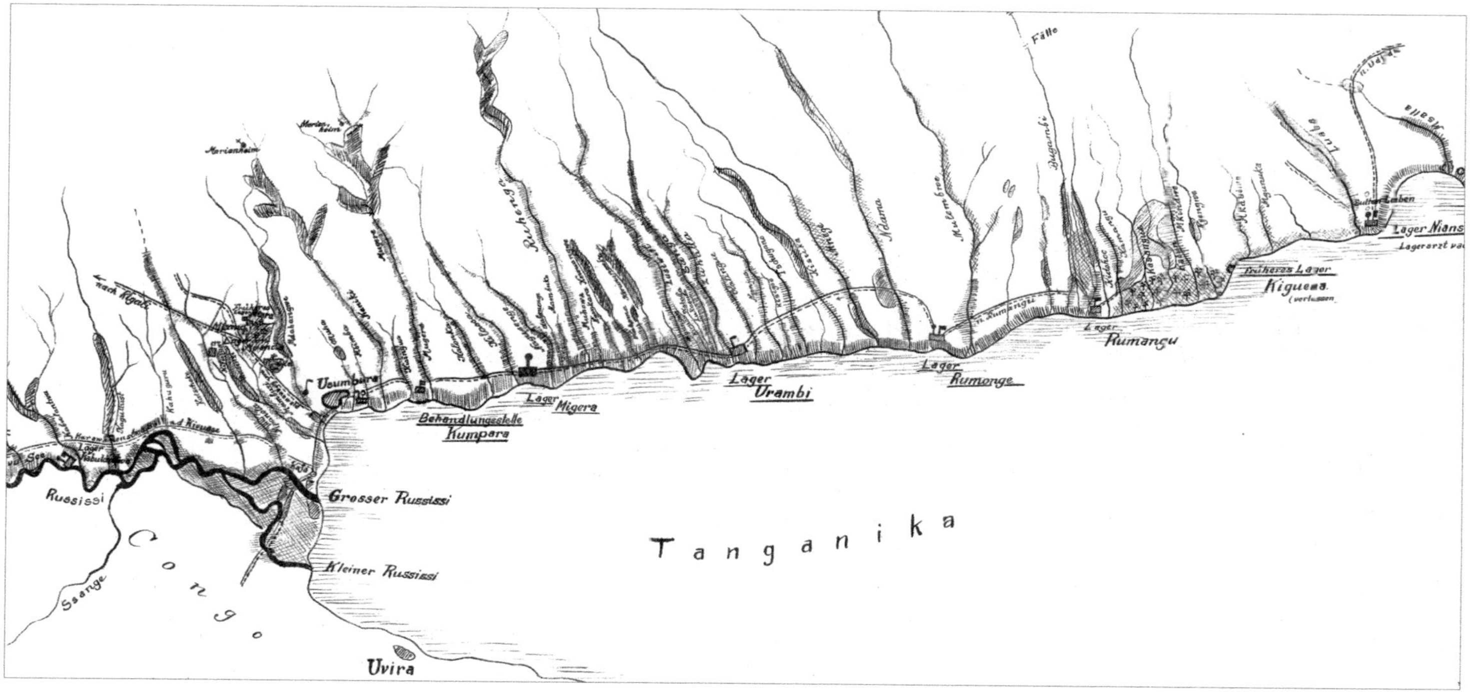
Tanganika
Congo
Russissi
Ssange
Grosser Russissi
Kleiner Russissi
Uvira
Usumbura
Behandlungsstelle
Kumpara
Lager
Migera
Ruhenga
Lager
Urambi
Ndama
Mulembwe
Lager
Rumonge
Bugambi
Fälle
Lager
Rumangu
Früheres Lager
Kiguena
(verlassen)
Luaba
Lager Nians
Lagerarzt v.

But if one should shift from these distant, bird's-eye views to instead land on the ground, making an imagined, swinging pivot from a map hanging on a wall to the terrain itself where the everyday activities of a public health intervention occurred, clarity all but disappears. The camps are isolated outposts, set apart from established villages, colonial administrative stations, and lakeshore trading towns alike. They share no particular consistency in elevation, terrain, or vegetation, as contemporary ideas connecting climate and disease might have dictated—even their proximity to the lakeshore is irregular. Some are near to concentrations of sick people, others are not. They might be surrounded by dense forest, intensively farmed land, or wide swampland. Situated within local geographies rather than imperial perspectives, policymakers' decisions about siting and location are not evidently intuitive, efficient, or rational. Rather, the siting of sleeping sickness camps was contingent, perplexing, and jarringly unique. Interrogating these maps produces a series of questions: Why did colonial attention focus here or there, then, and not elsewhere? Why put a sleeping sickness isolation camp in one place, and not in another nearby? Why did a camp focus on certain communities, and not on their neighbors? What was here, or there, before a camp was built?

These questions lead to still others that animate my broader inquiry into the history of politics and health in the Great Lakes region. How did the colonial choice to site an intervention at one place or another interact with extant meanings and uses of that place by the people living nearby? Did a camp's location overlap, conflict, or establish some kind of congruence with extant sites of healing, political power, or economic production? Did the pasts of these places impact how the targeted populations—sought after as patients, carriers, or suspicious cases—went to colonial sites and under what circumstances? Did *where* and *how* an intervention was located affect *how* people availed themselves of the treatments offered there? Sleeping sickness camps did not, of course, simply drop from the sky and slot neatly and smoothly into open, empty land. They resulted from strategic decisions by researchers, doctors, and administrators and often from negotiations with nearby political authorities. Where colonial officials located a sleeping sickness camp had meaning for people nearby, particularly in a cultural milieu such as the Great Lakes region, where place-centered healing practices had a deep history and where management of land was a fulcrum of political power.[21]

More broadly, thinking about where a public health intervention makes its home attunes us to its fundamental social and political contexts. Imagine the specificity of a new, dedicated building with fresh construction, a room inside a church or school with other uses during the week, an established government dispensary in a small town, or an urban hospital's busy ward.[22] An intervention site's context has ramifications for how (or whether) people use it; these ramifications derive from the experiences and judgements of its target populations regarding its cost, its efficacy, or its legitimacy, but also its emplacement. Yet research on colonial public health, and health interventions in history more broadly, largely leaves the siting, location, and development of interventions uninterrogated and the consequential implications for public health unexplored. By approaching the locations of public health interventions effectively as a fait accompli, we reify the logic of past practitioners as the principal way of understanding an intervention. For practitioners who worked in settings such as colonial eastern Africa, racialized ideas of cultural difference were fundamental to their logic, ethics, and strategies. Thinking critically about the places where colonial public health and research occurred allows us to reveal their blind spots and expose their intellectual biases in order to understand the lives and motivations of those people most affected.

Throughout this book, I argue for a reconsideration of sleeping sickness control efforts that understands historical local contexts to be fundamentally important to how people used the camps and how trajectories of colonial public health changed over time. In the societies that provide my case studies—the Ssese Islands of the Buganda kingdom, the Haya kingdom of Kiziba, and the southern Imbo lowlands of the Rundi kingdom—locally oriented political, social, and therapeutic traditions shaped how and where people lived.[23] There and more broadly in the interlacustrine region, politics, social life, and healing had long been embedded in particular places or kinds of spaces.[24] How people lived within or moved through particular places, and how they understood the implications of inhabiting, using, or traveling through them, were matters grounded in historic efforts to carve out a prosperous, healthy life.[25] These efforts manifested in the organization of domestic spaces, in agricultural practices, in patterns of trade and migration, and in strategies to heal or avoid illness. Situated, localized knowledge was paramount. Colonial anti–sleeping sickness measures were profoundly affected by the embeddedness of interlacustrine populations' experiences and intellectual worlds. Processes of negotiation and engagement between African elites, European doctors and administrators, and wider populations, for example, determined where sleeping sickness camps were located and how and when they were built. The past and present uses and meanings of those places shaped how people utilized the camps situated within them and, by extension, had implications for the efficacy of sleeping sickness research, prevention, and control measures.

This book engages with crucial questions of health and politics by looking to the processes through which African and European actors refined their definitions of illness and its causes, contextualized widespread illness and misfortune, set the political and social parameters for their amelioration, and reconciled colonial public health campaigns with the circumstances of daily life. Case studies from colonial Uganda, Tanzania, and Burundi explore the potential magnitude of the rupture presented by sleeping sickness specifically, as well as the continuities evident in African responses to other forms of illness and misfortune during this era. Epidemic sleeping sickness and broad-ranging interventions may have been novel in the early twentieth century, but they were not without precedent. African political authorities' historic responsibility to maintain the health of their kingdom and populations influenced their interest in engaging with sleeping sickness interventions, as did the new dynamics of political power that colonial incursion brought. By examining how knowledge, strategies, and tactics regarding widespread illness related over time in this interlacustrine, intercolonial milieu, we see clearly how African engagement, situated within extant political, economic, and therapeutic systems, fundamentally shaped ambitious and wide-ranging colonial public health programs.

PERSPECTIVES ON THE HISTORY OF SLEEPING SICKNESS

In both historical and medical literature, sleeping sickness epidemics in the early twentieth century are a singular sort of disaster in eastern and central Africa, vast in scope and unprecedented in the scale of human death. Concomitant with colonial incursion and subsequent economic and political imperatives, widespread illness and death from epidemic sleeping sickness loom large—a crisis that constituted a great rupture in the lives of populations in the Great Lakes region. But if we are to focus our analysis on the people affected by these epidemics, rather than the imperial panic they triggered, we must query the nature of the disaster and the extent of the rupture, asking not only how serious was sleeping sickness to interlacustrine societies, but also how it fit into or departed from known points of reference and comparison.

Historical epidemiological research has begun this important work, looking back at records from the Uganda epidemic to understand how and why mortality was so explosive in the early twentieth century. These multidisciplinary studies point to the importance of considering climate, food security, disease ecology, and epidemiology in assessing the disease's impacts in the early twentieth century, and make reference to more recent outbreaks as well.[26] Their findings are provocative. One set of research examines case records and mortality rates to conclude that the Lake Victoria epidemics were due to a novel exposure to a different parasite (*T. b. rhodesiense* rather than *T. b. gambiense*) triggered by aggressive cattle restocking efforts that caused acute, fast-moving infections and higher mortality rates. The introduction of a non-endemic parasite was, in this research, the epidemic's spark.[27] Other studies likewise use historical climate data and colonial health statistics before and during the Uganda epidemic to assess the impact of climatic variation—specifically several consecutive years of unreliable rainfall and drought—and colonial rule on food security, people's use of tsetse habitats, and human vulnerability to parasitic infection. Here, sleeping sickness mortality rates actually masked more widespread misery and hunger, exacerbated by both colonial policies and crop failures that made populations more vulnerable to trypanosome parasites.[28] Such work has an intellectual affinity to path-breaking work on sleeping sickness in the Belgian Congo that established clear links between the advent of "the colonial disease" and forced labor, rubber collection, and mobility into and out of tsetse habitats generated by the Belgian regime.[29] Broadly, this vein of research makes clear the devastating impact of sleeping sickness on vulnerable populations, but is equally insistent that scholars not understand sleeping sickness as a "natural" phenomenon inherent to African environments. Scholars thus refute colonial arguments of the coincidental or epiphenomenal nature of outbreaks of sleeping sickness, while also acknowledging the complexity of identifying what or who precisely touched off these epidemics and how. The extent of the crisis for affected communities was significant, to be sure, but disease dynamics were not natural or inevitable phenomena.

Studies that have sought to understand how sleeping sickness mortality changed over time at both the small and large scale in Africa situate sleeping sickness in different possible, immediate contexts, such as climatic variation, pathogenic variation or virulence, labor regimes, or food security and human vulnerability. Drawing on their insights interrogating the nature and scale of an epidemic at a population level, this book pursues related concerns: If we can get a sense of what caused disease dynamics to change in the past, what can we yet learn about how people understood these changing experiences of illness and death within their own frames of reference? What did they do in response? How did their actions affect colonial interventions? Studies of sleeping sickness and colonial public health have, by and large, not focused on these issues. Instead, understanding particular historical dynamics of morbidity and mortality, as well as the catalysts of past epidemics, has taken center stage. This book, by contrast, teases out the place of sleeping sickness among wider disruptions around Lake Victoria and Lake Tanganyika and fits this episode of illness into other experiences of illness and misfortune that provided intellectual points of reference and a toolkit of practical strategies for affected communities. It argues that African populations understood sleepy, wasting forms of illness with reference to previous forms of serious or widespread illness and death, particularly recent outbreaks of kaumpuli or *rubunga* on the northern and western shores of Lake Victoria, as well as pox-causing illness more widely. This book incorporates and expands disease-specific histories of bubonic plague, cholera, and smallpox in eastern and central Africa that have not previously been placed into dialogue with the history of sleeping sickness.[30] Seeing important continuities in both intellectual approaches and practical strategies taken by affected people, it also shows that people took measures against sleeping sickness that had historical precedent: they consulted known and proven healing resources, reoriented domestic and social spaces, and made claims on political authorities. For some communities, such as those forced to abandon homes and farms and move into "fly free" areas in Uganda, sleeping sickness arguably caused a significant rupture in everyday life and livelihood; the Ssese Islands archipelago, one of my case studies and part of the Buganda kingdom, was effectively depopulated for most of the first half of the twentieth century.[31] In other areas, mortality catalyzed deep and durable change. But focusing only on the singularity of the disaster of epidemic sleeping sickness erases the intellectual, therapeutic, and political work that many people put into living through it. Focusing instead on that work illuminates durable continuities across the nineteenth and twentieth centuries.

The particular interlacustrine cultural context of this book is crucial to understanding the variety of intellectual and practical resources available to affected populations by the late nineteenth century. My research on the responses and efforts of affected communities at the center of this book builds on robust studies of the social and political development of interlacustrine societies ranging over the past millennium.[32] Studies of developments in deep historical time provide the basis for my engagement with linguistic and intellectual innovations amid epidemic illness, as well as my approach to long-standing political, social, and therapeutic resources oriented around clans, healing societies, and spirit mediumship.[33] These earlier histories of interlacustrine politics and society are also in dialogue with analyses of political legitimacy, health, and prosperity as conditions changed with the advent of colonial incursion in eastern-central Africa in the nineteenth century. Foundational work on the relationship between political legitimacy and health—understood in terms of fertility, prosperity, and/or the absence of serious illness, among others—provides a key register within which I analyze reactions to epidemic illness on the Ssese Islands and in Kiziba.[34] Scholarship on eastern-central African societies has encouraged me to be particularly attentive to the politics and meanings of specific places and kinds of spaces, as well as people's movements within them.

This book brings the insights of studies of health, politics, and healing into dialogue with studies of the technologies and tactics of colonial disease prevention. I focus on the emplacement, development, and ongoing work of colonial sleeping sickness camps and the situated intellectual, therapeutic, and political worlds of the people that the camps targeted. Here, my approach to the early colonial era in the Great Lakes region is also guided by studies of late colonial and post-colonial health and illness. These accounts view efforts to define disease, healing and medical practices, and treatment-seeking as both embedded within and evidence of broader changes. This scholarship has shown that individuals and communities navigated illness or misfortune in creative, generative ways and tried to achieve health and prosperity amid a rapidly changing world through evolving and complex practices.[35] The histories I offer here restore a sense of the messy, negotiated, and deeply contingent nature of early sleeping sickness research and prevention efforts; they underscore how profoundly these efforts were shaped by local experiences. This is especially important when, in time, anti–sleeping sickness campaigns have come to be understood as rigid and draconian manifestations of colonial power, and, further, when medical and scientific literature continues to either obliquely or directly credit colonial campaigns as effectively reducing sleeping sickness mortality.[36]

My work builds on scholarship that established clear connections between the political, social, and ecological disruptions of colonial incursion and the spread of trypanosomiasis (among other maladies), and that has shown how sleeping sickness was intimately linked with new, extractive economic processes such as mining or rubber collection.[37] Histories of sleeping sickness that explore these connections generally keep within the confines of the nation-state and its colonial predecessor, emphasizing the singular approaches of the different European imperial powers and colonial administrations to controlling and preventing sleeping sickness.[38] Some have focused on the experiences of a specific region; others have addressed entire colonial programs to understand their implications for later national histories.[39] While histories of research emphasize the transnational and intercolonial nature of past scientific and medical efforts, and Africanist studies of labor and migration have long traversed colonial-national boundaries, this book is the first study to consider sleeping sickness prevention and control within a transnational and intercolonial frame.[40]

The particular circumstances of Lake Victoria and Lake Tanganyika, where multiple colonial states divided the lakes' shores and engaged directly with one another around the problem sleeping sickness posed, encourages this approach. But the lived experiences of littoral populations, where mobility around the lake and connection with other societies were central, make it an intellectual necessity.[41] Reflecting shifts in historical scholarship toward transnational and comparative methodologies, and, equally importantly, recognizing that the lives and experiences of Africans and Europeans alike were shaped by the vigorous mobility of people, goods, diseases, and ideas around the lakes, this book frames the problem of sleeping sickness within the ecologies and landscapes around Lake Victoria and Lake Tanganyika. This reframing of sleeping sickness not simply as a Ugandan, Tanzanian, or Burundian concern foregrounds the connections between populations that preceded partition and endured despite the advent of the colonial state. Considering the phenomena of sleeping sickness mortality, prevention, and control within an interlacustrine world—a world defined by historic states and tributary kingdoms, complex economies of land and labor, and the lakeshores' ecosystems—rather than in colonial-national units allows me to focus on the vitality of African mobility and interchange. This interlacustrine and intercolonial frame also allows me to pay particular attention to polities and societies for which colonial borders were a new imposition and one with varying significance for daily life. By virtue of their location at or near colonial borderlands, these populations had distinctive experiences of mobility and sleeping sickness. They were marginal to the centers of power in the region: distant from commercial and political hubs of the Indian Ocean coast and peripheral to the capitals of interlacustrine kingdoms. But they held an important place in colonial prevention and control campaigns and were central to managing the spread of disease in a new era of public health surveillance. Often, sleeping sickness research, surveillance, and prevention were African populations' earliest and most consistent engagement with Europeans or the colonial state, and the book's interlacustrine and intercolonial framing illuminates similarities and divergences in their experiences.

I also approach my three areas of focus—the Ssese Islands, the kingdom of Kiziba, and the Imbo lowlands—with time in mind, concentrating on a particular moment when sleeping sickness had a high impact for colonial and African authorities alike. The early 1900s were a moment of uncertainty: neither African authorities nor healers nor European scientists nor colonial bureaucrats had a firm grip on where the sleeping sickness epidemic came from, how precisely it spread, or what measures should be taken to control it. This productive uncertainty shows how simultaneous intellectual, political, and practical efforts of European and African actors mingled and conflicted in generative ways.[42] I show that accretions of new information and processes of scientific change in tropical medicine and public health more broadly did not occur solely based on Europeans' intellectual orientations and experiences—the "eureka!" moments of white researchers in a remote, humid laboratory or a dusty field site. Rather, new ideas and strategies that manifested in colonial sleeping sickness policies—such as the atoxyl-focused sleeping sickness camp—had their origins in interactions with and adaptations to the political, social, and environmental dynamics of Ssese islanders, Ziba royal authorities and their subjects, or Bwari and Rundi people in Imbo. Researchers, doctors, and colonial public health officials immersed in sleeping sickness work also absorbed elements of the intellectual worlds, morality, and political ideologies of their African interlocutors, even if these Europeans at the time saw those African people primarily as patients to dose, bodies to study, or people to target.

Sleeping sickness proves a particularly apt tool for prying open the discrete eras of modern African history—divided by colonial rule, the world wars, or political independence—to facilitate considerations of historical continuity in public health.[43] This stems from the persistence of sleeping sickness as a health concern in rural Africa and relates to the nature of human African trypanosomiasis itself. It is a focal disease, its transmission limited to particular places: environments where its fly vector thrives and where human and animal hosts of the parasite live or transit. People, parasites, and flies have to be in specific places, together, within a particular span of time, in order for the trypanosome parasite to undergo development in both its host and vector and to survive successfully. Break the chain of contact at any point—secure people from fly bites or prevent flies from ingesting parasites as part of their blood meal—and transmission ceases. And so, biomedical approaches to trypanosomal infections, beginning in the period of my study, developed an environmental and ecological orientation that spatialized the disease and the potential for epidemic outbreaks around "fly zones."[44] Work on sleeping sickness and other such ecologically specific, vector-borne diseases has since persistently prioritized the environmental dimensions of health and illness in identifying at-risk populations or ideal targets for vector control campaigns, generally limiting work by the climatic range or ecological niches of their disease vectors.[45] This has meant that, across the twentieth century, vulnerable and affected African populations have seen successive interventions by different regimes, states, and nongovernmental organizations, each oriented around that spatialized, environmental logic of sleeping sickness control and each building on precedents in particular ways. The book's three case studies in the Ssese Islands, Kiziba, and the southern Imbo highlight the kinds of complex relationships that often accompanied and shaped public health interventions historically and continued to inform subsequent interventions after World War I. I prioritize people's experiences to understand meaningful points of reference and resonance that impacted their engagement with, and therefore also the efficacy of, health interventions.

After decades of centrality in imperial research agendas, eastern and interlacustrine Africa emerged as hubs of global health activity after the 1990s. Global health programs have frequently come to supplant the core health-related functions of the state and often altered citizens' engagement with national governments.[46] Programs in Africa (as elsewhere) frequently focus on the strategic deployment of specific pharmaceutical or medical goods and emphasize community participation, sustainability, and capacity-building. Yet, while these programs are sometimes flummoxed by local complexity or unpredictability with historical roots, consideration of their contexts by global health practitioners tends to be strongly present-oriented. Programs and interventions often proceed without a sense of history, failing to reckon with historical precedents in specific contexts or lacking full perspective on comparable programs that have sought to solve the same or similar problems.

Nuanced appraisals of the successes and failures of historic public health campaigns should provide an expanded framework within which we can evaluate modern programs' practical tactics as well as their ethical implications. A rich, new vein of scholarship, relying on diverse scientific, medical, and political archives alongside ethnographic and oral history research, has built a narrative of health and politics stitching together the impacts of a long twentieth century.[47] This work has begun to carve out space for new, interdisciplinary dialogues, advancing productive conversations between health policymakers and scholars of health programs and informing the conceptualization of future programs. While my case studies examine the period a century ago, they highlight the historical contexts in which particular environmental approaches to vector-borne disease control proceeded or faltered—still pressing matters for modern campaigns around sleeping sickness, as well as onchocerciasis and schistosomiasis, among others.

Current sleeping sickness programs fit with other global health programs aiming at the elimination or eradication of diseases that predominate in rural African communities, consonant with current trends toward the "scaling up" of health programs and the pursuit of ambitious global agendas.[48] This study of colonial sleeping sickness camps around Lake Victoria and Lake Tanganyika shows that scaling up has a longer history, one rooted in colonial desires for widely applicable public health schemes and economic efficiency. Likewise, it broadens the history of the paradigm of treatment-as-prevention so relevant for HIV/AIDS that was, as Guillaume Lachenal argues, truly pioneered in French colonial campaigns against sleeping sickness.[49] This book's case studies demonstrate that targeted populations readily confounded public health policymakers' and practitioners' designs to operate at the level of entire territories or kinds of environments. This book also periodically considers mobility, in particular, to reveal the disconnects between plans and circumstances on the ground, exploring it as an epidemiological factor, a lens through which public health interventions came into focus, and an element of popular treatment-seeking strategies. This book thus speaks directly to the persistent challenges of surveilling, reaching, and monitoring access to interventions in the target populations of public health.[50] Considering how and why people have historically availed themselves of treatments and what factors shape those activities has implications for understanding the dynamics and difficulties of public health practice in the present day.[51] I hope, then, that this book will raise questions about the nature of participation in public health interventions, about the importance of historical precedents and experience, and about the factors affecting the sustainability of interventions—questions that project organizers and planners might ask initially in order to achieve their wider goals.

AN ORIENTATION TO THE GREAT LAKES REGION

This book centers on the kingdoms and scattered peoples of Africa's Great Lakes (interlacustrine) region. This region is distinctive on the African continent for its geography, containing Africa's highest mountains and largest bodies of water. The book's three case studies—the Ssese Islands, Kiziba, and the southern Imbo—focus our attention on the northern interlacustrine region, an area bounded by Tanzania's Malagarasi River in the south and the hinterland of Lake Mwitanzige (Lake Albert) in the north. The distinctive climate and environment of the interlacustrine region differentiate it

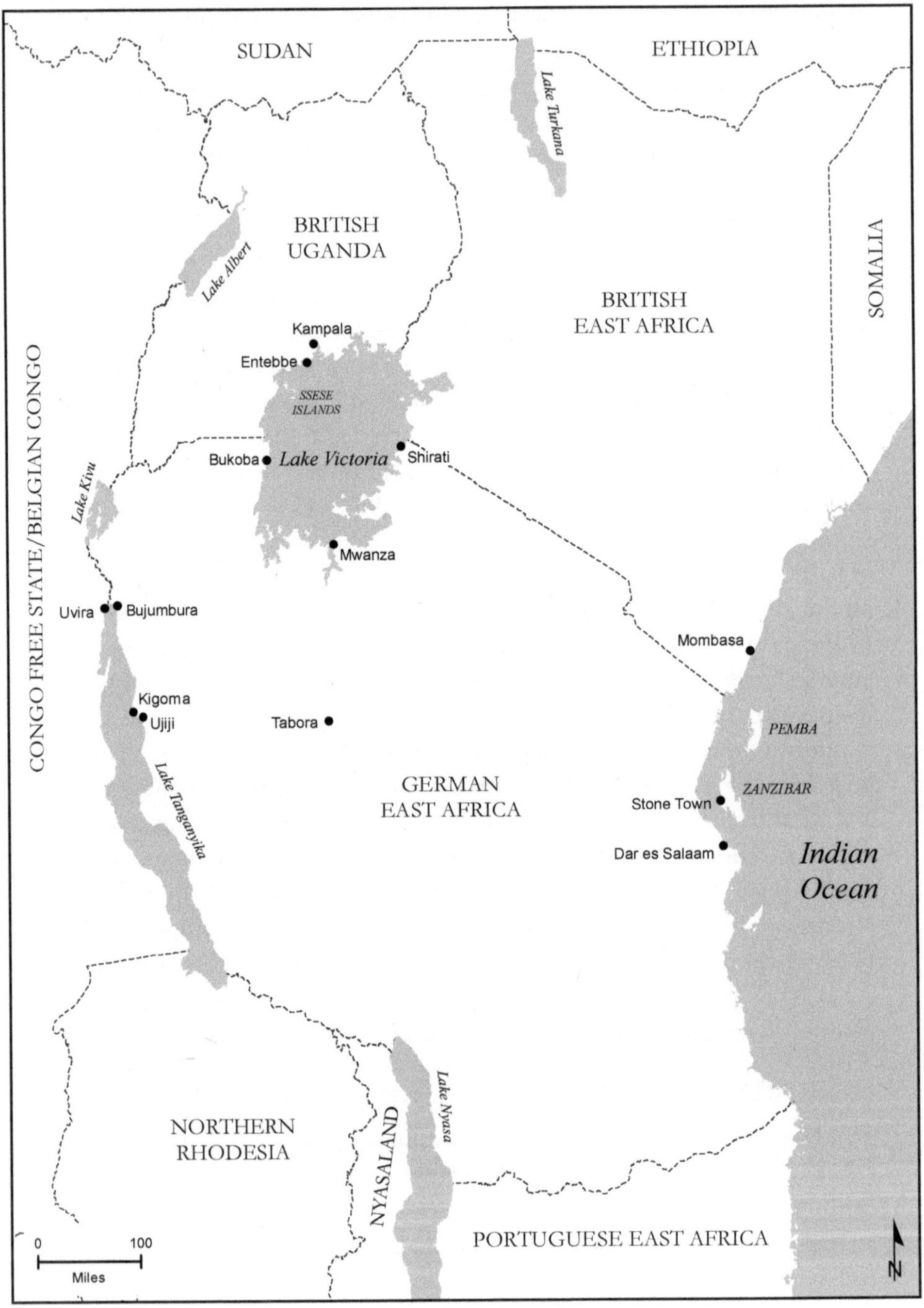

MAP 1.1. The Great Lakes Region. Map by Brian Edward Balsley, GISP.

from nearby plains, semi-arid savannahs, or river basins in eastern-central Africa; the cultural innovations of its populations have given the region an enduring analytical coherence for scholars. Continuities in general cultural forms such as sacred kingship, patrilineal clans, or spirit mediumship stretched across this large territory and helped define the intellectual worlds of its inhabitants. Such continuities, borne of a connected past, meant that people across the region understood political power, causes of illness, and possible steps for its remediation from a similar perspective.

This section introduces several of the root consistencies and broad continuities found across the region's societies historically. It offers readers—particularly those less specialized in African history or less familiar with the Great Lakes region—an orientation to the central aspects of interlacustrine societies that shaped life and livelihood for people living in the late nineteenth century. It illuminates the important political and social institutions, as well as economic and environmental trends, that shaped daily life. These central elements and key trends in the region's history provide a foundation for understanding the local variations and specific political and therapeutic frameworks that influenced how people in the Ssese Islands, Kiziba, and the Imbo lowlands managed illness and sought health and prosperity in the early twentieth century. This, in turn, allows us to see with greater precision how and why people affected by sleeping sickness interacted with colonial disease prevention campaigns as they did and to understand the broader consequences of these interactions for colonial public health.

Environmental, Social, and Political Dynamics of Interlacustrine Societies

Over the centuries after 1000 CE, populations in the Great Lakes region innovated political and social frameworks that would continue to influence the intellectual worlds and everyday lives of people living in the nineteenth century. Some of these innovations created structures that defined political power and governed land tenure and use, while others generated identities that bound together wide networks of kin and fictive kin groups. Still others provided ways of understanding connections between people, their environment, and wider cosmological forces and defined actions that could ensure health and prosperity.[52] Throughout, geography, climate, and environment historically played a role in shaping agriculture, economic activity, and social organization in the region.[53]

Agricultural sophistication and diverse food production were central to regional populations' prosperity.[54] On the highlands and near the lakeshores of the region, early populations in the first millennium CE farmed endemic crops such as eleusine (finger millet), squash, and sorghum. The availability of different crops at different times of year within seasonal rainfall regimes provided security for populations, as staggered harvests of annuals combined with perennial crops to safeguard against famine.[55] Uptake of non-endemic plants and their cultivation as food staples characterized ongoing, gradual agricultural innovation spanning several centuries before 1500 CE. Such innovation resulted from connectivity with other parts of the continent, as well as the circulation of people and goods around the western Indian Ocean. Alongside endemic sorghum, squash, and eleusine, people cultivated new arrivals from Asia, such as peas, taro, and banana, and then subsequent imports from the Americas, such as sweet potatoes, cassava, and new species of beans.[56] In some areas, such as the Burundian shore of Lake Tanganyika, people also cultivated oil palms, a tree species originating in western Africa.[57]

Generally gendered labor regimes emerged. Clearing work (such as that needed to prepare a field for yams) was typically done by men, while the daily tending of fields and crops fell to women.[58] Cattle-keeping further augmented agricultural production and food security in multiple ways, and cattle clientage bolstered political authority and stitched together individuals and households or compounds.[59] Fishing also flourished along the lakes and rivers, relying on sophisticated technical and labor inputs, fitting into agricultural production, and augmenting food security.[60] Specialized production of valued trade goods such as iron hoes, salt, dried fish, palm oil, and barkcloth occurred alongside the circulation of foodstuffs and livestock produced within households, driving patterns of trade that connected different ecological zones.[61] Scattered deposits of iron and salt throughout the region led to hubs of smelting and salt production among interlacustrine societies and catalyzed trade in hoe blades and salt; production of pottery, as well as barkcloth from ficus trees, was also widespread.[62] Diverse agricultural production, herding of cattle and small ruminants, and exploitation of heterogeneous natural resources facilitated the growth of populations from roughly 1600 CE onward. Local and regional trade connected these growing polities.

The intellectual resources available to the populations who would ultimately contend with epidemics in the late nineteenth century were rooted in pivotal political, economic, and social changes that occurred in the region between 1500 and 1900. In this era, monarchies rose and expanded, clans and healing cults evolved and spread, caravan routes stitched the lake and coastal littorals together, and people and goods circulated with unprecedented vigor and range.[63] Institutions of kingship and chiefship that emerged in the region were generally patrilineal, structuring and consolidating power within royal family lines. Political power and social prestige cohered around the royal or chiefly house and its expansive network of dependents and kin. A king's residence and its associated court functioned as a political hub, with large royal households comprised of adult monarchs and their wives and children, as well as important senior relatives, alongside countless laborers and people who filled particular ritual roles.

Political authority rested in a ruler's ability to ensure enduring prosperity for his followers or subjects. This involved strategic decision-making—waging war, levying tax or tribute, managing production and access to land, distributing surplus resources—within a framework of mutual obligation. Successful kings also acted to mediate the power of ancestral and other spirits upon their people through maintenance of rituals that kept society and ecology in balance; chiefship came to blend the political, ritual, and material.[64] While political structures and institutions were heterogeneous and took on locally specific forms, some root consistencies were also distributed over a wide geographic area, such as the institution of sacred kingship in the form of the Rwandan and Urundian *mwami*, the Bugandan *kabaka*, and the *mukama* in Bunyoro and Buhaya. From palace to province to district to chiefdom to village, relations of mutual obligation knitted together administrative structures across increasingly large territories. Interlacustrine royal and chiefly power from around 1000 CE had intertwined with that of clan leaders and healers to set the rhythms of daily life and keep them in tempo with spiritual forces—to mark, for example, when to begin cultivation, how to seal a new alliance or relationship, when and how to make war, or what measures to take to avoid widespread illness.[65]

Clans that provided social connection and cohesion, sometimes in counterpoint to royal political ideologies, also flourished in this era. Clans bound people to one another locally and sometimes regionally.[66] As a hierarchical, patrilineal kinship relation, clan affinity manifested through the common association with particular totems, most often an animal or plant that had played a historic role in a first-comer or ancestor's life, as well as through taboos observed, such as the common avoidance of particular foods.[67] Clans linked people in familial and fictive kin relations to a sense of place and space, tying people to land and giving sites meaning and significance within local cosmologies and social worlds. Relations between clans, and thus clan members, defined a person's social world by determining patterns of marriage and access to material and spiritual resources, while also locating individuals and families within durable social groups.[68] Locally, clan elders and senior family members controlled the allocation of land. Clan elders also maintained shrines for and spirit mediums of important ancestors to connect the worlds of the dead and the living, thus ensuring access to powerful healing resources fixed to specific sites around the interlacustrine region.[69]

Cults of healing and mediumship created ways to access social, spiritual, and material resources for many people, whether on the margins of political and social power or deeply connected to royal and clan networks in the kingdoms.[70] Healing of serious ailments and resolution of persistent problems focused around skilled healers and mediums, people who used gifts of connection with diverse spirits to identify causes of misfortune or illness and set a path toward health and prosperity. Some mediums connected people to powerful figures of a society's past—kings, gods and goddesses, or clan ancestors—or to deities in its present cosmology. Their intercessionary work often reinforced the powers of divine kingship or clan connections. Other forms of mediumship connected people to powers outside of royal and clan ideologies: to territorial "nature" spirits (*misambwa*) and to spirits of ancestors within a family or household (*mizumu*).[71] As well, *kubándwa* spirit possession, an ancient tradition that formalized into an institution of possession, mediumship, and initiation early in the second millennium CE, became centrally important for efforts targeted toward healing and prosperity in the region. In the ensuing centuries, the *cwezi-kubándwa* healing complex had developed and covered most of the region. It combined established traditions of kubándwa spirit mediumship with the exceptional powers of *abacwezi* spirits, deities often associated with particular places and/or environments who were also connected by ancient lineages to the ruling dynasties of the interlacustrine kingdoms. Two such deities, Kaumpuli and Mukasa, still influenced people's experiences of illness and health around Lake Victoria in the late nineteenth century. Cwezi-kubándwa deities and their mediums had particular territorial ranges, representing the system's grafting onto older, place-oriented misambwa spirits, but also focused around sites of particular power where major shrines were typically located. Healing powers concentrated at major shrines, where resident mediums acted as intercessors between treatment-seekers and spirits or deities, but also could be accessed at other, minor shrines as well as through mediums who lived in a community.[72]

Nineteenth-Century Transitions in Interlacustrine Life and Livelihood

By the nineteenth century, durable social and political institutions with deep historical roots shaped the everyday lives and intellectual worlds of the people on the Ssese Islands, in Kiziba, and in the southern Imbo who are the central subjects of this study. But also significant was the impact of changing regional dynamics, particularly those animated by the powerful, expansionist states that had emerged in the recent past. The rise of four interlacustrine kingdoms—Buganda, Urundi, Rwanda, and Bunyoro—shaped and was shaped by wider regional changes, particularly as new connections with the Indian Ocean coast commenced in the mid-nineteenth century. Understanding the nature of those new influences and forces in the wider region requires us to pivot away from the fertile littorals, highlands, and grasslands of the lakes region and look eastward to the Indian Ocean coast as well.

Newly prominent kingdoms like Buganda, Urundi, Rwanda, and Bunyoro were aggressively expansionist in their orientation, consolidating power in a territorial core and co-opting or subduing their unruly peripheries into tributary roles.[73] Two key factors enabled their territorial expansion in the eighteenth and nineteenth centuries.[74] Political centralization, undergirded by familial and clan ties as well as generational social groups, created stronger states. Entrepreneurial economic activities underwrote and facilitated territorial expansion. Political authorities in these kingdoms also successfully utilized the political and material resources provided by historic chiefship and clanship, alongside innovations in infrastructure, military organization, food production, and trade.[75] Growing cohesion and power at the centers of state—the capitals that grew up around royal palace complexes—were achieved by the accumulation of labor and of trade goods, as well as through the growth of bureaucratic institutions and infrastructure that expanded the reach of the state.[76] Long-standing trade around and between the lakes had connected communities that produced valuable goods such as iron hoes or copper and the products of agricultural labor, fishing work, and local craftspeople. Increasingly robust connectivity and the demands of expansionist kingdoms moved goods overland and via canoes across and around the lakes, linking people directly to markets in the kingdoms' distant urbanizing centers and to new hubs along paths of transit as well.[77] These powerful states existed alongside many, many smaller domains and, across the nineteenth century, forced these smaller polities into subordinate, tributary relationships of alliance or defense as they expanded—this was the case with the Ssese Islands and Kiziba and Buganda, less so with people in the southern Imbo and Urundi. Wars driven by expansion generated displacement and insecurity for the populations in their path. Violence disrupted food production and raids produced captives, dependents, and slaves.[78] New mobilities catalyzed by war and trade also meant that many interlacustrine populations experienced hunger, insecurity, and epidemic disease in new ways.[79]

In the early nineteenth century, wider regional flows and influences became more important as interactions between the interlacustrine kingdoms and peoples of the savannas and coast began to define a larger territorial—and indeed subcontinental—arena of exchange and engagement. Shifting political and economic trends in the western Indian Ocean interacted with entrepreneurial political ideologies further inland to produce new interventions into the politics and economies of eastern African populations.[80] By 1800, port cities dotting the Indian Ocean coast from Mogadishu (in modern Somalia) to Sofala (in modern Mozambique) had seen Arab, Indian, and African influences blend into a distinctive Swahili coastal culture over several centuries. Europeans, too, came and went, attracted by thriving trade in everything from precious metals to exotic spices and driven by competition for geopolitical primacy. Independent Swahili city-states had used their strategic position—good harbors, gateways to rich hinterland regions—to great advantage, while astutely engaging with new powers in the Indian Ocean world to try to preserve their autonomy.[81] Centrally important in the increasing connectivity between the coast and the Great Lakes region, and to new possibilities for insecurity and prosperity for interlacustrine populations, were two parallel developments in the 1820s and 1830s: Omani migrants established a sultanate in Zanzibar and claimed suzerainty in Swahili coastal ports; and large, organized caravans began to travel between the coast and lakes. The presence of coastal newcomers—read as "Arab" or Swahili by local populations—would only increase in the hinterlands by midcentury, as would African involvement in moving goods and people to the coast.[82] After 1840, Zanzibar became the hub of a transcontinental, Afro-Arabian state; Omani plantation agriculture, especially clove production, soon exhausted supplies of free labor and generated demand for unfree labor in the form of slaves from the mainland.[83] Concurrently, historic demand from the Indian subcontinent for East African ivory was joined with growing desire for ivory luxury and status items in Europe and the United States. The recently established caravan routes connecting various coastal cities to the Great Lakes became conduits for the movement of slaves and ivory out of the African hinterland, and of printed cloth, weapons, beads, and other manufactured goods into internal markets.[84] The caravan trade brought cultural change to the Great Lakes region as well, as coastal goods (such as printed cloth) came to connote prestige and worldliness and Islamicized, Arab-Swahili identities also traveled.[85]

This new slave-ivory nexus catalyzed significant, if uneven, changes in the politics and economies of the interlacustrine kingdoms between the 1840s and 1880s. Expansionist states, through wars and raiding, had long generated captives; such captives had historically been integrated into internal production as slaves, but now also became valuable commodities for labor markets further afield.[86] Arab and Swahili traders concentrated in entrepôts on and near the lakes and in proximity to caravan routes: Mwanza and Kampala at Lake Victoria, Ujiji and Uvira at Lake Tanganyika, Tabora on the central plains, and Kasongo in the Congo River basin.[87] The most vigorous, direct engagement between traders and the major interlacustrine states happened in Buganda, where an entrepreneurial, aggressive regional imperialism made it an active partner in generating and moving slaves and ivory within the wider region.[88] The many smaller kingdoms and polities connected with the caravan trade were affected by the movement of slaves, ivory, and other commodities in the region, some as targets of raids for slaves, some through pressure to generate ivory for chiefs to trade, and some as sources of caravan porters. By contrast, the highland kingdoms of Urundi and Rwanda saw less direct engagement until the later nineteenth century, by virtue of their relative isolation from major caravan routes.[89]

Coastal populations in East Africa had had contact with European traders, particularly the Portuguese, in prior centuries. But with the expansion of northern European empires and increasing agitation against the slave trade in the nineteenth century, European interest in influencing both coastal and hinterland African lives grew more assertive. European travelers followed caravan routes to explore the geography of the interior, "discovering" the sources of the Nile and the chain of vast inland lakes in the 1850s and 1860s on expeditions facilitated by Arab and African translators and guides.[90] Christian missionaries were often at the leading edge of European engagement with African populations and both Catholic and Protestant missionaries moved inland to the lakes from coastal footholds after mid-century. Continental partition by European powers at the Berlin Conference in 1884 established the Belgian sphere of influence in the Congo Free State as extending to the western shores of Lakes Tanganyika and Kivu. Jockeying for control of the interlacustrine region, conceived as a "second shore" for German power in eastern Africa, fit into wider imperial calculations regarding the balance of power in Europe. Border negotiations ultimately established the British to the north of Lake Victoria along 1 degree southern latitude, and the Germans to its south.[91] But while Buganda, the Victoria hinterlands, and the central caravan route via Tabora to Ujiji had become the focus of European energies, it was not until the 1890s that the hinterlands of Lake Tanganyika were traversed and surveyed by colonial officials.[92] Thereafter, first through exploratory expeditions aimed at mapping the region and then through military campaigns aimed at securing German colonial power, Urundi and also Rwanda were integrated into the German colonial state; similarly, the eastern regions of the Congo Free State at Lake Tanganyika were drawn more tightly into the Belgian colonial regime.[93] Ultimately, the British claimed modern Uganda, Kenya, and Zanzibar, the Germans claimed modern mainland Tanzania, Rwanda, and Burundi, and King Leopold II of Belgium, the modern Democratic Republic of Congo.

As European colonial incursion both accelerated and broadened, the 1890s were a time of increasing stresses on health, peace, and prosperity, generating significant upheaval in nearly all aspects of African life. Waves of disruption and disaster in the form of drought and famine, cattle diseases, war, and human illness had widespread and profound impacts on people on the lakes' shores and hinterlands, sickening and killing many, driving new local mobilities, and creating new tensions around political legitimacy. Epidemic sleeping sickness followed these other forms of illness and misfortune. Amid rapid change and diverse challenges, the possible resources and strategies available to secure health and prosperity for interlacustrine societies also shifted, as political regimes changed and therapeutic diversity increased. And yet, as the history of controlling illness and securing health amid new epidemics in the region demonstrates, certain political, economic, and therapeutic relationships and institutions continued to be resonant for interlacustrine populations. Their meaning and importance in daily life throughout the era of increasing colonial intervention demonstrates their resilience, even as the grave consequences of late nineteenth-century disruptions make clear the complexity and vulnerability of the political, social, and ecological balances in place.

OVERVIEW OF THE BOOK

This book traces the experiences of health and illness in communities that were affected by sleeping sickness and were central to the development of colonial disease-prevention strategies. Three case studies ground the three parts of the book: Part I examines the Ssese Islands of colonial Uganda between 1890 and 1907, Part II examines the kingdom of Kiziba in colonial Tanzania between 1890 and 1914, and Part III examines the Imbo lowlands of colonial Burundi between 1890 and 1914. Each part follows accumulated knowledge, ideas, and practices that changed as the German colonial anti–sleeping sickness campaign unfolded from research on the Ssese Islands to the first sleeping sickness camp in Kiziba to the implementation of a scaled-up campaign in the southern Imbo. Part I focuses on the Ssese Islands as the site of research and treatment development efforts spearheaded by eminent German scientist Robert Koch; these efforts culminated in the development of the *Schlafkrankheitslager* (sleeping sickness camp) as a model for German colonial strategies. Part II examines how the camp model fit into both precedents of managing widespread illness and changing political dynamics in the kingdom of Kiziba. In Parts I and II, I explore continuities across the late nineteenth and early twentieth centuries in an effort to understand the points of reference and intergenerational touchstones that allowed people to make sense of sleeping sickness upon its arrival. Part III pivots to follow how ideas of sleeping sickness prevention that developed out of these experiences at Lake Victoria, which explicitly focused on location and mobility, played out in the particular context of northern Lake Tanganyika. Each part, then, also traces German colonial efforts as connected to those of British and Belgian regimes nearby.

Part I begins with a brief orientation to the history of the Ssese Islands and Buganda in the late nineteenth century. The Ssese Islands, an archipelago in northwestern Lake Victoria and part of the Buganda kingdom, were widely understood to be a hotspot of sleeping sickness in the early twentieth century. This orientation introduces the reader to specific social, political, and environmental aspects of life on the Ssese Islands and the northern rim of Lake Victoria, setting the scene for chapters focused on healing, mobility, and the interaction between established and new ways of addressing illness. Chapter 1, "Finding Sleeping Sickness on the Ssese Islands," argues that important continuities existed between historic responses to widespread illness and those to seemingly new forms of misfortune at the turn of the twentieth century. It first establishes the general contours of islanders' lives and livelihoods in the late nineteenth century, focusing on the islands' political and ritual importance within Buganda and also situating Ssese mobility and livelihood within the islands' distinctive environments. These elements of life would be fundamental to Ssese experiences of serious illness to come. This chapter then examines the range of responses that islanders and lakeshore populations historically employed in times of illness and misfortune, looking at kaumpuli to highlight responses that involved new mobilities or reorientations to domestic spaces and surrounding environments. With the serious illness kaumpuli as a key point of reference, chapter 1 closes with an exploration of changing responses to an increasingly common form of wasting death, mongota, which European researchers would come to translate as sleeping sickness. It demonstrates that Ssese islanders made important moves to mitigate this new form of illness and death, drawing upon both established strategies for healing and avoiding illness as well as the resources of increasing therapeutic diversity on the islands.

Chapter 2, "Healing *Mongota*, Treating Trypanosomiasis: Research on the Ssese Islands," continues to follow the history of mongota on the archipelago, anchored by a research expedition on the Ssese Islands led by German scientist Robert Koch in 1906–7. This chapter argues that Ssese islanders' experiences of previous misfortune and illness and the diverse therapeutic landscape they inhabited shaped their engagement with entrepreneurial German scientists. Ssese islanders' engagement with the German expedition's diagnostic techniques and therapeutic regimens influenced both practical research techniques and theories of disease control that would be exported throughout German East Africa and define the German anti–sleeping sickness campaign. In particular, the historic importance of the Ssese Islands and recent uses of specific sites on Bugala Island, where the German research site was located, significantly impacted Ssese engagement. I examine the advent of the Schlafkrankheitslager, or sleeping sickness camp, and Koch's attempts to suppress sleeping sickness through the use of months-long regimens of atoxyl, an arsenic-derived drug.

Research on the Ssese Islands led colonial scientists to historic relationships—epidemiological, economic, and social—that connected the islanders to the Haya kingdoms of the western lakeshore, specifically to the kingdom of Kiziba. Kiziba would ultimately become the key site in the region for German anti–sleeping sickness measures. Part II begins with a brief orientation to important social, political, and environmental aspects of life in Kiziba, one of the eight kingdoms of Buhaya in modern northwestern Tanzania, in the late nineteenth century. It offers deeper historical detail on the local factors that shaped royal power and the political economy of land and labor, elements that would shape Ziba and colonial efforts to mitigate the impact of illness. Chapter 3, "The Prince and the Plague: Politics, Public Health, and *Rubunga* in Kiziba," argues that the political and social dynamics of sleeping sickness must be understood in the context of another illness, rubunga. This chapter explores the social, political, and environmental factors that shaped Ziba society and wider Haya approaches to illness, healing, and misfortune, including approaches to both rubunga and sleeping sickness. I analyze outbreaks of rubunga (usually translated as bubonic plague) in the 1880s and 1890s to uncover intersections of tactics to prevent disease and mitigate misfortune by Ziba royal authorities and the newly arrived German colonial regime. I argue that rubunga served as a foundational experience for the implementation of both Ziba and German understandings of disease prevention in the early colonial era, during a time of significant change in many aspects of Ziba life. Rubunga provided a practical model of how health and politics could intertwine in the early colonial era, one that shaped subsequent responses to widespread death and disease.

Directly on the heels of rubunga came another widespread illness and, with it, further colonial public health interventions. Chapter 4, "Gland-Feelers, Elusive Patients, and the Kigarama Camp," explores the creation of the flagship German intervention at Lake Victoria: the sleeping sickness camp at Kigarama. Focusing on the local economies of land and labor that shaped the location and trajectory of the camp, it examines the engagement of the Ziba kingdom's young monarch, Mutahangarwa, with German colonial officials. This chapter illuminates the factors that shaped how and why people sought or rejected the treatments offered at Kigarama, pointing to the importance of clan-based land distribution, seasonal labor, and shifting royal power. I argue that Haya practices of land allocation overlapped with place-centered traditions of royal authority to make Kigarama a space imbued with Ziba political power as well as a site for the acquisition of material resources and access to colonial therapies. This chapter also follows the fortunes of a cohort of new colonial auxiliaries, *Drüsenfühlern* (gland-feelers), whose work to search for hidden cases of sleeping sickness reveals the complex interplay between royal prerogatives, colonial desires, and individual interest in the thick of the public health campaign. Here, I offer new readings of the spaces and tactics of colonial public health in order to interrogate local meanings alongside colonial intentions and understand the Kigarama camp within Ziba geographies and economies.

Though the Lake Victoria epidemic commanded colonial attention firmly and quickly in the first decade of the twentieth century, German energies in eastern-central Africa soon turned to Lake Tanganyika, where sleeping sickness appeared to spread unchecked. Part III shifts to focus on the littoral of Lake Tanganyika, where German sleeping sickness interventions had begun in parallel to those at Lake Victoria. Part II begins, like parts I and II, with a brief orientation to important social, political, and environmental aspects of life on the coastal lowlands of Lake Tanganyika known as Imbo and areas on the western shore with connections to those lowlands. I focus particularly on contexts useful for understanding the particular dynamics of mobility and illness that shaped anti–sleeping sickness work in the region. Chapter 5, "Mobility, Illness, and Colonial Public Health on the Tanganyika Littoral," examines mobility between the opposite shores of Lake Tanganyika—the lowlands of the southern Imbo region in German Urundi and the Ubwari peninsula of the Congo Free State/Belgian Congo, areas connected by vigorous trade and migration. I show how lacustrine mobilities and their routes and hubs contributed to the spread of sleeping sickness and came to define the emplacement and scope of subsequent colonial prevention efforts. I piece together the importance of historic mobilities across the lake for life, livelihood, and experiences of illness for linked Rundi and Bwari (and other Congolese) populations. I argue that the parameters and constraints of colonial interventions, particularly bush-clearing work aimed at destroying tsetse habitats, resulted from the vigorous mobilities, distinctive environmental conditions, and heterogeneous populations in the southern Imbo. The book concludes with a discussion of how histories of sleeping sickness and its control help us understand current global health challenges.

A NOTE ON LANGUAGES AND CONVENTIONS

This book relies upon source materials created by speakers of English, French, German, Oluhaya, Kiswahili, Kirundi, and Luganda, some of whom also used distinctive dialects within those major languages. I have maintained German special characters or spelling in use in the early twentieth century, but have standardized German translations of African place-names and terms to reflect modern standardized spelling in the relevant African languages; for example, the word *Schauri* (German) is written as *shauri* (Kiswahili); the place Kiguena (German) is written as Kigwena (Kirundi). I follow orthography of the historical languages of the Great Lakes region from David Lee Schoenbrun's *The Historical Reconstruction of Great Lakes Bantu Cultural Vocabulary: Etymologies and Distributions* (Cologne: Rüdiger Köppe, 1997) and modern conventions of the International African Institute for all African languages.

PART I

The Ssese Islands, c. 1890–1907

The Ssese Islands, c. 1890

An Overview

LOCAL DYNAMICS on the Ssese Islands in the late nineteenth century played a central role in shaping the epidemiology of sleeping sickness in the early twentieth century and also influenced the nature and trajectory of early research and control efforts based there. Alongside the overview of regional and interlacustrine history provided in the Introduction, I here offer an orientation specific to the Ssese Islands in order to highlight distinctive aspects of Ssese politics, society, environments, and economies that impacted experiences of illness and misfortune as well as efforts to heal and prosper.

A hilly and dispersed archipelago of eighty-four islands in the northwestern corner of Lake Victoria, the Ssese Islands were a distinctive feature of the northern part of the lake.[1] The islands' location and topography shaped islanders' social and political worlds and livelihoods in the late nineteenth century. Islanders lived within diverse ecosystems: dense forest, mixed grassland with scattered trees, reed-choked swamps, and wide, open beaches. The centers of the islands were drier than the margins nearer to the lake, and covered by grasslands and small clumps of trees as well as and denser forest. Those dense forests and open, grassy areas sloped down toward a shoreline irregularly cut by deep coves and bays. While Ssese populations utilized forest crops and resources, forests were also spaces apart from homes, fields, and grazing lands—sites of burial and therefore places of ancestral spirits, for instance. On some of the islands' bays and coves, a sandy beach offered a good access to the lake, while in others the shoreline was a thick mass of reeds or stretched into a swamp. The hilly,

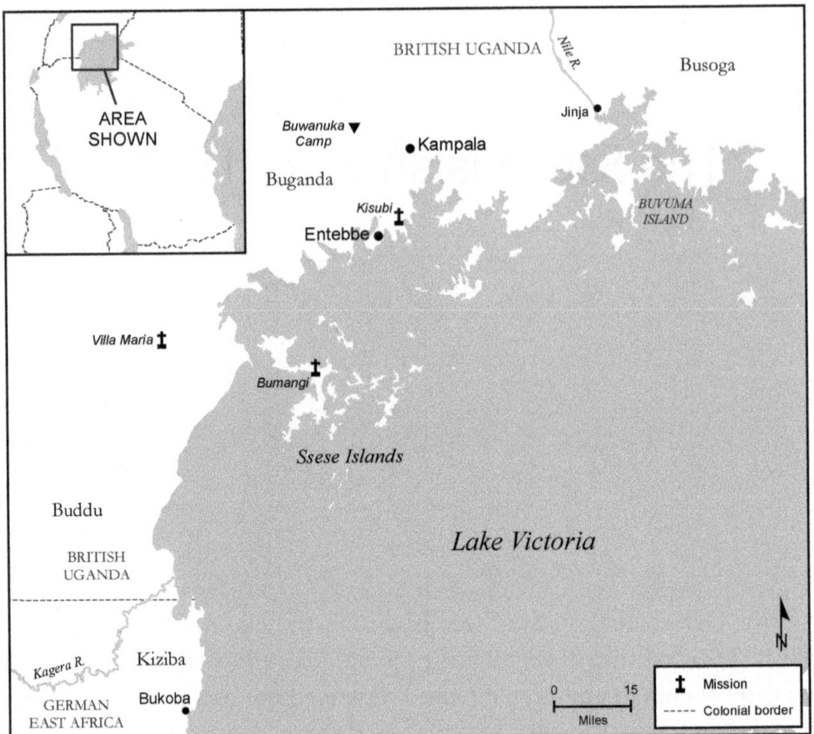

MAP 1.1. Northern Littoral of Lake Victoria. Map by Brian Edward Balsley, GISP.

grassy central areas of the islands provided grazing lands and were often bounded toward the lakeshore by a belt of trees.[2] The islands had their highest elevation at their most central points, with elevations sloping down toward the lakeshore. The lake, then, lay below homes and villages, separated from them first by grasslands and then by forest or swamp.

Ssese villages and their social geography were tailored to the islands' environment. Villages fit into the mixed forest-grassland ecosystems of the islands' interiors and homes appear often to have been located advantageously where forest and grassland met. In the mid- to late nineteenth century, Ssese homes were loosely grouped into non-nucleated villages connected by well-worn, meandering paths.[3] Typical homes of non-elites were circular, domed constructions with exterior walls of reeds covered in grass thatch, divided internally by barkcloth curtains or reed walls, and with a hearth for cooking inside. Chief's homes, by contrast, were larger, with multiple poles supporting a broader roof, exterior walls supporting the roof, and a larger interior space divided into separate rooms.[4] The typical home, regardless of status, was situated on flattened, cleared ground and set among numerous banana trees, with groves also kept clear of undergrowth to allow growing other crops.[5] Crops that later observers considered typical for the Sseses—plantain bananas, yams, and coffee—optimized the heavier rainfall regimes on the lake's shores and were cultivated alongside vegetable crops.[6] Rainfall on the islands was bimodal, with rainy seasons lasting for two months, typically beginning in March–April and September–October, and among the heaviest on the Buganda littoral.[7] Within the rainy and dry seasons, cultivation of annual crops and management of perennial tree crops shaped labor demands and dictated bursts of activity. Missionary diaries from the late 1890s indicate that November and December were busy months for planting, as well as a season of relative dearth as final stores were used up and new plants had not yet matured.[8] Alongside farming, islanders also kept sheep, goats, and cattle.[9] Fishing and fishing-related work such as the construction of nets and traps were also a central component of Ssese livelihood in the late nineteenth century, as islanders actively exploited the lake's fisheries, both for their own consumption and to market dried fish elsewhere around the lake. Fishing was "critical to regional diet and the local economy" along the lakeshore.[10] Ssese fishing at the turn of the century was sophisticated and complex, involving spearfishing, setting woven traps, and hook-and-line fishing, with gendered and generational specialization.[11]

Canoe-building and producing the tools of fishing knitted together forest, household, and lake among Ssese communities, as trees became planks for canoes and rowers' oars, raffia and other fibers caulked the canoes' gaps and was fashioned into nets and basket traps, and fish provided sustenance and income. These activities involved men and women, young and old, and tied islanders intimately to the lake and lakeshore environments. Large, sewn canoes made from planks of specially selected wood defined long-distance travel on Lake Victoria in the nineteenth century and Ssese expertise was key to their widespread use.[12] Contemporary Ganda and European narratives provide a sense of the impressive vessels afloat: they were regularly upwards of forty feet long and four feet wide and manned by dozens of rowers; the prow of some canoes extended with a battering ram or had animal horns fixed to it.[13] By the late 1880s, Ssese expertise in boatcraft and rowing was well known to outsiders, as islanders created and manned many of the vessels circulating on the lake—their expertise was, by all accounts, unmatched. Canoes connected the Ssese with their Ganda and Soga neighbors, but also with populations further afield. Sophisticated, impressive Ssese vessels were also imbued with ritual and spiritual significance that connected rowers to the wider powers of the great lake and its deity, Mukasa.[14]

Ssese islanders had close political, economic, ritual, and military ties with the powerful kingdom of Buganda on the mainland some twenty miles immediately to the north.[15] The islands' intimate links with the Ganda state grew out of political and economic changes in the nineteenth century and made a deep impression on political and economic life on the islands. Over the course of the early and mid-nineteenth century during the reigns of *kabaka* Suna II (reign c. 1830–57) and kabaka Mutesa (reign c. 1857–84), the Ganda kingdom emerged as a local power and regional empire, incorporating the Sseses and other areas into its sphere of influence.[16] Essential to the state's emergence as a regional power was the development of a navy that could engage local rivals, raid for valuable captives, and incorporate smaller kingdoms around Lake Victoria as tributaries, as well as fleets that captured trade with coastal caravans they met on the lake's southern shore.[17] Plying the lake required diverse and complex knowledge that Ssese islanders, as well as other lakeshore populations, had developed and retained over generations, but Ssese expertise in rowing, navigation, and boatcraft now became instrumental in the growth of the Ganda fleet of war canoes in the mid- to late nineteenth century.[18] Canoes and Ssese boatcraft had long linked the islands to communities around the lake, but now served to connect nodes within a wider, regional Ganda sphere of political and commercial influence. Ssese oarsmen played a crucial role in long-distance lake trade and Ganda warfare at the expense of economic stability for island communities.[19] Dense populations in Buganda were sustained by production in outlying estates, with the Sseses among them. Thus, by the late nineteenth century, the Ssese Islands were a "province" of Buganda, with Bugala Island (the archipelago's largest) serving as the anchor of Ganda governance.

Politics oriented around the largest and most populous island, Bugala, as well as the large, easternmost island of Bukasa, while ritual power centered on the powerful *lubaale* Mukasa's principal shrine on nearby Bubembe Island. Bugala, also referred to simply as "Sese" by Europeans, is an irregularly shaped island that curves southwest away from the Buddu shore before angling sharply to the south; a wider northern section joins a wider southern section at the narrow neck at Bumangi. Directly east of the southern half of Bugala is Bubembe Island, and beyond it, the second-largest island of Bukasa. The islands cluster close to one another and many lay less than a day's row apart, such that travel between the islands was frequent in the nineteenth century.[20] The miles-wide expanse of the lake defined the islands' orientation to Buganda and Busoga on the mainland to the east and north, but the mainland lay closer on the west, with the islands separated from Buddu by only a narrow channel. Within the archipelago, canoe travel was by no means easy or reliable, however. Changeable weather and squalls—as well as occasional encounters with hippopotamuses—made the journey uncertain and dangerous for rowers on both large and small vessels.[21] By the late nineteenth century, islanders could connect to mainland markets and communities and Europeans wishing to travel could generally find rowers to make the journey.[22]

Ssese integration into the Ganda state required grafting centralized, elite royal politics onto the necessarily dispersed governance of the islands. Within the Ganda bureaucracy, province-level authorities held sway over rulers of districts of the larger islands or of individual, smaller islands; *nnamasole*, the Ganda queen mother, traditionally held influence over land and politics in the Sseses, constituting another connection between the palace and islands.[23] Political authority on the Sseses focused at the village level, within district and island hierarchies, such that powerful chiefs ruled the larger inhabited islands. Particular to the individual islands were hereditary chiefs, powerful political figures whose titles and political-ritual roles were linked to control of particular areas.[24] These men were the primary interlocutors of missionaries and colonial authorities, and, alongside other key chiefs, would have appointed village-level leaders.[25] Clan affinities also structured relations between islands and mainland. The historic leader of the powerful lungfish clan, titled *gabunga*, held symbolic and practical power on the islands as a consequence of his central position between Ganda political power and Ssese-based ritual power.[26] Gabunga held the role of "admiral" or "head of canoes" of the Ganda fleets, reinforcing links between Ssese rowing and the Ganda state (a related role, titled *kweba*, served as provincial chief). Lungfish clan members also mediated worship at the lake deity Mukasa's principal shrine on Bubembe Island, stitching together clan, palace, and the islands. Other clans with connections to rowing and boatcraft also shaped political life on the islands.[27]

Dramatic changes in the 1880s and 1890s redounded to alter life on the Sseses. The arrival of Arab-Swahili traders at Lake Victoria and in Buganda in the mid-nineteenth century, followed by the arrival of increasing numbers of Christian missionaries in subsequent decades, made for vigorous economic and cultural exchange around the lake.[28] At the same time, rivalry within the Ganda palace made for increasing volatility toward the end of Mutesa's reign in the 1880s, turmoil that only increased as his son Mwanga took power in 1884. Muslim, Protestant, and Catholic factions had developed in the palace, particularly among young elites, early in Mwanga's reign, the kabaka's sense of threat from external forces, particularly

Christian converts and missionaries, led to escalating violence.[29] Roughly simultaneously, outbreaks of both bovine pleuropneumonia and rinderpest (diseases affecting cattle and other ruminants) flowed through the region in the 1880s and 1890s, with serious, although not necessarily uniform, mortality among cattle-keeping societies' herds.[30] Epidemics of serious human illness overlapped with cattle diseases and, along with crop failure and famine, devastated individuals and households. Importantly, this convergence of misfortunes "undermin[ed] one of the key functions of the kingship as a key point of patronage and distribution"—spelling political trouble for the new kabaka.[31] Amid these wider disturbances, ongoing violence toward different religious communities, and Mwanga's increasingly onerous demands for taxes and labor, the kabaka was overthrown in a palace coup in 1888. British imperial interests, drawn in in the 1880s amid an outcry over religious persecution of Anglicans and attracted by the region's economic potential, became involved in palace affairs first under the auspices of a chartered company. Uganda then fell into the British sphere of influence in East Africa mandated by the Anglo-German treaty of 1890. A formal protectorate followed in 1894. Under the new regime, the child Daudi Chwa was installed as kabaka and the power of the Ganda ruling council (the *lukiiko*) was preserved. Land that had previously been the kabaka's prerogative to disperse to clients and allies was dramatically reduced, and all other "unclaimed" land was now owned by the British Crown.[32] Buganda would remain centrally important to British rule of the larger, multiethnic Uganda Protectorate, but the Sseses receded to the periphery of Ganda, and therefore also colonial, politics.

The 1880s and 1890s were decades of significant change around the Lake Victoria littoral and, indeed, in central and eastern Africa more generally. While we cannot presume a wholesale disarray in Ganda or Ssese society amid the overlapping crises of war, cattle disease, human illness, and famine, it is also clear that death, illness, and insecurity changed daily life—sometimes in temporary responses, sometimes in permanent reorientations. In wider perspective, the potential causes of trouble and insecurity in the 1890s for Ssese islanders and those on the lakeshore were legion. The religious wars that tore Buganda apart in the late 1880s and early 1890s had material consequences, destroying some Ssese households, villages, and boats and leaving some homes looted of livestock and household goods.[33] These wars removed able-bodied Ssese men from other work on the island, as they served as rowers on the Ganda fleet.[34] The diminishment of agricultural and economic manpower and resources to such a wide and lengthy extent would have made many households more vulnerable to the ravages of infectious diseases. In some cases, new local mobility resulted as people sought temporary assistance from nearby missions or, perhaps, migrated to areas of greater relative security where family or clan connections might offer support. Throughout, people assertively sought healing and amelioration of misfortune within the range of historic strategies and in an increasingly diverse therapeutic marketplace.

1 Finding Sleeping Sickness on the Ssese Islands

GEOGRAPHY AND environment—the lake and the islands—oriented political and ritual power, while also shaping the types of labor and production that the Sseses provided within the growing Ganda empire. This chapter examines the island contexts within which the illness called *mongota* appeared and on which colonial energies would ultimately focus in order to illuminate the social, political, and environmental dynamics of widespread illness. A singular aspect of the Sseses that shaped political, economic, and treatment-seeking activity around Lake Victoria was the presence of the *lubaale* Mukasa's principal shrine, situated among other sites of healing or cosmological power. I open this chapter with a discussion of the political and ritual dynamics around Mukasa's shrine and the powers that mediated contact between the Sseses and Buganda before turning to an exploration of the historic sources of illness and misfortune that struck island and littoral communities. I then examine the illness kaumpuli to consider how Ganda ideas about illness as well as practices of managing widespread illness accommodated the rapid change and emerging therapeutic diversity of the late nineteenth century.

Both the nature of misfortune and the possibilities for healing and relief were changing near the turn of the century, with the emergence of a new illness, mongota, that caused people to nod or sleep markedly, as a pivotal moment. This chapter closes with an exploration of changing Ssese responses to mongota. I argue that the advent of mongota catalyzed the deployment of diverse strategies to cope with the illness and death it caused, as well as shifting engagement with European missionaries on the islands. Focusing on how such strategies changed over time, I look particularly at historic precedents of place-centered responses to widespread illness, such as new mobilities or reorientations to domestic spaces and surrounding environments. Ssese islanders made important moves to mitigate the impact of illness and death, and their actions demonstrate historical continuities in responses to widespread illness during an exceptionally disruptive and tumultuous era in littoral life.

The matter of human African trypanosomiasis—of sleeping sickness—will emerge as centrally important to Ssese lives in chapter 2. Colonial and tropical medicine attentions fell on the islands with increasing scrutiny after 1905. Ssese and Ganda ideas and practices around illness, which I examine in depth in this chapter, shaped the field upon which those later sleeping sickness research and treatment efforts would occur. The Sseses proved a particularly influential and productive zone for the articulation of the intellectual and strategic foundations of the German sleeping sickness campaign, ranging from particular ideas about African mobility to specific dosage regimens with atoxyl. The sleeping sickness camp that German researchers ultimately founded on the Sseses provided a springboard for the wider campaign in affected areas of German East Africa, shaping colonial public health strategies elsewhere on Lake Victoria and further afield at Lake Tanganyika. However, in this chapter, I explicitly do not work with sleeping sickness in a European biomedical framework of disease classification or causation, hewing instead to Ganda and Ssese categorizations of illness.

THE SSESE ISLANDS: POLITICS, PROSPERITY, AND DANGER WITHIN THE GANDA EMPIRE

By the mid-nineteenth century, the Ssese Islands were remarkable for their position at the nexus of ritual, military, and therapeutic power within Buganda and the lake's northern shores. Important for elites and ordinary people alike were Ssese shrines to the powerful lake deity Mukasa, whose presence made the islands a space of power, healing, and potential danger amid the vast lake.[1] The intertwining of royal, clan, and ritual power in previous centuries had made Ssese ritualists central to the *kabaka*'s maintenance of legitimate rule within Ganda royal cosmologies, as well as in other kingdoms around the lake. While the majority of the population occupied a subordinated position within the centralized Ganda state, powers located on the Sseses remained indispensable for efforts to secure prosperity and restore health and made the islands a singular site within littoral political, ritual, and therapeutic dynamics. These social and political worlds of Ssese islanders—created by both long-standing political and ritual relationships as well as recent adaptations to new conditions of life—shaped how people responded to colonial impositions, including those around illness and health.

Within regional cosmologies, the lake and the Ssese Islands were places of significant power.[2] Powerful forces of fertility, prosperity, and nature circulated around Mukasa, the lubaale or "national spirit" of the lake, who was associated with fish, rain, winds, children, and especially twin children.[3] Mukasa's main *kubándwa* shrine sat on the island of Bubembe, nestled in the center of the archipelago; tradition held that the lubaale had been born on the nearby island of Bukasa.[4] Ganda and Ssese people seeking healing had diverse sources to consult, but Mukasa's significant power offered the lubaale's mediums and shrines a corresponding potency to resolve challenging matters and ensure prosperity. Roscoe and Kagwa's extensive explanations of Mukasa and his powers attributed "benign" force to the lubaale, characterizing him as a "god of plenty" who "gave the people an increase of food, cattle, and children" and "sought to heal the bodies and minds of men."[5] In addition to desire for successful voyages or productive fishing by people frequenting the lake, others also sought amelioration of illness and misfortune from mediums who, through the lubaale, could identify the cause of trouble and offer direction to address it. Mukasa and *balubaale* were "indispensable to the common people" as "providers of health and fertility."[6] Childless women sought fertility from Mukasa and his companion Nalwanga; his local mediums would have offered solutions for maladies alongside a family elder or nearby herbalist or healer, perhaps when persistence or complexity suggested additional resources were needed.[7] Importantly, Schoenbrun argues, Mukasa and the Ssese Islands provided sources of "information, creativity, and fertility" to ordinary people through practices of gathering and supplication at the deity's shrine there; they offered resources for realizing self-sufficiency, prosperity, and "respectable adult belonging" to people living on the littoral from the eighteenth century onward.[8] People historically accessed these resources at shrines scattered around the lakeshore and at the main Bubembe shrine, assembling in gatherings both large and small. Large gatherings to consult Mukasa at a shrine might occur regularly and rhythmically, focused on the lunar cycle and the timing of the new moon and spaced every three months, or, less predictably, in response to other triggers: a "public calamity" such as widespread illness or famine, or instances of royal consultation.[9]

Mukasa determined the temper of the lake and therefore the fate of people traveling on it—whether weather would be calm or the waters rough, how rowers would fare in their collective work, whether canoes would be threatened by hippos or meet with unseen rocks.[10] Veneration and propitiation of Mukasa was a key aspect of the labor and experience of traveling the lake for Ssese rowers especially. Rowers or fishermen might make an offering for safety at a local, minor shrine or as they worked on the lake.[11] A loose network of mediums around the littoral maintained those minor shrines that were scattered around the lakeshore.[12] In the early 1880s, Fr. Léveseque, transiting the lake from Buganda to a mission on the southern edge of the lake by canoe convoy, recounted Mukasa serving as a touchstone for rowers in the daily experience of rowing—explaining a hippo's growl or a slow journey—as well as a means of seeking intercession in difficult circumstances. Ssese rowers offered the lubaale ripe bananas before setting out across the open lake, to "feed" Mukasa, employing rituals that maintained connections between their terrestrial farms and home life and their work on the water.[13]

Mukasa's power was formidable and his reach was wide, extending deep into Buganda and far beyond the Sseses.[14] The legitimacy of interlacustrine chiefly authority depended on the health and prosperity of a ruler's population and good relations with the lubaale Mukasa were important to Ganda royal power. Veneration of the powerful god by the Ganda kabaka maintained a connection between the islands and the Ganda royal court.[15] The kabaka's veneration of Mukasa, through offerings of people, fowl, livestock, barkcloth, and cowries at his shrine, and reciprocal gifts of fish from Mukasa's priests to the kabaka, directed the exchange of symbolic goods between the Ganda court and the Ssese shrine within regular, ritualized festivals.[16] Mukasa's shrines and mediums also offered resources to Ganda and other leaders in dire circumstances. According to early twentieth century ethnographies of the Ganda, just as an individual might seek the cause and remediation of a challenging or serious illness from one of Mukasa's shrines, so, too, might the kabaka seek counsel with Mukasa's Ssese medium "if any plague began to rage in the country."[17] In times of difficulty, such as a famine that struck the region in late 1880, the kabaka might send frequent gifts to the central shrine's medium on the Sseses. In the same era, missionaries reported, kabaka Mutesa relied on consultation with the medium to determine when to make war or travel on the lake.[18] Mukasa's medium also famously brought life in the capital, Kampala, to a halt during efforts to define and address the kabaka Mutesa's illness in 1879: no one could trade.[19] Intimately linked, Ganda rulers and lubaale "were involved in a continuously negotiated relationship" mediated through such exchanges of symbolic goods, maintaining independent "realms of action" as well as reciprocal, mutual obligation.[20] On the Sseses and in mainland Buganda, these connections between broader chiefly authority and Mukasa's power were reinforced historically through connection to specific sites from which the most significant healing power emanated. The person titled *Sewoya* was, as chief of a major area on the Sseses, historically involved in the ceremonies that annually reconstituted Mukasa's shrine as a sacred and powerful space, along with the men bearing the names *Semagala*, *Kaganda*, and *Gugu*.[21] Thus, many positions of authority on the Sseses related to chiefly mediation of the powers of particular lubaale or of activities around Mukasa's principal shrines.[22] Mukasa's importance for safe and secure life around the lake—as protector of fishermen and rowers and controller of weather and good catches—meant that the lubaale's influence also extended beyond Buganda. Royal responsibility for maintaining health and prosperity by mediating cosmological forces also connected other regional kings to Mukasa and to his Ssese priests. In one example, ceremonial "fire" from the Sseses was required for the installation of the *bakama* of Kiziba, to the southwest of the islands, where it was ceremonially used to cook the king's food and heat his person; it was kept burning in the palace hearth until a *mukama* died.[23]

Historic ritual and political connections between Ssese shrines and the Ganda court provided one logic for interaction between the islands and the mainland, structuring the engagement of political elites and ritual experts around the mediation of cosmological power centered on the islands. The relations that connected aquatic and terrestrial worlds were experienced differently by the people who worked the lake, however, for whom engagement with Mukasa affected life, health, and prosperity in everyday ways. Rowers' ready invocations of Mukasa as they labored, moving people and goods across and around the lake, point to the lubaale's role in shaping mobility and security for littoral populations. The lakeshore and the forested fringes of the Ssese Islands were spaces that islanders had to traverse as they went about their days—to fish, to collect water, to travel, to work, or to bring livestock to drink. Gendered but comprehensive use of the lake meant widespread activity at the islands' edges; the lakeshore was rich with vegetation and fish, a place of fecundity and potential. But nearness to the lakeshore had implications for health and security.[24] On the Sseses as elsewhere in the era of Ganda military campaigns and raiding for captives, it was also a space of vulnerability to violence and force for people living on the islands: living permanently near the lake occurred only in extraordinary circumstances, as we will see. The lubaale's powers mediated those mobilities and potentials for island and littoral communities, linking terrestrial and aquatic worlds through places where the lubaale Mukasa's powers could be accessed in widespread small, local shrines. These social and environmental aspects of Ssese life and livelihood would influence how islanders and littoral communities engaged with new kinds of illness in the coming decades.

MALADIES, MISFORTUNE, AND HEALING ON THE SSESE ISLANDS AND NORTHERN VICTORIA NYANZA, C. 1890

The circumstances of cosmological power and its mediation would shape how Ssese islanders dealt with widespread illness in the early twentieth century, and, ultimately, the emplacement of colonial public health. Previous experience and extant practice are also central to understanding dynamics of treatment and mitigation of sickness and misfortune that would later emerge around new outbreaks of widespread illness. I here examine therapeutic practices and the intellectual worlds within which littoral communities in Buganda understood illness. By the late 1800s, Ssese islanders and others living along the lakeshore had diverse resources to manage the precarity and prosperity of life and cope with illness and misfortune. Generally, they availed themselves of an expanding, pluralistic set of therapeutic and medical resources that included family members, healers, local shrines, and missions. Within Ganda therapeutic and etiological frameworks, healing an individual's illness depended on the mediation of spiritual forces and also addressed a wider set of relations: between a person and ancestral spirits; between a person and his or her family, kin, or clan relations; between a person and deities in Ganda cosmologies. Mission healing, too, drew upon a framework that integrated spiritual and material causes of and treatments for illness.

Consulting knowledgeable family members likely provided a first possibility for the sick. Elder members of one's kin or clan networks might direct a person to locally available botanical remedies, perhaps with the assistance of herbalist healers with specialized knowledge of healing plants.[25] More persistent or worrisome ailments drove the sick and their relations to seek further specialized assistance. A healing expert—who might act as a diviner, herbalist, kubándwa medium, or utilize these skills in combination—determined the cause of the illness and set a course for its remedy.[26] Treatment-seeking was a social endeavor and healing accounted for a person's relationships; the involvement of household and kin and attention to social relationships corresponds with the public nature of healing and its overarching emphasis on ensuring collective health and prosperity.[27] Healers attended to specific physical or temperamental signs but connected changes to the body and temperament to a wider social and spiritual world. A person's actions or behaviors, when they contradicted taboos or proscriptions, might cause illness or pain; likewise, behavior or activity that disrupted good relations with ancestors could also cause harm.

Healers might also act on the sufferer's body, addressing physical manifestations of illness. Detailed descriptions of Ganda healing in practice are rare; historical and ethnographic narratives offer only a selective view of the therapies, techniques, and medicines used by healers around the turn of the century.[28] Nevertheless, sources discuss that a healer's interaction with a sick person might involve applying botanical materials on the skin, preparing medicines to ingest, letting blood, or directing smoke or steam onto or into the body, for instance. Ganda healers attended to specific foci of pain, swelling, or irregularity as well as more generalized weakness or malaise, even as they engaged with spiritual or socially grounded woes.[29] Healers employed horn cups to draw blood from incisions on affected parts of the body, for example—on either side of the head, in case of headache. A healer might apply also a heated iron implement to cause blisters in order to draw out sources of pain from a particular place deep in the body, or use a blistering agent on sites of swelling.[30] A healer might also recommend a regimen of repeated washing of the sufferer's body with mixtures of particular plants and the drinking of plant, animal, and other tinctures.[31] Therapies were also applied to ill bodies: Roscoe approvingly recounted fevered patients sitting under barkcloth steam tents, applying an immersive, surrounding remedy for a particular symptom. Healers might require a sick person or family member to procure powerful substances—bones, saliva, urine—which were used to make medicines or amulets, as well as animals used for augury.[32] The occupation, social status, and clan of a sufferer were important to healing, as they positioned a person with regard to taboos, protective deities, or possible transgressions. Ganda etiologies, fitting within a wider moral economy, further accounted for the gender and age of the sick.[33] The sick person and his or her family also likely pursued overlapping therapies, shifting strategies if an illness persisted, in pursuit of effective cure.[34] These interventions, particularly if the healer was not a medium, occurred in tandem with propitiation of spiritual forces—efforts to properly maintain ancestral graves or shrines, offering gifts of food, livestock, or goods to a medium—as well as complying with a healer's instructions to restore balance and health to relationships. Alongside efforts to address an individual's manifestation of illness, healers and mediums also worked to resolve misfortunes on a different scale when illness became more widespread or frequent in Ganda society. Serious or widespread illnesses—those that redounded to impact a household, family, or larger population—might be attributed to correspondingly serious disruptions in relations between people and a lubaale or, conversely, to the behavior of the kabaka, who himself held secure the health of the kingdom's population.[35]

Some maladies might resolve with the use of medicines a healer made; others might have historically required consultation with a lubaale or ancestral spirit for resolution. In such cases, islanders would have sought solutions for illness and misfortune from kubándwa mediums, including those of the lubaale Mukasa. Appeals for healing to deities other than Mukasa are less well documented, but Cohen notes that the Ssese Islands were "a veritable hive of deities," with sixty deities associated with the islands in diverse interlacustrine traditions.[36] Several of the powerful balubaale with connections to the Sseses also had lineage links to Mukasa. Musisi, lubaale of earthquakes and progenitor of Mukasa, had one of his two principal temples on Fumve Island, where offerings might be made to keep the earth calm. The lubaale's powers also extended to affect fecundity and pregnancy, in both invocations of his potential to impact pregnant women and in amulets bearing the same name that were associated with fecundity and childbearing.[37] Wanema, father of Mukasa and another powerful deity, historically had his temple on nearby Bukasa Island, which was also renowned as Mukasa's birthplace.[38] Buswa forest on Bugala was sacred to Mukasa's son Mirimu, a lubaale with implications for victory in battle.[39] Kagwa's research tallied another six minor deities located on Bugala Island, another specifically on Bugala's Buninga peninsula, and seven more on Bukasa Island.[40] These other deities, alongside appeals to other prominent balubaale such as Kaumpuli, who had a mainland shrine, could have offered islanders connections to other kubándwa mediums in addition to those oriented to Mukasa when seeking relief or healing.[41] Deep associations with potent, generative forces were woven into island names: Bugala Island's name resonated with the root *-gàlá*, which glossed "physical force of life" and also connoted fertility on one's maternal side.[42] All told, particularly for adults during mongota whose grandparents would have had direct experience engaging with Mukasa and other balubaale as part of life on the Sseses or elsewhere around the lake earlier in the nineteenth century, this constellation of historically important and potent forces made island sites places that people had gone to and could still go to for relief or aid.[43] Such processes of treatment seeking occurred in a dense social field. Given the prominence of Mukasa's shrine on the islands and the prominent role of Ssese clans and political authorities in shrine activities, people affiliated with Mukasa's shrine might have also been family or clan members of the supplicant. Further, seeking healing required utilizing social connections to marshal necessary resources. Appeals to balubaale required material goods—contribution of foodstuffs or of livestock, for instance—that signaled veneration and acknowledged a medium's intercessionary powers, and thus also potentially required tapping into wider networks of family or affinity for resources.

By the late nineteenth century, Ssese therapeutic resources were diverse. Mediums or healers, on the one side, and missionaries, on the other, both sought to provide healing within systems that linked material and spiritual etiologies and treatments. The arrival of Arab-Swahili traders at Lake Victoria and in Buganda in the mid-nineteenth century, followed by the arrival of increasing numbers of Christian missionaries in subsequent decades, made for vigorous cultural exchange around the lake that introduced Islamic and European diagnostic systems and therapeutic practices and added to the healing resources available to Ganda and Ssese populations at the time. By the 1890s, mission medicine had become available to many Ssese islanders, as well as to populations elsewhere around the northern shores of the lake.[44]

The acceleration of both Protestant and Catholic missionizing in the late 1880s meant that many on the Lake Victoria littoral lived within a day's journey of a Christian mission or community of converts. Two groups—the Catholic Society of the Missionaries of Africa (White Fathers) and the Anglican Church Missionary Society (CMS)—were of particular relevance to the Sseses. Amid the religious and civil wars of the late 1880s and early 1890s, the White Fathers founded missions first at Bugoma, on the westernmost point of Bugala Island near the Buddu shore, and then at Bumangi in the island's center.[45] The Sseses, like most of Buganda, were a contested field of evangelization, and after years of religious unrest and occasional confrontation, British authorities intervened to "divide" the islands into Protestant and Catholic spheres in 1891, much to the chagrin of the White Fathers.[46] By 1898, Anglican missionaries had built a station on Bukasa, a southeastern island facing out into the lake, ruled at the time by a Protestant chief, Danieli Kaganda. From Bukasa, missionaries supervised a few dozen small congregations; they began building a church on Bugala in 1902, providing them a base on the eastern end of the island. But the CMS generally encouraged the growth of small churches in the villages under Ssese readers (as they called converts) rather than worship at a central station.[47] Both Catholic and Protestant missionaries relied on established missions on the Buganda mainland as springboards for their Ssese outposts. By all accounts, mission staff traveled frequently between islands and mainland, and the Ssese missions also served as way stations for confreres dependent on canoe transport and lakeside routes in traveling elsewhere around the lake.[48]

Healing was ideologically central to both Catholic and Protestant missions' ministries, although missionaries' capabilities and expertise sometimes differed as much as their approaches to converting and saving souls.[49] Missionaries often went into the field with some basic medical training, allowing them both to manage ailments that might affect Europeans with no alternative for treatment and to offer treatment and care to those they wished to convert.[50] For Ssese islanders, the White Fathers missions at Bugoma and Bumangi on Bugala Island and the CMS missions on Bukasa and Bugala Islands presented an additional source of healing and means to ameliorate misfortune—whether or not it involved the kinds of "genuine" conversions that missionaries sought. Locally on the Ssese Islands, CMS missions made medical care and treatment less of a priority than their mainland Buganda counterparts, while the White Fathers gradually sought to formalize and expand their capacity for medical care.

CMS missionaries offered no formal clinic or hospital to their Ssese parishioners and medical resources for acute crisis were limited.[51] Anglican missionaries sent people with complicated or persistent illness to the CMS hospital at Mengo, near Kampala, and would request that a doctor visit the islands when necessary.[52] The Ssese CMS missionaries, in comparison to those on the Buganda mainland, did not prioritize medical work in their evangelizing and did not establish sites of formal, regular medical treatment. The structure and nature of the CMS Ssese mission may have hindered it from serving as a resource for healing or medicine, regardless of missionaries' training or goals. Its early years saw frequent turnover of personnel due to illness, and necessary staff itinerations between posts on large islands and dispersed daughter churches meant that the men who led the mission were often away as much as they were at home. Missionaries' engagement with their readers did sometimes involve matters of health and illness, however, and was especially focused on women missionaries, women readers, and their children.[53]

By contrast, the White Fathers on the Sseses actively integrated medical treatment into their mission life and, over time, increased their capacity to do so, dispensing remedies and offering care in hospitals and hospices. By 1895, the White Fathers mission at Bumangi included a school and a small hospital with a few dozen beds, serving an estimated population of fifteen thousand on Bugala Island.[54] Priests regularly cared for a few dozen people in the hospital, assisted by local catechists.[55] Of note for responses to widespread illness, and ultimately for epidemic sleeping sickness, was the White Fathers' ready provision of medicines to their Ssese and Ganda charges. They dispensed a variety of available remedies, many typical for the era: a variety of purgatives and emetics, drugs presumed to affect the circulation, and drugs to relieve pain. Priests treated one another, and sometimes African patients, with calomel (mercury chloride) as a purgative, saltpeter (potassium nitrate) for rheumatism, "calaya" for hematuria (blood in the urine) or blackwater fever, citric acid to calm vomiting, and brandy.[56] They administered quinine for a wide variety of complaints, including but not limited to diverse manifestations of fever, and also dispensed laudanum. Some of these remedies were also given to their catechists and nearby families.[57] For the illness called kaumpuli (which they equated with bubonic plague), in the 1890s, for instance, priests gave their Ganda patients, variously, aloe as an emetic, "acide phénique" (phenol, carbolic acid), quinine, and cantharides, an ancient treatment for edema that could be used to produce blisters on the skin.[58] In the main, the White Fathers mission and hospital, despite some staff turnover, gained a strong foothold as a hub for healing, utilized regularly by catechists and their relations as well as by nearby communities more broadly in times of intensifying crisis such as outbreaks of widespread illness.[59]

Missions on the Sseses, as elsewhere, functioned as points of exchange and distribution of valued goods alongside and sometimes overlapping with medical interventions.[60] On the Ssese Islands on the whole, and Bugala Island foremost, Christian missions provided an important precedent for colonial interventions and institutions focused later on addressing epidemic sleeping sickness. The missions would later offer tropical medicine researchers a springboard to launch their work: social connections would facilitate relationships within which experimental treatment and control measures were arranged and make available the physical spaces within which these measures would play out. In parallel to these material and social resources were experiential points of reference for dealing with widespread or disseminated instances of sickness and death. Chief among the causes of those was kaumpuli.

KAUMPULI: INTELLECTUAL WORLDS AND STRATEGIES OF AMELIORATING MISFORTUNE

Experiences of illness, particularly of what appear to be epidemics that sickened and killed many, surface in diverse sources created around Lake

Victoria in the late nineteenth and early twentieth century: early colonial reports, missionary letters and diaries, accounts of the occasional traveler making his way through the region, and oral histories and traditions. One of those causes of illness and misfortune, kaumpuli, illustrates how historic Ganda ideas about illness and strategies for mitigating or avoiding it were connected to practices of doing so in the late nineteenth century. Outbreaks of widespread illnesses could and did move into and out of the framing of kaumpuli—it was not a universally applicable etiology. But kaumpuli provided a coherent, meaningful, and capacious means of understanding sudden and serious illness in Buganda by the late nineteenth century. Moreover, kaumpuli could catalyze mobilities and reorientations to domestic spaces and evinces the kinds of intellectual and pragmatic resources available for people faced with outbreaks of illness. Discussions of kaumpuli and cholera in missionary texts from the 1880s and 1890s open up space to consider central elements in Ganda nosologies as well as strategies of seeking treatment and healing in the late nineteenth century. Focusing on illness categorized as kaumpuli in the period between roughly 1880 and 1905 underscores the flexibility and expansiveness of Ganda etiologies and nosologies and discourses of illness and causation. It also proves a complex, multilayered problem that is good to think with. Considering kaumpuli allows us to apprehend the simultaneity of intellectual work in different but intersecting systems, situating Ganda ideas of illness and wellness within an era of widespread social, political, and epidemiological change, while also exploring the mutability of European biomedical models in the same era.

By the late nineteenth century, two specific balubaale were associated with certain kinds of illness and death that struck Ganda populations. The minor lubaale (Ndaula/Ndahura) Kawali was associated with irruptions on the skin, while the better-known Kaumpuli, a deity born of ancient transgression and misfortune, brought "plague" into people's lives.[61] While Kawali seems to have been associated with a particular type of illness—one which caused raised bumps or lesions on the skin—the lubaale Kaumpuli could have diverse impacts on human health. Important for epidemics to come was how his power registered in widespread illness in Ganda communities, striking people with disease and driving them from their homes.[62]

Between the 1880s and early 1900s, illnesses causing wasting, vomiting, and/or diarrhea fit into the etiology of kaumpuli, as did illness causing fever, pain in the chest, inflammation in the armpits, groin, and glands.[63] These categorizations, gleaned from mission diaries and contemporary ethnographies, varied over time. In the 1880s, for example, missionaries equated kaumpuli with cholera, based on conversations with their African interlocutors and observations of a few sick people, suggesting that signs of Kaumpuli's power could involve weakness or rapid wasting, diarrhea, and vomiting, as well as fever and changes to skin tone or appearance (e.g., bluish or darkened lips, sunken eyes). More often than not, its end was death.[64] While contemporary definitions provide an extensive terminology covering pain in the belly, vomiting, and diarrhea, among others, the symptoms missionaries recorded at the time as signs of cholera had no specific Luganda gloss, other than an association with kaumpuli, underscoring both the novelty and severity of this way of ailing.[65] In the early 1890s, another missionary clearly equated kaumpuli with an illness vaguely defined as "plague," describing it as "a disease attended generally with swelling of the glands, and pain in the chest," and noting further that "it is very prevalent after the rains."[66] By the late 1890s, however, Europeans around Lake Victoria firmly understood kaumpuli to be bubonic plague (Fr., *peste bubonique*), a disease characterized by dramatic swelling of glands in the armpits and groin (buboes), fever, weakness, blackening or suppuration of the skin around the buboes, and death. This particular iteration of plague, well known in European history, had by the late 1890s also become associated with an identifiable germ.[67] Kaumpuli, therefore, could align with the presence of *Yersinia pestis* in the body. But within another few years, further diversity was fitted into kaumpuli. In 1902 to 1903, an itinerant British scientist reported that Ssese islanders named as kaumpuli an illness associated with fever, swelling of the face and areas of the neck, swelling of the glands, wasting, sleepiness, and death. Severe diarrheal disease also remained an aspect of other cases of kaumpuli simultaneously.[68] In each situation, missionaries or scientists used the Ganda word *kaumpuli* to describe specific, widespread illness around them, both reporting the presence of epidemic disease and disseminating a "local" name for it. The term *kaumpuli*'s utility, for missionaries, was in facilitating translation and communication, and they used the term freely, if sporadically, over two decades as an equivalent for illnesses they defined variously as cholera, plague, and sleeping sickness.

A rich body of historical epidemiological scholarship considers, broadly, what killed people in the past based on historical narratives, archaeological data, or genetic research; indeed, epidemiological analysis of historical sources often illuminates connections between populations, or relationships between climate, food production, and disease, that might have otherwise fallen away from political or social histories.[69] The era I examine—the late nineteenth and early twentieth centuries—has received significant and diverse scholarly attention owing to the depth of the crises that occurred to challenge regional health and prosperity at the time, and for the implications of these early crises on long-term epidemiological and social change in the region.[70] Scholars of this region and era benefit from a rich set of sources on causes of illness and death on the Ssese Islands and the Lake Victoria littoral, particularly as Anglophone and Francophone clergy, colonial officials, and Ganda narrators sought to fix morbidity and mortality to specific and consistent modern biomedical causes. At first glance, their accounts confirm several key milestones in the global history of disease: that the disease known today as bubonic plague (caused by the bacterium *Yersinia pestis*) killed many in the 1890s, that the disease modern readers would recognize as smallpox (caused by the virus *Variola major* or *V. intermedius*) periodically devastated the region in the nineteenth century, and that an illness correlating with symptoms of cholera (caused by the bacterium *Vibrio cholerae*) struck populations in the latter third of the same century.[71] These milestones allow us to link eastern-central African disease histories to changes in migration, commerce, or climate in Eurasia and Africa or the Indian Ocean littoral. Such historical epidemiological scholarship is centrally concerned with positively isolating and identifying causative agents of past epidemics. As such, it is oriented around discovering the possibilities of what, in biomedical and microbiological terms, historic vernacular illnesses were biomedically and how this information might illuminate the related histories of migration, environmental change, or politics, for instance. Here, I am not concerned with which presently known pathogens can be equated with episodes of kaumpuli in the past. A preoccupation with what a historic illness actually was in modern biomedical terms first obscures how people understood or experienced disease at the time, and, second, privileges microbiological and biomedical logics of explanation over those in use at the time (which would, in the case of late nineteenth-century Buganda, be anachronistic). The equivalence or non-equivalence of bubonic plague and kaumpuli or cholera and kaumpuli is not at issue.

Indeed, the complexities of African, and particularly interlacustrine, cosmologies, nosologies, and healing practices are flattened in European accounts that sought to associate a given term with a suite of symptoms and outcomes based on European biomedical concepts. European missionaries' growing confidence in associating a "local" (meaning African-language) name to a specific biomedical entity paralleled scientific and medical practitioners' efforts in the same era to deploy the growing consensus around germ theory to fix pathogens, etiologies, symptoms, and, ideally, prevention measures or treatments.[72] Particularly in colonial contexts, these processes of defining, glossing, and equating illness and disease subordinated extant etiologies and nosologies—and their related intellectual worlds—to those of colonizers (or, at times, of the missionaries that preceded them).[73] Further, in the long view of history, the productive uncertainty of the early colonial period—years of interplay, of mutual observation, of discussion and engagement, of contention, of violence—is then lost in the shadow of teleologies of scientific sophistication and biomedical precision globally. Thus, the concern about defining an illness like kaumpuli biomedically reorients inquiry toward present-day knowledge and intellectual worlds. Instead, I emphasize that kaumpuli demonstrates how thoroughly the late nineteenth century was an era of fundamental contingency and uncertainty for both African and European populations in the Great Lakes region when, in some cases, extant nosologies and etiologies were strengthened, rather than weakened, by irruptions of novelty and unpredictability.

I place these three consecutive accounts of illnesses under the rubric of kaumpuli to consider several different implications for understanding populations' historical experiences. One suggests an epidemiological telltale: that kaumpuli referred not to a specific suite of symptoms or changes in a person's body, and to a specific, individualized etiology of illness, but rather to any grave, serious illness, perhaps especially one that could spread and kill more widely. Kaumpuli was thus used to identify epidemics of cholera, bubonic plague, and sleeping sickness/human African trypanosomiasis that struck littoral populations in succession between roughly 1880 and 1905.[74] But sources also indicate that Kaumpuli was well known to missionary observers as the name of the lubaale of "plague," who brought illness into people's lives. Its powers affected people at least three times in as many decades, covering multiple generations. And so, another reading of these texts: "kaumpuli," a word gathered by European interlocutors and fixed to particular signs of illness, referred not just to the body's changes, but also established etiology, naming the external, spiritual force which acted on human lives and bodies with increasing frequency in the late nineteenth century. *Kaumpuli*, here, would not describe the disease alone, because this isolation of physiology from cosmology was not a conceptual or practical reality for sufferers at the time.[75] Rather, *kaumpuli* could describe certain changes to the body, but also named the unseen, but very present, forces that determined which individuals or communities suffered, when relations between deities and people fell out of balance, and whose mediums and shrines could promise intercession and resolution if honored appropriately. *Kaumpuli* could serve to signify Ganda taxonomic and etiological thinking that located a particular species of disaster, misunderstood by missionaries as a name for a particular illness, but still serving to signal belief in cosmological forces.

Attribution of diverse illnesses to Kaumpuli's power signals an expansiveness in Ganda nosologies that would have facilitated the incorporation of new threats to health and prosperity into extant systems, and that also would have allowed experienced healers and/or powerful mediums to claim continued power to intercede as the world changed around Ganda populations. Consider the differing presentations but widespread devastation that unchecked diarrheal disease, suppurating buboes and overheated bodies, or weakness, wasting, and uncontrollable sleep might have on a given community: each visited disaster upon the population, but in diverse ways and timeframes. Increasing severity or difficulty could shift the nosology of an illness into the realm of kaumpuli, a disaster visited by its namesake lubaale. These differing identifications of kaumpuli were not, then, a conflation of diseases or symptoms, but rather evidence of the work kaumpuli could do as a capacious categorization of an illness and attribution of its causes. As a context for later epidemics, Kaumpuli's malign powers would confound European efforts to seek equivalencies in Ganda and European names and definitions of disease, as they attempted to pin in place a set of signs and problems that were more complex, variable, and contingent.

An outbreak of illness linked with kaumpuli in the 1880s offers insight into the intellectual precedents and strategies in circulation on the Buganda shores of the lake. Catholic missionary diaries reported that an illness causing diarrhea, vomiting, and frequently a quick death—in between one and four days—gripped the city of Kampala by mid-April 1881. People around the city called the illness *kaumpuli*, missionaries reported; the kabaka Mutesa referred to it as *lumbe*, glossed in the 1890s as sickness, disease, or death.[76] The epidemic generated panic and disruption as people attempted to evade the illness, safeguard their families, and stem the tide of wider misfortune around them. It also generated engagement with kubándwa mediums—as well as Catholic missionaries—as people sought tools and strategies to do so.[77]

Intertwined with the use of kubándwa mediums or missionary doctors for healing were other responsive, preventive practices. Identifying an illness as of Kaumpuli facilitated collective responses that temporarily redefined everyday life in different ways. Reactions to kaumpuli indicate that people changed the rhythms of daily life and oriented differently to the environment and people around them in response to widespread illness: by isolating the sick, by moving away from homes and villages, or by suspending typical social obligations. Sources indicate that, for instance, the Ganda responded to kaumpuli in early 1881 by isolating those with signs of the illness in separate rooms or dwellings.[78] Kagwa and Roscoe noted respectively at the turn of the century that arrival of this illness sometimes necessitated that Ganda people abandon their homes, without differentiating between the sick and the well.[79] Kaumpuli's mediums here played an important role in structuring such movement. Mediums might establish that death associated with the lubaale was attached to a person or household, triggering movement away. Kaumpuli's medium also ritually welcomed people to return to their homes and farms following an outbreak of illness (glossed by Kagwa and Roscoe as plague), and received goods and gifts in return.[80] The lubaale's intervention was spatial as well as spiritual, with implications for the long and short terms. People might temporarily leave their homes and farms to flee places of illness or cease the collective labor of planting or harvest.[81] When they returned, along with the lubaale's blessing, beer was brewed and offered, restoring social bonds. Departures from homes and farms were sometimes precipitated by animals, particularly rats, sickening or dying, both Roscoe and Kagwa assert, a pattern corroborated elsewhere in the region at the same time period.[82] Kagwa's ethnographic notes from the late nineteenth and early twentieth century recalled that "if a person had swollen glands it was said that this god had done it," and noted that the "plague" that Kaumpuli's name connoted still struck fear into populations because of the many deaths it could cause.[83] Several years later, in 1908, the principal medical officer of Uganda would note a case "of a disease called by the Baganda Kaumpuli, which is associated with the occurrence of bubo."[84] Médard asserts that Kaumpuli and his mediums mediated the return of survivors; the possessions of the sick belonged to the lubaale upon their illness, and the lubaale and medium facilitated their return (and thus the return to normal activity) when widespread illness had abated.[85] As ever, widespread disease also had implications for the stability of the state, safeguarded by political authorities, clan elders, and healers. Practices that worked to ensure population health also worked in the interest of political leaders and the ritual leaders and kubándwa mediums with whom they engaged.[86]

Experiences of kaumpuli also offer insight into different collective strategies for addressing widespread illness that would prove relevant as epidemic sleeping sickness came to the region. But unlike the serious, periodic illnesses that people survived alongside everyday injuries and ailments, the next widespread disease that communities around the lake encountered would significantly and permanently alter social and political relations. Efforts to prevent it would change the geography of daily life. As *bamongota*, "those who are drowsy," appeared more and more frequently on the Sseses and word spread of increasing mortality, Ssese people marshalled social resources and directed diverse strategies to cope with new manifestations of illness and misfortune. Their efforts would ultimately become entangled with those of colonial authorities and itinerant researchers, as illness brought new outsiders into the orbit of Ssese islanders and others living near Lake Victoria.

MONGOTA: THE SSESE ISLANDS IN A GATHERING STORM

European reports of a new and strange disease first emerged from Buganda in 1901 via doctors Albert Cook and J. Howard Cook, brothers and CMS missionary physicians at work in the hospital at Mengo.[87] By late 1901, White Fathers missionaries on the Sseses noted that their parishioners suffered from "the sleeping sickness" as well. In his annual report for 1901–2, Fr. Ramond at Our Lady of Good Comfort at Bumangi reported that, spiritually, the mission's fortunes were fine, but the mission found itself in grave circumstances otherwise: illness took its toll on the islanders to whom the Fathers tried to minister, creating a population weakened by disease and death. Kaumpuli—here understood as bubonic plague—he reported as a familiar threat. Newer and less predictable was the illness "called sleeping sickness," which was "very terrible and very murderous for all."[88] Of a population on Bugala estimated at twenty-six thousand, Fr. Ramond reported that the mission had treated six thousand people in 1902.[89] Elsewhere on the same island, Anglican missionary H. T. C. (Henry) Weatherhead noted that "nearly everywhere [he] met with murmurings with regard to the sleeping-sickness" on his itinerations around the archipelago and that "the death-rate on Sese [Bugala Island], hitherto not very high is, we now fear, increasing."[90]

Soon known to Europeans throughout the Uganda Protectorate and eastern Africa colloquially as sleeping sickness, the illness was also called *mongota* in Luganda, translating as "one who sleeps" and deriving from the verb glossed contemporarily as "to nod" or "to be drowsy."[91] Centered on unpredictable sleepiness or disrupted sleep to define the malady, news of mongota peppered the diaries and notes of European observers around Lake Victoria in the few dozen months between the epidemic's outbreak and the identification of the causative pathogen and vector in 1902. After that point, Europeans generally referred to a specific biomedical entity, sleeping sickness/*maladie du sommeil*, or sometimes trypanosomiasis after the parasite believed to cause the disease.[92] Despite the dangers of illnesses brought about by Kaumpuli, pox-causing diseases, respiratory infections, and the ever-present potential precariousness of rural agricultural life, mongota appeared to be different in how people deteriorated and died, causing severe degenerative changes in a person's body as well as their temperament. Grave illness was far from unknown, and misfortune in its many forms struck islanders, but this illness slowly and inexorably brought weakness, thinness, unpredictable behavior, and impenetrable sleep: death in a new form.

Missionaries' reports of early sleeping sickness patients and subsequent government reports of the wider epidemic focus on the abject misery and illness of the afflicted, but also, markedly, on a fatalism in their Ganda and Soga interlocutors.[93] Colonial-era sources frequently characterized African responses to epidemic sleeping sickness around Lake Victoria as a combination of fatalistic, brutal, and primitive: the sick were cast out completely by their families, chiefs sent away the sick to suffer "in the bush," people killed and consumed all of their household's livestock at once, the dying were discovered in squalor and alone. These early accounts, while likely capturing the physical and psychological toll of illness and death, have also had the consequence of creating a durable narrative where shock and inaction characterize the African response. Particularly in descriptions of the "pagan" or "heathen" practice of casting out the sick rather than abiding with them, tropes of Christian charity were at the forefront of missionaries' writings about the disease, serving strategically to emphasize the ongoing need for evangelization and resources to their readers (and funders) back home and to underscore the missionaries' commitment to particular communities or individuals amid the epidemic. Though Kuhanen and others have noted that Ganda authorities attempted to respond proactively to control the spread of disease, Ssese sources flesh out how these elite mainland moves were paralleled by energetic activity in affected households and villages on the islands.[94] Further, this close reading, alongside a rough sketch of the known historical epidemiology of the disease, offers a sense of the practical strategies that Ssese populations used and underscores that their approaches shifted as the epidemic unfolded. Examining these shifts not only illuminates the changing nature of the epidemic and its widening impact, but also helps us understand why islanders engaged as they did with different therapeutic possibilities, first with missionaries who remained as the islands' population declined, and subsequently with the German sleeping sickness research expedition that arrived in 1906. Ssese islanders fit their experiences with mongota into wider political changes commensurate with their position amid Ganda and British imperial spheres as well as within existing intellectual and therapeutic frameworks and experiences of illness in previous generations. Mongota generated diverse responses on the islands, particularly given its overlap with British colonial encroachment and the widening availability of mission-centered medicine for people living on the lake's shores and islands.

An initial response to mongota was elemental: islanders spoke of it, named it, and discussed it with travelers or those in wider clan or kinship circles. Generally, across the Great Lakes region, distinctive, locally specific names for an illness associated with nodding were in use at the time: mongota in Luganda and *isimagira* in Oluhaya, for instance.[95] This novelty and relative simultaneity suggests that people in nearby communities were contemporarily categorizing a set of changes to the body and temperament—here, a nodding sleepiness—as a single illness and differentiating this from others.[96] Informants in the 1910s underlined the initial novelty of mongota on the islands, for example, though other illnesses causing fever and sleepiness had been known.[97] CMS missionary George Pilkington's Luganda-English dictionary (one of the earliest made) glossed *bongota* and *simagira* as "to nod" or "to be drowsy" in the 1890s, with distinct words—*tulo* and *ebaka*—glossing "to sleep." But the connotations of the words *mongota* or *isimagira* seem to have changed over time amid the early epidemic, shifting from an association with nodding or drowsiness in the 1890s and early 1900s to a firmer connotation of sleep in subsequent years. As nodding gave way to sleeping or unconsciousness in "those who were drowsy," mongota became an illness of sleeping. The meanings of *bongota*, correspondingly, seem to have cohered around sleeping rather than sleepiness or nodding amid and after the burgeoning epidemic.[98] By 1902, Ssese islanders also called the illness *tulo*, a word that glossed sleep, but not nodding, even as missionaries referred to it as mongota.[99] As well, at some point before 1904 people around Lake Victoria came to associate peculiar swelling on the body—it is unclear whether exclusively on the back of the neck, or more widespread on the body—with illness that ended in sleeping and death.[100] Some also attributed the illness to the lubaale Kaumpuli, generating the talk that led European observers to report connections between sleeping sickness and that extant nosology of serious and widespread illness.[101]

As they discussed and defined mongota, Luganda-speaking Ssese islanders fit it within their experience of concurrent political change, making sense of illness and death in relation to the potential consequences of dwindling communities. H. T. C. Weatherhead recounted how, by 1904, missionaries "hear[d] it said that the English have brought this sickness by the 'magic,' that they may 'eat' the land. 'Has not the Government made a law that all uninhabited land shall belong to the English?' Therefore, they want to kill the people off the coast lands and islands that they may claim them."[102] Land tenure at the time oriented around paternal and agnatic relationships and social reproduction depended on access to land. Islanders knew that deaths among them and on the mainland on the scale that mongota caused could disrupt land tenure and fundamentally change such durable arrangements. With "the English" as the new players in the region claiming "uninhabited land," Weatherhead's Ssese informants reasoned that depopulation could only redound to the benefit of the British colonial regime.[103] That Weatherhead's informants also explained mongota as a sickness brought by "magic" by the "English" indicates that people fitted colonizers into cosmological and nosological systems where human malevolence wreaked widely felt havoc. The association between English presence and widespread mortality also underscores that people categorized mongota as something new—or, at least, significantly different in its scope and impact—and intimately linked with experiences of recent British arrival in the region.[104]

Within households and villages, people reconfigured life around mongota in several ways.[105] Illness and death triggered changes in mobility that we might compare to earlier responses to kaumpuli on the mainland, encouraging circulation away from areas where people were sick and perhaps also movement to consult lubaale shrines and powerful kubándwa mediums. On the Sseses, mongota began to erode remaining islanders' prosperity and livelihoods as it sickened fishermen and farmers, men and women, across the archipelago. Locally, people deployed strategies to mitigate mongota, setting the sick apart from the well, but also settling sick people together. Some might have acted similarly to nearby Ganda communities that in 1902 isolated the sick, avoiding smoking from the same pipe or eating together.[106] Missionary sources recount early recognition of the illness on the Ssese Islands in 1902–3 and particular steps taken to isolate, but also care for, the sick, such as settling a group of sick people together or lodging a sick person away from other homes and providing a caretaker. In November 1902, for example, CMS missionary Aileen Weatherhead wrote in her journal of a house that had recently been built around twenty minutes' walk away from the Bugala mission. This house was a space for the sick, not a preexisting home for particular people, and was notable enough that the Weatherheads took British researcher

Cuthbert Christy there directly when he journeyed through the area in search of cases of sleeping sickness.[107]

The case of a young man named Isaya, employed as a servant in the Weatherhead household, further illustrates how these processes of isolation and care might unfold. News of Isaya's illness came to the Weatherheads from other mission youth, who raised the alarm with a story of Isaya putting a pot on to boil and inexplicably falling asleep. Recognizing the signs of sleeping sickness, the Weatherheads sent Isaya away to his relatives. Soon, both Henry and Aileen Weatherhead reported, Isaya's relations on Bugala Island had "built a little house on an open space near the shore where others who have the disease live," and had designated an elder female relation to care for him.[108] This arrangement lasted for some time. Aileen Weatherhead journaled to her relations in England that they had sent Isaya a book to write in and some fishing line, that he might stay occupied; Henry Weatherhead later noted, "it took him six months to die."[109]

Ssese communities moved sick people out of households, Isaya's case suggests, relying on familial responsibility for each individual. Parallel sources on Bugala also indicate that efforts to avoid the spread of the illness coordinated at the village level as well. Fr. Ramond of the White Fathers Bumangi mission noted in May 1903 that "each of the major villages has an average of ten patients set apart to prevent contagion. Each patient has his separate hut where he was treated and fed by his relatives during the long months that the disease lasts until inevitable death comes to end his miseries. . . . During the last months of his painful existence the patient seems to lose the use of his faculties—he vegetates rather than thinks."[110] Ramond's account corroborates other contemporary accounts of the epidemic's initial demographic impact on younger members of the population, whose parents or relatives might yet have survived to help care for them. The villages that he and the Weatherheads described had apparently become a commonplace around Bugala Island at the time. At Buninga on the island's northern peninsula in the summer of 1903, the White Fathers' Bumangi diarist recounted that there were a number of such villages where "the *bamongota* were placed a little apart; everywhere they [the Bassese] built huts outside the villages."[111] Coordinated efforts to isolate the sick at the village level were likely the consequence of regulations issued by the kabaka's powerful regents in May 1902, who ordered chiefs to

> gather together all sick people. . . . Take them away to a place half an hour away from their house and build a shed on high ground to put the sick men in and set fire to the scrub near the

> house where the sickness was, one hundred yards on each side. . . . Food and water is to be taken to the sick people. . . . You, the chiefs must build the houses for the sick people to go in. Every chief is to see that someone gets to look after the sick. . . . Don't eat fish.[112]

This regulation from Kampala preceded British scientists' confirmation of the causative parasite and fly vector of sleeping sickness to colonial officials in April 1903, as well as concurrent suggestions to gather the *katikiro* (Luganda, chief minister) and principal chiefs to disseminate information to affected populations.[113] It significantly predated British colonial efforts to institute widespread bush clearance measures, depopulate fly areas, or control travel on the lake.[114] It provides, then, a sense of how Ganda authorities located the spreading epidemic within the existing political and public health landscape, with overlapping colonial, missionary, and Ganda responses to matters of health. The 1902 regulations asserted particular chiefly powers and obligations to maintain and care for the sick, balancing the management of those ill with the protection of those still well. Placing responsibility for providing food and water to the sick onto political authorities suggests that the regents recognized that chiefs might need to step in to ensure resources for sick people whose families could no longer provide for them, or whose social world had been changed by their illness. Regulations also speak to a sense of the spatial dimensions of the epidemic: where people lived, how they managed the environment around them, what spaces demanded attention, and what measures might be undertaken in place to impact the spread of disease. We gain, here, a sense of the practical distance that authorities could place between the sick and the well—a half-hour's walk away—and of connections made between the growth of bush and scrub around homes and the health of people living within them. The regulations fit within the historic aspects of chiefship to safeguard the kingdom's wider health, as well as within the prerogatives of the kabaka and chiefs to allocate labor and the use of land. Pertinent to the Ssese Islands, as we will see, was the injunction to move the sick to "high ground." This, along with a prohibition against eating fish also included in the regulations, targeted chiefly attention to people living around the lakeshore or along waterways. Further, the injunction against eating fish—which would effectively have had the impact of keeping people away from riverbanks and lakeshores—would have constituted a significant burden for Ssese islanders in both food security and economic activity.

Ssese approaches to mongota changed over time, particularly in the initial years of the epidemic, and both drew upon and expanded from historic precedents for mitigating illness. Initial accounts also indicate that Ssese populations, as with elsewhere in Buganda and the lake littoral, addressed mongota within frameworks defined by experience with other serious illnesses. Strategic separation from the sick was one aspect of Ganda approaches to those stricken with the spreading, swollen lesions and open wounds of *bigenge*, for instance.[115] During an outbreak of *kiddukano* (a diarrheal illness) in late 1904, affected people left their houses for the forest and markedly avoided the Bumangi mission and its sick people.[116] Distancing the well from the sick echoes how people had historically left places of illness temporarily during a visitation of Kaumpuli's power. But resituating bamongota, as occurred on the Ssese Islands, was not congruent with recorded responses to other widespread illnesses, suggesting innovation amid its widening impact. Strategies similar to those that might have arrested bigenge or kaumpuli ultimately would shift to more drastic measures as mongota continued to spread in the early twentieth century.

Let us take the Weatherheads' descriptions of how the relations of the sick on Bugala Island ultimately settled the sick near to one another, but also nearer to the lakeshore, as a starting point. Many Ssese islanders spent time on the shore regularly and men may have had shelters to use while fishing or drying their catch there—indeed, the lakeshore's ideal tsetse habitat of abundant moisture and thick vegetation had likely exposed many to fly bites and thus the disease's causative parasite. But permanent homes were typically in the islands' interior, on higher ground.[117] To locate the sick in smaller homes nearer the lake was to set them apart, but not to maroon them without access to basic necessities like food and water. Indeed, the designation of an elderly relation to care for the sick boy Isaya immediately signals recognition of diminishing capacity and the need for sustained care and indicates that families or kinship groups addressed the degenerative progress of the illness as they shifted allocations of time and labor that their sick kin now needed. The grouping of several "little houses" together might have allowed kin to share time, labor, and resources as they managed the needs of the sick or enabled people in different stages of the disease to assist one another. But, importantly, these "little houses" were places apart from more permanent homes. A photograph from 1906 of a "camp of the sick near Bugala" matches missionary descriptions of the kinds of habitations that Ssese islanders built for the sick.[118] Compared with contemporary photographs and descriptions of typical homes around Lake Victoria, these "little houses"—later marked as a "camp" by German scientist Robert Koch—differed markedly in their layout and emplacement from a typical family home.[119] While the exact location of this small settlement is unknown, several aspects suggest its remove from social and domestic spaces in Ssese society. Firstly, the houses are clustered tightly together and some are constructed roughly, of differing sizes; materials used to build them are scattered in front of their doorways. Piles of brush and low trees or shrubs appear to circle the group of houses and a well-worn path crosses in front of it. The settlement sits at the margins of clumps of trees and grassland, with ground rising away in the background in one direction; in the other, the lakeshore is also visible. Accounts of Ssese isolation practices are not consistent with regard to the distance that people might be set apart, nor do they discuss the meanings or implications of that distance, but this camp near Bugala appears to fit the instructions of the kabaka's regents to the topography and environment of Bugala village, and appears also to accord with past approaches to illness that affected many members of a community. Its remove from the settled geographies of village life sought to keep illness from affecting others. But its exposed location and its temporary materials also signal its unsustainability as a place of durable social life. This little camp was not a place where people could tend a vegetable garden, keep small livestock, or cultivate banana trees. Rather, it was a place to rest and to shelter as death came.

The early responses of isolation and separation that missionaries noted would have followed months of accumulated experience in Ssese communities. Here, the specificity of mongota must remain central: though drastic, it was not a fast-moving disease like, for example, *lubyamira*, a widespread illness that had circulated a decade prior.[120] Mongota made people nod or sleep, in a gradual decline, whereas lubyamira literally laid people (and cattle) down swiftly. Progress of trypanosomal infection—how fast signs like disrupted sleep, mania, or coma might emerge—are and were variable from one person to another. Levels of stress and fatigue, how regular and nutritious one's diet is, or whether a person experiences multiple exposures to a parasite (i.e., multiple bites from infected flies) are several factors that scientists assert can impact a person's immune response to the parasitic infection and the efficacy of that response.[121] A case like that of Isaya, a young man and a domestic laborer likely mobile and active around the mission's vicinity and through fly vector habitats, suggests that he would have been exposed to the parasite and ailing for many weeks, if not a few months, before he fell asleep while he was supposed to be minding a boiling pot. Settling people with particular symptoms in a particular

FIGURE 1.1. **Camp of the Sick near Bugala.** *Courtesy of the Robert Koch Institute, Berlin.* This photograph from Robert Koch's expedition photograph album shows dwellings of the sick on Bugala Island. The area's elevation and vegetation indicate that the "camp" sat nearer to the shore of Lake Victoria and at a distance from Bugalla village, similar to the "little houses" set aside for *bamongota* by Ssese islanders. Source: Robert Koch Institute Archives, Fotoarchiv 6105, Fotoalbum Koch in Sese, 6105036 (1906-07).

space shows that affected households and villages had generated collective responses to the illness as more severe signs appeared with greater frequency. It is very likely that this move was mediated by political and ritual authorities—chiefs, clan heads, perhaps healers or kubándwa mediums—given frameworks where elder kin and clan or village members were responsible for decisions with bearing on productivity and prosperity.[122]

After these early moves to gather and isolate people showing signs of mongota, approaches and capacities to deal with mongota began to shift. Fr. Reynès, journaling his July 1904 itineration around Bugala Island between Bumangi and Bugoma, walked past village upon village filled with the sick, visiting some in their homes; the disease, he found, was widespread.[123] Reflecting on the fourth year of the epidemic in 1905, Reynès noted that people preferred to be at home, and could find devoted care even among distant relatives; though the mission provided patients with salt, fish, and sometimes meat, patients would forego such "little treats" to be in their home and among kin.[124] For many, then, care concentrated in the home, with family and networks of kin in established domestic spaces. Caring for stricken relatives initially corresponded with gender and age. Patterns of early infection suggest that men, particularly younger and more mobile men, were first affected, followed by adult women. Thus, missionaries reported women caring for both a spouse and male relative, children caring for older siblings and fathers, and, ultimately, entire families coping with illness among adults and children.[125] Mongota's effects cascaded to touch more and more of the Ssese population, as the disease struck ever more adults whose livelihoods and social roles (as fishermen aiming for trade, or as women fishing to provide a household with food, for example) exposed them to the parasite's fly carrier.[126] As more people sickened, too, fewer were available to care for the sick, to cultivate crops, or to produce food. Missionaries at Bumangi, where five "improvised nurses" cared for patients in the hospital, reflected, "The disease still raging with the same intensity, we look with dread to the moment coming, sooner or later, where the survivors who are still healthy will not be able to feed and care for the sick, at least if God does not end this scourge soon."[127] Missionaries on the Sseses noted that the islands' population had gradually diminished, either because of mortality from the disease or people fleeing from it to the mainland. Fields around Bugala Island lay fallow by late 1905, though not wholly abandoned; months-long devotion to caring for increasing numbers of sick drew labor away from preparing fields and cultivating crops, necessary work to sustain households into the future. The islands, one priest observed, "resembled a great battlefield after a long struggle."[128]

Place-centered ideas about illness and health shaped new relocations and local mobilities on the islands. On an itineration around Bugala, Fr. Reynès found a man affected with sleeping sickness caring for his elderly mother, aunt, and wife (who was apparently also affected by smallpox), also sick, sheltered under a large tree.[129] Reynès assailed the local village chief's "inhumane" expulsion of a sick family out of their village, but I read here a shift in tactics and evidence of a sharp narrowing of the possibilities for coping with the sick within existing social relationships. For this family, separation from their village meant not a "little house" attended by an elderly relative among others similarly affected, but removal from hearth and home. To be sent away from home and village and into the forest in such a manner may have reflected a chief's efforts or a medium's advice to safeguard remaining villagers by encouraging abandonment of a home to which death had come—consistent with responses to visitations of kaumpuli in past generations. Removal also suggests a heightened gravity of the disease; the priest's presumption about the drastic measures of the village chief begs the question of whether anyone remained to care for the ailing family. Within lived experience of other grave diseases in the area, particularly those attributed to Kaumpuli and for which abandoning home was a practiced strategy, the possibility also remains that the family, too, played a role in leaving their home in an attempt, however desperate and futile, to evade illness and death that was understood to attach to domestic spaces.[130] Indeed, as illness became widespread, more radical moves occurred. A Bumangi priest noted that "those who it has spared have fled toward the beaches, thinking, as they do, them to be less murderous."[131] The Ssese abandoned homes and farms temporarily and perhaps permanently as the crisis widened.

Amid the worsening illness, both Protestant and Catholic missionaries continued to offer some assistance to the sick. But in 1905 Anglican missionaries evacuated to the mainland and the White Fathers alone remained at Bumangi and on the islands. Like their colleagues on the mainland, the White Fathers on Bugala built hospices, which they called hospitals, to house both early and advanced cases of sleeping sickness. One priest described how the people who came to the mission needed much from the men and women who cared for them there, requiring "everything that a mother would do for her child."[132] On Bumangi, appealing to the mission for care was for many a last resort: "If we had here all of the unfortunates seriously stricken with this terrible disease on the island of Sese only, their number would increase to five hundred at least; but we receive only those who are without relatives who are able to care for them at home, or *even worse*, the poor madmen who are no longer wanted by anyone and who are chased away pitilessly. The others stay at their places, receiving the necessary care, be it from a wife, be it from a mother, be it from some other relative."[133] Priests could only offer palliative care through the months-long course of the infection.

We do not know what specific therapeutic recourse Ssese islanders sought from healers or mediums, as it is collective responses, rather than individual efforts, that were most visible to observers who documented the early years of mongota. Assembly at one of Mukasa's major shrines, or offerings at smaller local shrines, would have featured in islanders' possibilities for relief and healing. Tracing responses to illness in the epidemic's first years, individuals and families initially separated the sick, but kept them nearby; worsening mortality would have decreased the resources and labor available to sustain this. As resources diminished and concern deepened, and as priests made known that their hospital would care for "sleepers," as they called bamongota, people came to the Bumangi mission. Some came because they had no surviving relatives to care for them, while others were brought by relatives perhaps unable to care for them further.[134] Some catechists apparently also expressed a desire not to be far from a priest on their deathbed. In all cases, the mission functioned not as a clinic, but as a hospice, because no treatment for sleeping sickness was available. Within only a few years of the disease's arrival at Ssese, these situations of last resort likely confronted the islanders more and more frequently.

And so, by 1904, a researcher on the British Sleeping Sickness Commission based at Entebbe wrote that "sleeping sickness is very bad on the islands," noting that the population on the lakeshore had been "almost completely wiped out."[135] As months went by at the larger White Fathers hospice at Kisubi on the Buganda mainland and the smaller, thirty-six-cot shelter at Bumangi, however, rooms full of the sick began to slowly empty as people recognized that they and their relatives could not be healed, and as those sick continued to die.[136] The White Fathers had only rudimentary medicines, with palliative care focused on keeping people comfortable and perhaps better nourished. The therapeutic resources of Ganda communities, too, might keep a person from suffering more intensely, but could not cure. The missionaries stood by, watching together with the Ssese as death made "dry islands" of the archipelago and as the population continued to diminish.[137] Fr. Reynés somberly reflected in a 1905 report, "No one here is under any illusion about the outcome of the disease; as it now stands, for our Basésé, to be sick is to be already dead. They no longer say: 'I am going to care for father, my . . . sick brother';—but rather: 'I am going to bury my father, sick with sleep!'"[138]

Flight from the missions was paralleled by other mobility, too, as illness became widespread. Migration, alongside mortality, diminished island populations. Responses varied by a person's social position or political connections, as well as over time, particularly as mortality increased: many who could leave did, traveling to the Buganda mainland. CMS missionary Hugh Savile noted in 1905 that Bugala's "influential Natives" left their "island properties" for the mainland, fearing the disease.[139] This may have particularly applied to prominent chiefs, whose political roles connected them more closely to Buganda.[140] Such mobility reminds us that elites, particularly those with familial and clan connections to the mainland—or perhaps with origins on the mainland particularly—had a greater range of possibilities to cope with times of dearth or illness. For most islanders, especially as capacities and resources dwindled amid prolonged illnesses in a household, marshalling resources to cross the lake for the mainland would have been extremely difficult.

This chapter has foregrounded the important social, political, and economic factors that shaped life and livelihood on the Ssese Islands in the late nineteenth century, paying special attention to the political and ritual dynamics around Mukasa's shrine and the powers that mediated contact between the Sseses and Buganda. Historical experiences of illness, misfortune, and healing offer a sense of the intellectual and strategic toolkit available to island and littoral populations. The examples of kaumpuli in the late nineteenth century illustrate how intellectual and social worlds accommodated changing dynamics and experiences of illness, as well as the emerging therapeutic diversity of the time. Central to dealing with illnesses that affected people around Lake Victoria were place-centered responses to widespread illness, such as new mobilities or reorientations to domestic spaces and surrounding environments. The advent of mongota catalyzed the deployment of diverse strategies, based on past experiences, to cope with the illness and death it caused. Initially, these took the form of broad social interventions that separated the sick from the well amid spreading sleepiness and weakness, focusing on the living spaces of affected and unaffected people. These shifts and departures were important actions to mitigate the impact of illness and death and demonstrate continuities in responses to widespread illness and death during an exceptionally disruptive and tumultuous era in littoral life. Such measures bear consistencies with those utilized in situations where an illness attributed to Kaumpuli appeared to spread unchecked. Later moves shifted the nexus of care back to homes and villages and also involved shifting engagement with European missionaries on the islands. The toolkit available to islanders was diverse and full.

But as kin and clan resources were exhausted, as the relief that missionaries could provide grew less attractive, and perhaps as amelioration offered by kubándwa mediums proved fruitless, activity bustled in another arena, as tropical medicine researchers across Africa trained their focus on sleeping sickness, its causative parasite, and its fly carrier. Intensive research had been ongoing since 1902 in a British-supported laboratory near Entebbe, where the pathogen and vector had been confirmed. In the next few years, as the situation on the Sseses grew dire, scientific familiarity with the disease grew more robust and research efforts expanded—as did competition and collaboration among tropical medicine researchers at work in the colonial sphere. Sleeping sickness, after all, was for European scientists the newest puzzle to solve, and its potential impact on the wide swath of Africa where tsetse flies could live meant that diverse colonial and administrative resources could be diverted to help solve it. And so, by 1906, as the Sseses became "dry islands," with widespread consensus about the cause and mode of transmission of this new scientific entity, human African trypanosomiasis, colonial and scientific energies shifted to the problem of prevention and control. It was at this point that a new set of figures entered the scene on the Ssese Islands, driven there by peculiar ambitions and bringing with them wholly novel demands and potential solutions for the dire circumstances there: the second German sleeping sickness expedition, led by esteemed bacteriologist Robert Koch. Through the expedition's work, Ssese populations would engage directly and deliberately with formal medical research for the first time and, in turn, shape the goals and parameters of that research.

2 Healing *Mongota*, Treating Trypanosomiasis

Research on the Ssese Islands

The shores of the great inland sea became the focus of international scientific concern with news of epidemic sleeping sickness in 1901. Scientists at a new British laboratory at Entebbe generated research on the pathogen and its vector, but stayed largely out of the fray when it came to developing disease control measures.[1] But British interests were not the only ones threatened by the epidemic around Lake Victoria—German officials would also grow worried for their East African empire and fund an expedition led by prominent bacteriologist Robert Koch. Diverse epidemiological and practical considerations drove Koch to locate his research expedition in British Uganda rather than German East Africa. As a result, the German colonial anti–sleeping sickness campaign had its origins on the Ssese Islands in 1906–7, rather than in German possessions. The German expedition's work on the Sseses at Bumangi and Bugalla sought to find a cure for sleeping sickness and develop a general, transferrable model for the isolation, treatment, and management of sleeping sickness patients. Though it would only marginally achieve some of these goals and fail completely at others, Koch's Ssese work set the agenda for the formal German *Schlafkrankheitsbekämpfung* (sleeping sickness campaign), which ran from 1907 until the outbreak of World War I. Paralleling the links between communities on the Sseses and other parts of the Lake Victoria littoral, research on the Sseses also stitched together more tightly the lives and deaths of people living on either side of the new Anglo-German colonial borders on the lake as their mobility became central to research practice and ideas about disease prevention. Koch's tropical medicine research on the Sseses and his system of "sleeping sickness camps" served as a catalyst and model for German colonial interventions around Lake Victoria and Lake Tanganyika.

I locate Koch's tropical medicine research in a therapeutic framework with mission medicine and Ganda healing to explore the nature of engagement from Ssese islanders and regional populations and its ultimate impacts. Contrasting previous work that focuses on avoidance of British colonial impositions in the region, this chapter examines initial engagement with this early, formal intervention and treatment-seeking by littoral populations in an effort to situate the camp in its distinctive context and understand Ssese perspectives on the crisis of *mongota*. It identifies new therapeutic mobilities, focused on camp-based treatments, among Ssese populations and others around the lake's northern shore. It situates the rapid influx of several thousand people to Bugalla and a sister site at the nearby Bumangi mission hospice in the context of mission medicine, regional mobilities, and the powerful healing forces located on the Sseses discussed in chapter 1. The Sseses were an ideal research site, but work there served as a poor model for a portable intervention: the historic experiences of Ssese and Ganda populations and dynamics on the Sseses facilitated the German expedition's work, but were both place-centered and deeply rooted in local politics and cosmologies. My examination of Koch's sleeping sickness research and the changing nature of Ssese engagement with his work shows how Ssese islanders shaped both the process of producing knowledge about sleeping sickness and the design and trajectory of subsequent German interventions.

TROPICAL MEDICINE AND TRYPANOSOMIASIS

In the early twentieth century, sleeping sickness was an urgent research priority for the young discipline of tropical medicine.[2] The British Sleeping Sickness Commission's successful identification of the trypanosome parasite as the cause of sleeping sickness and the tsetse fly as its vector established their laboratory at Entebbe as a hub for trypanosomiasis research after 1902.[3] The epidemic drew other scientists to the region, too, including German researcher and physician Robert Koch. Koch, the most prominent living scientist in Europe, would ultimately be a defining figure in efforts to prevent and control sleeping sickness in German East Africa around Lake Victoria and Lake Tanganyika, influencing wider debates within colonial research and public health efforts. Several contexts for Koch's work on the Sseses influenced the character of his research: his desire to engage with the field of tropical medicine, his enthusiasm for research in colonial Africa and his connections in German East Africa, and his interest in making an impact on drug development and therapy. Known for his pioneering intellectual work on the etiologies of anthrax, cholera, and tuberculosis in the late nineteenth century, Koch had also contributed significantly to the development of new research techniques and technologies in microbiology. Efforts to translate these achievements into therapeutic innovation, notably to develop a treatment for tuberculosis in the 1890s, however, had ended in embarrassment.[4] Participating in the young field of tropical medicine, for which sleeping sickness was the next big challenge, was a large part of what drew Koch to East Africa. Sleeping sickness represented an opportunity to stay current and relevant in modern scientific research.[5] His interest in sleeping sickness dated from first reports of the disease, and he had undertaken a wide-ranging research expedition in East Africa in 1904–5 that brought him into firsthand contact with both people sick from the parasite in mission hospitals as well as with British researchers at work in Uganda.[6] The 1905 foray had laid the groundwork for an expedition dedicated to sleeping sickness in 1906–7 that would bring Koch to the Ssese Islands.[7]

Koch's expedition was part of a wider research agenda in tropical medicine focused on trypanosomes, animating research in European institutions and research forays into Africa. Between 1901 and 1909, eight different expeditions sponsored by European governments and tropical medicine institutes fanned out across Africa, each lasting at least a year and some resulting in permanent laboratories or research stations.[8] Ultimately, Koch's expedition would span nearly a year and a half and range throughout German and British colonial territory. The researchers first followed a counter-clockwise circuit around Lake Victoria, hopscotching by government steamer and small boat between consecutive coastal stations and larger settlements, including the British port of Port Florence (Kisumu), the German East African port of Mwanza, and Ukerewe and Kome Islands.[9] Within weeks of his arrival at Lake Victoria in 1906, Koch had already made arrangements with the British colonial government to travel to the Ssese Islands for research.[10] In Uganda, Koch could engage as needed with British colleagues on the Royal Society's Sleeping Sickness Commission and in the colonial medical service but also work autonomously. Koch's work came after the basic etiology of sleeping sickness had been established but before prevention or treatment measures had been standardized.[11] Developing strategies and tools to treat active cases and prevent the disease's further spread were at the top of most agendas at the time, and particularly prominent for German authorities concerned about its economic impact on colonial development. For Koch, this meant developing widely applicable drug therapies. Immediately prior to settling in on the Ssese Islands, Koch spent a few days in early August 1906 with British colonial officials and sleeping sickness researchers at Entebbe talking about different options for sleeping sickness prevention. Entebbe lay a short distance north of the Ssese Islands, in an area itself hit hard by the epidemic. Records of the visit indicate that British medical officers, researchers, and administrators had also not yet come to a consensus about sleeping sickness prevention in Uganda—options included the evacuation of the lakeshore and moving the administrative capital away from Entebbe—and that Koch was aware of the fluidity of the situation there.[12] He visited the brand-new and well-equipped laboratory at Entebbe, built for the Sleeping Sickness Commission but as yet unused, but bypassed it in favor of the promise of "*Krankenmaterial*" on Ssese: "only [there] did the outlook exist to find sufficient medical material and at once the possibility to study the *Glossina palpalis* [tsetse fly]."[13] And although he insisted to then Ugandan Senior Medical Officer Aubrey D. P. Hodges that he had no hope for a drug that could treat or cure sleeping sickness, his subsequent work on the Ssese Islands soon became famous—and later infamous—for claiming to do precisely that.[14]

The primary focus of therapeutic experimentation for sleeping sickness at the time was arsenic-based chemicals, particularly a drug called atoxyl. Atoxyl was derived from organic arsenic, first promoted as a drug in 1902 and first used experimentally to kill trypanosomes in 1905.[15] Koch's expedition carried stores of atoxyl to East Africa; his experimental vision for atoxyl fit with research practices used by other scientists at work on sleeping sickness expeditions in Africa and in European research institutions, evaluating it as a durable cure and treatment. While its long-term efficacy and safety were still open questions, recent research suggested that atoxyl still had potential as a powerful pharmaceutical tool.[16] Within a brief period between 1905 and 1910, several European tropical medicine institutions, in cooperation with metropolitan governments, sent expeditions to the African "field"; each worked with the arsenic compounds and chemical dyes that dominated experimental chemotherapies at the time. Viewed in comparative perspective, many methodological, theoretical, technical, and ethical continuities emerged among researchers using atoxyl, despite variations in local colonial approaches to preventing sleeping sickness.[17] British scientists at Entebbe, for example, used a dye called "trypan red,"

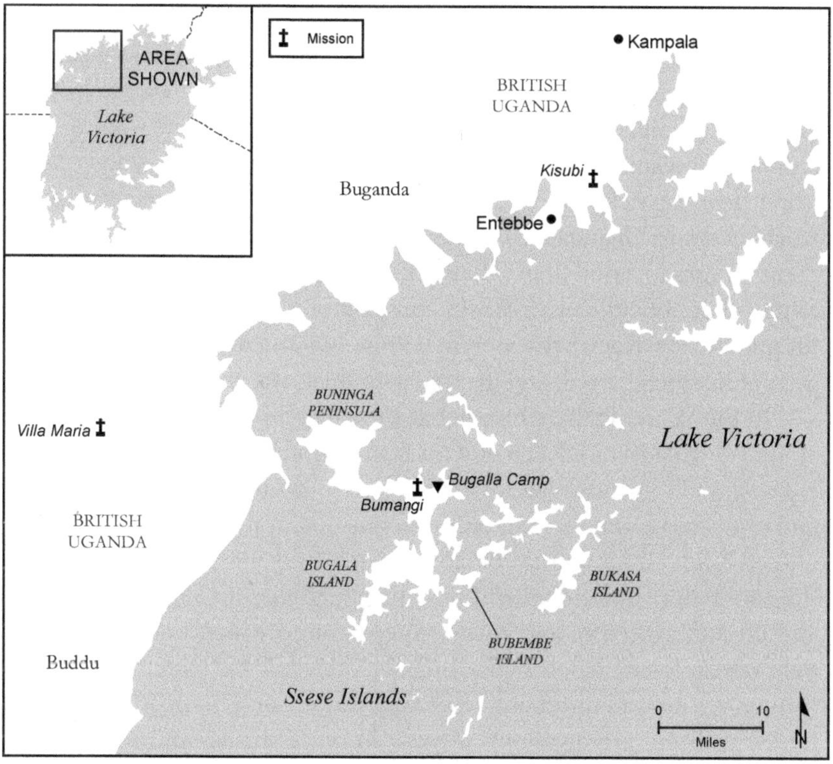

MAP 2.1. The Ssese Islands. Map by Brian Edward Balsley, GISP.

arsenic, and atoxyl on prisoners in Uganda regularly throughout 1904 and 1905.[18] The Liverpool School of Tropical Medicine furnished its doctors at work in West Africa with atoxyl to test its effects beginning in 1905; the French sleeping sickness expedition that worked in French Congo from 1906 to 1909 also tested atoxyl.[19] Belgian scientists at the time also used atoxyl to treat patients in Belgium and conduct research in the Belgian Congo.[20] Each used arsenicals and dyes in African patients, often without certainty of these chemicals' safety from animal experimentation.[21] While some experiments were promising, the trypanosome parasite at which the majority of doctors aimed seemed as yet impossible to kill permanently in its human hosts. But the potential of atoxyl, and the model of drug prophylaxis and cure, would remain compelling to Koch and define the nature of his engagement with Ssese populations.

THE BUGALLA CAMP

Robert Koch and his research expedition arrived on Bugala Island, the Ssese Islands' largest, in August 1906. Koch's sleeping sickness expedition—now numbering six German men and several "Baswahili" (coastal African men) who worked as laboratory assistants, guards, porters, cooks, and servants—found a grim situation that was, for the purposes of their research, ideal. The expedition's research on Bugala Island focused on a camp built around the abandoned Anglican mission at Bugalla village and the existing sleeping sickness hospice at the White Fathers Bumangi mission. Koch's work in Uganda also placed the expedition in the thick of African communities in a period of intensifying crisis and ongoing, extensive mortality.[22] Scientific documents and images produced during the expedition illuminate how Koch's research fit within, but also diverged from, existing ways of healing the sick and dealing with the crisis of mongota.

In the White Fathers hospital at Bumangi and in the scattered farms around Bugala Island, Koch and his team found scores of people in different stages of trypanosomal infection. Two mission sites, situated at either end of a "broad straight road cut from one end of the island to the other," offered Koch buildings suitable for research work and lodging.[23] Here was the abundant *Krankenmaterial* that the aging scientist so ardently sought and the infrastructure and relationships around which he could build a field research site. Koch chose Bugalla village as his base, assembling a camp there around the abandoned Anglican mission. At nearby Bumangi, he found some arrangements already in place for *bamongota*, here described by a priest: "This hospital consists, first, of one large earthen house, big enough to contain 36 sleeping places, a building reserved for the sick whose strength is still sufficient to walk around a little bit; 2nd, a dozen huts of reeds where we find those who have arrived at the final period of the sickness; 3rd, a hangar 20 meters long, which serves as a chapel; 4th, two houses where the male and female nurses live."[24] The expedition staff quickly set to work. Dividing the sick people they found into early and advanced cases, Koch and the team split up, with two men working among advanced cases in the Catholic mission at Bumangi and Koch and the others working with ambulant cases in the old buildings of the Anglican mission at Bugalla village. They soon had nearly a thousand people located in or around the two sites; by late October 1906, two hundred and fifty people stayed at the Bumangi mission and another eight hundred were in treatment at Bugalla.[25] German researchers worked in tandem at the two sites, less than an hour's walk apart.[26] The arrival of the German expedition and the establishment of their "Bugalla camp" led to a redistribution of the sick, according to the expedition's research priorities and the capacities of the White Fathers mission at Bumangi. Generally, most advanced cases were under care of the missionaries and less advanced, more ambulatory cases were at Bugalla. The Bumangi hospice, which Koch referred to as a *lazaret* (infectious disease hospital), housed advanced cases for "clinical study," while at Bugalla he could closely monitor the interaction between trypanosomes and experimental drug doses (and the impact of medicine on less-sick people for whom it might register more obviously).[27] The Bugalla camp's primary function, as a site of research, was to evaluate potential chemotherapies, particularly the use of atoxyl. Organizing and ordering of people in the Bugalla camp enabled Koch and his team to conduct experiments on the efficacy of several drugs and chemical dyes in killing trypanosomes in the human body, fitting within a growing contemporary tradition of colonial pharmaceutical experimentation.[28]

The German expedition's engagement with the Bumangi missionaries paralleled and augmented an approach to mission medicine by the British colonial administration that relied on missions to provide care and expand medical infrastructure. Amid the sleeping sickness epidemic, both the Catholic and Protestant missions more broadly provided a crucial stopgap in the Uganda Protectorate government's ability to survey for and nurse the sick, later recognized by the government with small monetary assistance.[29] For their part, White Fathers missionaries understood collaboration with the German researchers to redound to the benefit of their parishioners and worked to maintain a smooth relationship. German reliance on the missionaries' existing relations with Ssese populations to facilitate their work was explicit. "They ask us," Father Delévaux explained to a superior in October 1906, "to use our influence with the natives to encourage them to come to have themselves treated and we have fully succeeded." Remunerations for the mission's cooperation were direct, he noted, with a "compensation of 100 Fr. per month for the sick" functioning as "rent" and viewed as an opportunity to provide better nutrition for the sick.[30] As the missionaries encouraged local islanders to utilize German doctors' proffered treatments, they grafted the expedition's research-oriented biomedicine onto their own traditions of treatment and care, expanding once more the therapeutic resources that Ssese people could use. Their successful "influence" and encouragement relied on the relationships and obligations that missionaries had mutually developed with the Ssese in years past, through trade and gift-giving, social interaction, healing, and the common survival of the ordeals of illness, famine, and the past decades' fractious religious wars. At Bumangi, the priests' connection to the German team was also clear and close by—they shared the mission's spaces with the scientists and provided them with lodgings and work rooms, frequently sharing special meals as well. Although German researchers generally relied upon both the Catholic and Anglican missions' histories and connections with the surrounding population, circumstances at Bugalla were quite different: expedition members directly replaced the recently departed occupants of the Anglican mission. When the expedition co-opted the Bugalla mission site, it also removed the imperative of care and treatment from the sole influence of nearby Catholic missionaries by building its own work rooms and lodging for the sick.[31] Grafting its research spaces onto the mission buildings at Bugalla and commandeering an extensive space around the site, the German expedition carved out new spaces for biomedical research.[32]

The Bugalla camp fit into a Ssese landscape layered with historical meaning: political power, social connectivity, and littoral mobility intersected there. Bugalla's village and mission sat astride a high ridge that ran the length of the Sseses' largest island, in an open, airy space with a clear vantage of the lakeshore.[33] Bugalla village was a center of political authority on the archipelago after the newly installed *kweba* (a title related to the Ganda *gabunga* and used by Europeans to name the provincial chief himself) built his seat there in 1900; with a British colonial station on a nearby spit of land near the lake as well, it soon also became a hub of Anglican energies.[34] The permanence of Ganda political authority and the missionary presence registered in Bugalla's built environment. The kweba's residence was a large stone home, with further reed-and-thatch buildings around it; a stone church dominated the mission site, with wood-and-thatch buildings dedicated as missionary homes and kitchens and a mission school nearby.[35] Between Bugalla and the lake lay a long and wide slope of tall grassland, broken occasionally by clumps of trees; the abundant grassland provided materials for roof thatching as well as the construction of smaller homes and buildings.[36]

Koch's map of the Bugalla camp from November 1906 delineates discrete spaces for the team's lodgings, scientific and clinical work, and living areas of the sick. The camp occupied a large, cleared space around

FIGURE 2.1. *(opposite)* **Sketch Map of the Bugalla Camp.** *Courtesy of the Geheimes Staatsarchiv-Preussisches Kulturbesitz, Berlin-Dahlem.* This map of the Bugalla camp accompanied Koch's "Bericht über die Tätigkeit der Schlafkrankheits-Expedition bis zum 25. November 1906," which was circulated to imperial and Prussian officials in Berlin. The former Anglican Church Missionary Society mission's buildings (labeled "Ev. Mission") were the core of the camp, which also included various homes and barracks for the sick and housing for researchers. The work tents and treatment area shown in figure 2.3 are at the map's center. Bugalla village was on map left; the slope toward Lake Victoria begins at map right. Source: Geheimes Staatsarchiv-Preussisches Kulturbesitz, I HA. Rep. 8, no. 4118, R. Koch, "Bericht über die Tätigkeit der Schlafkrankheits-Expedition bis zum 25. November 1906," 25 Nov. 1906, 129.

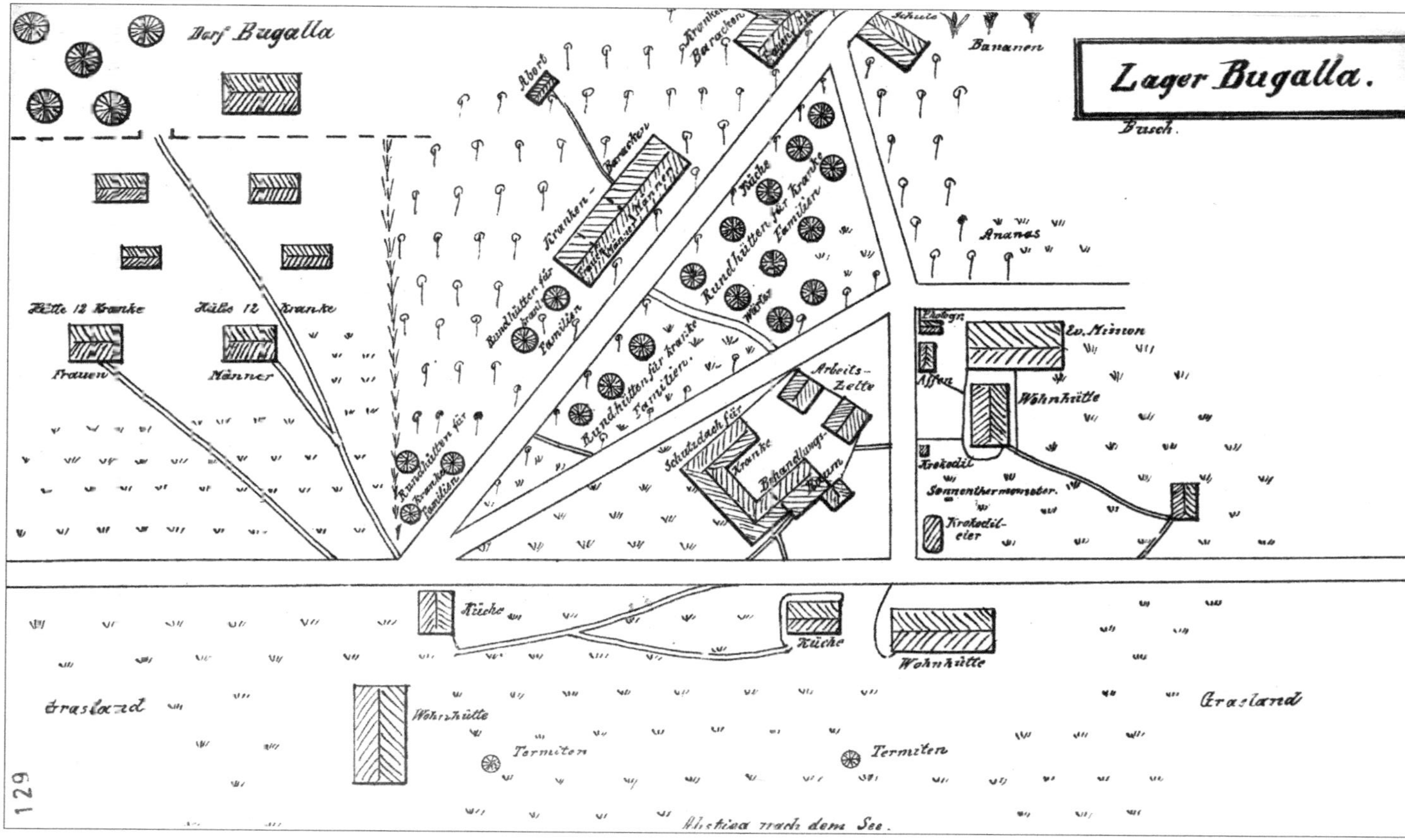
Lager Bugalla.
Busch.
Dorf Bugalla
Bananen
Ananas
Abort
Kranken-Baracken
Hütte 12 Kranke
Frauen
Männer
Rundhütten für kranke Familien
Küche
Rundhütten für Kranke Familien
Wärter
Arbeits-Zelte
Schutzdach für Kranke
Behandlungs-raum
Ev. Mission
Affen
Wohnhütte
Krokodil
Sonnenthermometer.
Krokodil-eier
Küche
Wohnhütte
Grasland
Termiten
nach dem See.

the former Anglican mission's buildings, which included the missionary's house, a church, and a school. It sat in the space between the mission, chiefly residences, and Bugalla village on open grassland above the lake. Local authorities—the kweba and the subordinate local chief, Semagala—facilitated, sometimes grudgingly, the delineation and construction of the camp's spaces.[37] The expedition pitched two "work tents" that formed the focus of the camp's clinical work, while the site's most secure building—the stone mission itself—housed the expedition's delicate research apparatus. Alongside the mission buildings, the German expedition staff apparently had the cooperation of the kweba to have round huts built nearby for families, as well as "a type of double barrack," to house individuals separated by gender.[38] The "large hospital operations" of the camp required significant construction to build the camp.[39] Near the mission-cum-laboratory, further outbuildings housed photographic equipment, research animals (primarily monkeys, juvenile crocodiles, and crocodile eggs), a thermometer, and lodging and kitchens for German staff. The two work tents faced a cleared area where patients awaited examination and treatment—an early photograph shows smoothly pounded dirt and a narrow, shallow trench around a square perimeter before the tents, around which people sat or laid.[40] In short order, this open space was fenced in fully and bounded on one end by a large, open-air banda that extended in a rough C-shape facing the work tents. The banda had been built soon after their arrival, Koch reported, to accommodate the hundreds of people who "streamed into the camp each day" and needed shelter from the frequent and heavy morning rains. Within the camp, but particularly under the open-air banda, Koch wrote, people could be "examined, registered, injected and punctured"—all the doctors' interactions with them under one roof.[41] These were orderly spaces, with scientists and patients each in their place and clear boundaries between them; the movement of patients was controlled by African orderlies who manned gates, assisted with research, and oversaw patients' lodgings. Apart from fenced-in areas around the treatment spaces and the mission, the camp's environs were open and people moved freely around.

The sick and their relatives began to settle near the camp, building temporary homes on the grasslands nearby, an activity that the expedition encouraged. People built small dwellings clustered together on land cleared of brush and grass, closer to the mission and village: a "provisional" camp, as the caption in Koch's expedition photo album reads.[42] The photograph's caption details the cleared location, ideal for the purposes of establishing distance between tsetse habitats and human populations, suggesting that the encampment may have been created concurrently with the early arrangement of the Bugalla camp, when expedition members became involved in locating the sick. While the "provisional camp" recalls the "little houses" near the lakeshore set up by the Ssese for their relatives in the early years of the epidemic, as well as a camp of sick people that Koch documented elsewhere on Bugala, it is also marked by some key differences from other sites.

Distinct from the other photograph (figure 1.1) of a settlement near the lakeshore, this camp has larger structures and shows many people of different ages. Here, the sick settled in a way that located them at a distance from homes and farms, in collective isolation from village life—the small, irregular homes of grass thatch are situated close together on the open grassland, with none of the typical features of Ssese domestic life in sight. But the photograph indicates that this larger cluster of homes was in proximity to the new therapeutic and material resources that Koch and his team might offer. This "provisional camp" is important for understanding not just the Bugalla camp's early days, but also its future life. It portends the new local mobilities that, fitting with extant Ssese responses to sleeping sickness, would shape the camp's fortunes. Where Protestant Ssese might have, in years past, sought assistance from Anglican missionaries by beckoning them to their own homes for advice or by circulating past the mission, people now relocated to the camp's perimeter. These local movements to the Bugalla camp and the Bumangi hospice were also accompanied by wider regional treatment-seeking mobilities on the northern littoral of the lake. Indeed, in late 1906 and early 1907, thousands of people circulated through German research sites on Bugala Island.

People's decisions to come to Bugalla or Bumangi involved material and economic considerations, some of which would ultimately affect the duration that people stayed to use German medicines. The crowds gathered at the Bugalla camp also represented the use of time, labor, and perhaps significant resources: people had to travel, sometimes great distances and on a potentially perilous journey, to get there. This travel fit into wider regional mobilities but was notable to observers at the time for its vigor and volume. Travelers from farther afield tapped into established long-distance routes overland on Buganda's main roads and across the lake. Those who came from greater distances—Buvuma Island, the Buddu coast, Kiziba—were transported across the lake on large canoes, perhaps conveyed partway in the smaller dugout canoes also in use.[17] Each journey would have represented the deployment of material resources (such as supplies of food for travelers and compensation for rowers) as well as the organizational energies of the sick and any family members who accompanied them. A group that arrived at the Bugalla camp from Kiziba, a kingdom in northwestern German East Africa with historic ties to the Sseses, suggests the scope of these efforts. In April 1907, fifty-eight Ziba people arrived at Bugalla from Kiziba, of whom twenty-eight were prospective patients and the remaining thirty were rowers and caretakers of the sick; the size of the group and number of rowers suggests travel on several larger canoes. Such a group would have required provisions for the sick and their caretakers, provisions and some form of compensation for the rowers, and organization and leadership of the canoes and their crews; the boats may have carried other goods or people wishing to travel to the Sseses as well.[44] Another large group of thirty people, six of whom were ill enough to not survive a return journey, traveled to Bumangi from Busoga around the same time.[45] Once on the islands, some people debilitated by sleeping sickness arrived at Bugalla in net hammocks slung between two people, or were helped along with the support of stronger relatives. Some may have traveled alone; some appear in photos to have traveled with their children or younger relations.[46] Coming to the camp required much from populations already in dire straits, but thousands made the journey.

FIGURE 2.2. **Bugalla: Provisional Camp.** *Courtesy of the Robert Koch Institute Archives, Berlin.* This image, taken in the fall of 1906, shows dwellings built by people gathered nearby the Anglican Church Missionary Society mission site in a "provisional camp" as Koch's expedition began its work at Bugalla. The open grasslands around the former mission at Bugalla and near Bugalla village, where Koch based his sleeping sickness expedition, soon became home to many people seeking treatment. Lake Victoria is in the background; a British fort was located on the peninsula at upper right. Source: Robert Koch Institute Archives, RKI Fotoarchiv 6105, Fotoalbum Koch in Sese, 6105037 (1906-07).

To identify suspected cases of sleeping sickness, Koch and his colleagues were particularly interested in patterns of sleep disruption and a generally healthy versus wasting, emaciated body. But they screened potential patients with an eye on more subtle signs: tremors in the tongue or limbs, elevated body temperature, unsteady gait, pallor and health of the skin, and the presence of swollen lymph glands on the back of the neck (Winterbottom's sign) and elsewhere. The researchers selected "particularly characteristic cases" to be admitted into the treatment regimen, from among the hundreds of people who initially appeared at the camp.[47] People who came to the camp were, in series, questioned, enrolled, palpated, examined, and triaged. Some were given shelter in or around the camp. Many had some area of their bodies punctured—a fingertip, an earlobe, the back of the neck—to extract blood or lymph for laboratory confirmation of the presence of trypanosome parasites. Some subsequently received injections or took medicine by mouth. Doctors took temperatures twice daily. The desired process of bringing people into the camp, from the perspective of German scientists, was a linear one. People would be integrated into the structure and order of laboratory-focused research and clinical observation and then followed through courses of experimental treatments for sleeping sickness.

Newly enrolled, punctured, and diagnosed people received injections of atoxyl, as culmination of a person's integration into Koch's research regimens on Bugala Island.[48] Koch had anticipated using a significant amount of the drug when planning the 1906 expedition, as he sought to test different doses of atoxyl to try its effectiveness as a treatment and develop a therapeutic protocol.[49] At Bugalla and Bumangi, Koch's team typically injected atoxyl between a person's shoulder blades; injections were painful and sites sometimes became infected.[50] The atoxyl regimen also involved monitoring the impact of atoxyl on trypanosomes, including the short-term efficacy of atoxyl, how quickly trypanosomes disappeared from lymph, and the long-term success of the drug in killing the parasite. Researchers sometimes changed injection sites between the back and closer to the inguinal lymph nodes (in the groin) and performed regular gland punctures at the respective sites to test whether atoxyl affected only the trypanosomes found in glands closest to the injection site or worked systemically.[51] Throughout, atoxyl injections were accompanied by further, regular cervical gland punctures and blood draws, as well as temperature measurements; painful and invasive procedures were regular. In October 1906, Koch recommended a typical treatment regimen with atoxyl on two

FIGURE 2.3 **Interior of the Bugalla Camp.** *Courtesy of the Robert Koch Institute Archives, Berlin.* This image from the Bugalla camp shows the primary area of interaction between researchers and patients, a cleared and fenced-in area adjacent to researchers' laboratory tents. People moved through this part of the camp during an initial evaluation for sleeping sickness and, subsequently, for atoxyl injections, gland punctures, and temperature checks. Accompanying text notes that patients in the foreground have late-stage sleeping sickness. They have thermometers inserted under their arms and are being held upright; the thin wooden blocks with printed numbers are their *vipande* (identification tags). Researchers' two work tents are in the background and a fence bounds the area's perimeter. Source: Robert Koch Institute Archives, Friedrich Karl Kleine Nachlass, Box 2.

consecutive days, with a seven-day pause between "double injections," continuing for at least two months.[52] He then argued that relatively higher doses given further apart—every ten to twelve days—would be more effective for killing the tricky trypanosomes. By early 1907, his regimen regularized to a ten-day pause between consecutive dosage days.[53] The regimen itself was subject to significant experimental variation; Koch varied the dosage of atoxyl given in each injection from an initial dose of 6 milligrams to a final dose of 500 milligrams, briefly ranging as high as 1 gram per injection.[54] The schedule of injections in a course of atoxyl, as well as the need to follow individuals throughout treatment, meant people had to cycle regularly through the camp's treatment spaces over a period of two to three months in total following their admission. Connection to the camp, then, continued over weeks and months.

"IT IS A FEVER": THERAPEUTIC MOBILITIES AND GERMAN TREATMENTS ON THE SSESE ISLANDS

Koch's reports and correspondence from the early months of work on Bugala Island are shot through with a heady enthusiasm for atoxyl's potential as a mass treatment for sleeping sickness. Atoxyl apparently had an initial impact on some of the Bugalla camp's patients, as the drug killed some of the parasites affecting the body's systems. After eight to ten weeks of injections, Koch reported, some people who had been suffering from disruptions in sleep, impaired mobility, weakness, and incontinence appeared to become more wakeful, move with greater ease, and regain their appetite.[55] Trypanosomes seemed to disappear from the bodies of people treated between September and November 1906 and those few parasites that remained, he asserted, seemed to have a "defective appearance" under the microscope.[56] Malignant side effects were not yet on researchers' radar and Koch claimed that these initial results came with no signs of adverse effects from atoxyl. Less promising were impacts of atoxyl in the severely ill, whom Koch surmised had extensive and irreversible changes to the central nervous system. The results of initial treatment with atoxyl encouraged the scientists, missionaries, and evidently the Ssese people in and around the camp—each had a material improvement that they could associate directly with treatment.[57]

Word of these first impacts of treatment spread among communities dealing with mongota around the Ssese archipelago and the nearby Buganda mainland, percolating through colonial networks as well. Koch remarked in late November 1906 that "more and more patients are streaming in from the Ssese Islands and also from a greater distance on the mainland." The Ugandan governor, Henry Hesketh-Bell, also reported to the Colonial Office that "natives suffering from sleeping sickness are flocking to [Koch] in hundreds."[58] It is tempting to read the governor's and the scientist's reports of mass movements in response to the German expedition's work as pure hyperbole. British authorities interpreted mobility directed at the Bugalla camp as enthusiasm for Koch's treatment, validating imperial rhetorics of European biomedical sophistication and therapeutic superiority and marrying these to the good optics of African popular embrace. Visible approval, in this vein, also injected a measure of optimism into an ongoing epidemic that otherwise looked exceedingly deadly and difficult to arrest. Positive colonial and medical reports of newly vigorous mobilities oriented toward Bugalla are congruent with independent accounts in missionary sources, however, indicating that circumstances of treatment-seeking were somehow changing in late 1906. While later narratives of sleeping sickness interventions around Lake Victoria focus on evasion and avoidance, diverse evidence shows that people engaged with urgency with the Europeans at Bugalla in this early moment.[59] These urgent and assertive mobilities demand consideration within Ssese and Ganda intellectual worlds and therapeutic logics in order to understand what animated them, how they changed, and what impact they had on Koch's research and thus on German colonial public health measures.

The Bugalla camp benefited from the local, intimate history of missionary medicine and the recent history of efforts to isolate bamongota to safeguard the wider population. Movement to the camp and mission and into the German treatment regime indicates that people folded camp-based and camp-associated activity into their understandings of illness and past experiences of healing and intervention. Overlapping with these locally focused ideas and activities on the Sseses were broader, regional historical dynamics of place-based healing and spirit mediumship. While British and German observers viewed their medicines as the attraction that drew so many people from the islands and mainland to the German camp, the Ssese Islands had been a center for healing long before the Germans arrived. These histories and contexts shaped engagement with the camp, and thus also influenced Koch's ideas about isolation camps as the portable model around which German anti–sleeping sickness measures could focus.

Records from Bugalla and Bumangi indicate that the initial arrival of people from the Ssese archipelago was accompanied by arrivals from "all around Buganda" and that total numbers of people awaiting treatment at the camp rose precipitously between September 1906 and January 1907.[60] Mission diaries, scientific reports, and personal correspondence give a clearer sense of the volume of people who arrived over time. Upon his arrival at Bumangi in August 1906, Koch found a mere dozen people in the White Fathers' hospice; several more people may have been living in isolated encampments around Bugalla or on the lakeshore. By mid-October, the expedition had enrolled 550 people at Bugalla, and, by month's end, 700 people, still not meeting the demand of 800 people Koch estimated had gathered daily around the camp. At Bumangi, another 250 circulated through a doctor's observation locally and 45 stayed in the hospice.[61] By late November, 79 people stayed at Bumangi and 907 were treated at Bugalla, and in December 1906 site totals for each were 300 and 1,200 patients respectively.[62] By January–February 1907, the Bumangi treatment tally had more than quadrupled in two months, to 400, along with the 1,500 people treated at Bugalla.[63] The uptick in admissions in November was in part due to people arriving from further afield: of the roughly 1,000 people then in the expedition's purview, half were from Bugala Island and nearby, 200 from elsewhere in the archipelago (chiefly Bukasa), and 300 from the Buganda mainland and more distant islands such as Buvuma and Kome.[64] Missionary diarists at Mitala-Mariya on the mainland to the northwest and the mission at Bumangi remarked on distinctive mobility among people affected by sleeping sickness or mongota that directed "sleepers" to the Ssese Islands in the same time frame as Koch's autumn reports and Hesketh-Bell's comments.[65] By April 1907, Koch's camp had also received a group of 58 people from Kiziba, in German East Africa; many Ziba had historically lived on the northern part of Bugala Island in fishing camps.[66] At their peak in early 1907, the Bugalla camp and Bumangi mission hospice counted nearly 2,000 people in their logs, with still others gathered nearby. Priests at Bumangi would later reflect on the large numbers of people gathered at Bugalla: some 2,500 "in the beginning," that is, in the autumn, and as high as 3,000, which would have clearly outstripped the expedition's atoxyl supply and the camp's carrying capacity.[67]

News of treatments available on Bugala Island (then called simply "Sésé") spread throughout the region and drove urgent mobilities. "Word has spread like wildfire that a cure for sleeping sickness was discovered on Sésé," the Bumangi diarist wrote, "and so a wretched mob descend[ed] on Sésé to go and drink the cure for the disease" during the month of November 1906.[68] As news of a new treatment around Bugala spread, a Bumangi priest wrote elsewhere, "from Buddu, Busoga, and all of Buganda, it is a mad dash of panicked people."[69] People also continued to come directly to the Bumangi mission and to the White Fathers hospice there, straining their capacities to organize meat, water, and shelter. "We receive far too many people, especially from outside," the diarist wrote on November 15, 1906, noting simply, "It is a fever."[70] Where Koch and Hayes-Sadler perceived enthusiasm, the priests read desperation and a departure from typical patterns of mobility that they knew from labor migration or trade. But the "fever" for treatment had a logic within patterns of treatment-seeking and traditions of coping with widespread illness in the region, where travel to Ssese Island shrines had historically offered access to powerful resources. The crush of people seeking healing served not as an endorsement for the German expedition's offerings above all other options, but as evidence of an aggressive utilization of novel regimes and technologies, which occurred in modalities with strong local resonance, among other options.

The focus of movements to Bugala Island was the different medicines and treatments available at the Bugalla camp and the Bumangi mission outstation. Medicines circulated through interactions with the German doctors and their Kiswahili-speaking assistants, also sometimes mediated by White Fathers missionaries. By mid-October, when the Germans had enrolled over five hundred people and had begun atoxyl injections, they had generated a growing cohort of people experiencing and observing the treatments they gave. Those treatments were diverse: alongside injections and related blood draws, gland punctures, and temperature checks they also, apparently, dispensed oral quinine regularly. While atoxyl's impacts were not consistent and were not durable, when positive, they were dramatic. Initial injections, as the drug killed some of the trypanosomes in a person's body, rendered some temporary improvement in the strength, mobility, or wakefulness of some patients. Thus, after a series of double injections—a point arrived at in October or November for the first patients enrolled in August and September—a difference in ability and temperament might have been evident to some patients and their relations. Missionary observers remarked frequently, if cautiously, about the "ameliorative" impacts of atoxyl in these first months, signaling a difference in patients after atoxyl injections.[71] The "fever" for treatments at Bugala in October and November was intuitive for them, too.

German tallies and initial evaluations provide a sense of processes of admission, examination, and treatment from the researchers' perspective. While the detailed personal stories of individual treatment-seekers are largely unavailable in extant records, I read scientific materials for implicit and explicit narratives of mobility and motivation, visible in reports of travel as well as accounts of an individual's provenance, occupation, and duration of illness. Taking treatment-seekers' mobilities seriously shows how German tropical medicine research occurred within interlacustrine intellectual worlds and diverse therapeutic practice. Movement to and from Bugala Island and Koch's Bugalla camp fit into historic, wider patterns of treatment-seeking in Buganda as well as activity specific to the Ssese Islands. But understanding individual experiences of Ganda intellectual worlds and their therapeutic frameworks, particularly amid the crisis of mongota on the Ssese Islands, is a particular challenge for scholars, given the forced depopulation of the islands after 1910 and its disruptions to social life and administrative records.[72] Ganda and Ssese narratives, which would help to clarify people's processes of examining and reinterpreting colonial biomedicine, are thin in archival records.[73] But placing accounts of patient responses in dialogue with records of mobility makes Ssese processes more visible in the ebb and flow of engagement with German researchers.

Ssese and Ganda patients understood the impacts of German treatments on their own terms, not necessarily as a direct correlation to specific injections, but associated with different elements of their encounters with the German expedition. Patients' intellectual work around camp-based therapies focused on two particular elements that illuminate this process: first, the drinking of medicine (versus having medicine injected), and second, the experience of gland punctures. Patients' historical experiences with Ganda and mission-based healing traditions involved extended social interactions, engagement with spiritual forces, and the use of specialized, powerful medicines and objects. Their experiences provided flexible and expansive definitions of healing and therapy within which people located aspects of camp-located "treatment" that led to a change in a person's illness. Thus, while German doctors noted that their Ssese patients requested cervical gland punctures and touted their embrace of atoxyl injections, local priests simultaneously noted an alternative focus of enthusiasm: people came to "drink the cure for the disease." Oral doses of bitter quinine and the twice-daily ritual of temperature checks in the armpit were healing them, people told the priests, and indicated that they felt "injections were secondary."[74] This represented the reality of German medicine that many experienced: drinking quinine was the most consistent element of therapy for some patients. Inconsistent access to sufficient amounts of atoxyl meant that doctors elected to inject only certain patients, according to their value for consistent scientific results, and offered regular drinks of quinine to the rest.[75] Initially, some patients had also expressed to Bumangi priests that they preferred to drink medicines rather than experience painful atoxyl injections.[76] Concurrently, patients also experienced gland punctures of the back of their necks, with and without also receiving injections in their backs. People then made arguments about the therapeutic value of these cuts and punctures on swollen areas. Koch noted that the "small incision" of a gland puncture had become an "integrated part of the treatment" of atoxyl injections.[77] Koch's reports also noted that people's focus on unusual swelling on the neck was twinned with a sense of serious illness that had preceded his own research; his patients had a more nuanced sense of etiology and related intervention than he understood.

How did people seeking treatment from the German expedition at Bugalla or Bumangi understand doctors' actions and therapies with regard to other experiences of healing interventions? Here scholars must work around lacunae in the historical record and in sources about healing and therapy, for although sleeping sickness and mongota caused widespread death, drove new mobilities, and motivated strategies of mitigating and coping with illness, contemporary sources on specific Ganda therapies that addressed this problem are muted. The dense accounts of Ganda life assembled by Kagwa and Roscoe, for instance, make no mention of mongota or how people understood or coped with sleeping sickness; intensive ethnographic research seems to have been put on hold amid the epidemic. Evidence of Ganda healing for other illnesses within the same era, however, suggests that people from around the lake could have readily placed German examinations and interventions into frameworks defined by their experiences of healing and found utility in them. Despite the absence of critical aspects of *kubándwa* healing practice, such as intercession with spiritual forces, interventions on the body by Ganda healers and German researchers might have held meaningful parallels amid the widespread and severe crisis that mongota constituted. Drinking a medicine to heal resonated with diverse extant practices. Healers prepared medicines as part of rituals imbuing them with power and then offered them to the afflicted to be drunk; herbalists' preparations were sometimes also ingested. Swallowing bitter quinine would have echoed ingesting these medicines, as well as the bad-tasting medicines offered sporadically by the White Fathers in the past. Abnormal swelling, pain, and widespread sickness fit into previous experience of illnesses, allowing people to nest early episodes of this illness within *kaumpuli* as an explanatory rubric as they narrated its early years; the emergence of sleepiness as a distinctive aspect then led some to identify mongota (or *tulo*) as something different. Regarding attention to cervical glands, people articulated a connection between particular kinds of inflammation and the appearance of wasting thinness and a nodding, sleeping illness.[78] At the time, Ganda healers directed their therapies at specific physical manifestations of illness and/or specific sites on the body—cupping and withdrawing blood from areas of the head for headache, raising blisters on the skin above painful or swollen parts of the body. Healers likely directed their attention to new, tender bumps under the skin that accompanied pain and changes to the temperament in some people suffering from mongota. Incisions or raising of blisters around such swelling would have fit within their treatment practice, and cutting perhaps extended to excising entirely what healers felt or found. Entebbe researchers Gray and Tulloch noted in 1904–5 that people proactively "advocated the removal of the enlarged glands as a means of cure," and suggested a long-standing association between this type of swelling and illness, even as their own research was only beginning to consider how gland inflammation might be a reliable early sign of disease.[79]

Further, chiefly use of, or promotion of, camp-based therapies would have explicitly connected the social worlds of Ganda ritual, cosmological and therapeutic power, and colonial medicine. The men who bore the names Semagala, Kaganda, and Gugu were all Ssese authorities with connections to ritual reconstitution of Mukasa's Ssese shrine. They surface in records from the early epidemic as leaders whose populations were affected by mongota and visited German sites on Bugala. Sewoya was the principal chief on the Buninga peninsula of Bugala Island, an area hard-hit by the epidemic. Sewoya himself was suspected of ailing from mongota in 1905, as well as from *kabotongo* (a condition with suppurating lesions on the skin)—both serious matters for a sense of wider crisis among his subjects. That chiefs themselves ailed was portentous for prosperity more generally; their search for new treatments offered by the Germans was a visible embrace of the promise of camp-based therapies.[80] Many of Sewoya's subjects had shown signs of mongota and also likely suffered concurrently from epidemics of smallpox and dysentery observed by Bumangi priests.[81] So, when Sewoya brought the Bumangi missionaries a young man "said to be cured after three injections" by the Germans shortly after they began their work, he signaled a chiefly power in commanding the mobility of the healed man, but also demonstrated his custodianship of this man and other subjects by going personally to the mission to discuss the treatments offered by the Germans.[82]

The influx of people to the camp indicates the activation of local treatment-seeking mobilities—perhaps wholly new, perhaps taking a new trajectory—that built on diverse therapeutic practices and traditions for Ssese and Ganda populations. The places where treatment and healing occurred were also significant, as some treatment-seeking mobilities were focused around the missions themselves. Let us once again ground ourselves in place and locate the German expedition's activities in the context of the mission sites that they occupied: the two rooms and several open-air structures dedicated to two doctors at the Bumangi mission and the abandoned Anglican mission buildings near Bugalla village. Both missions had cooperated with initial investigations by the Royal Society's Sleeping Sickness Commission in years past, directing researchers to known cases.[83] Attention to mongota within mission spaces, then, had been continuous for the preceding five years. The White Fathers at Bumangi had been attending to seriously ill bamongota ("those who are drowsy") who came to the mission and visiting those living in the island's villages since noting the illness's arrival in early 1902. Bamongota in the area circulated through and around the Bumangi mission, frequently suffering from other illnesses as well—*kawali*, diarrheal illnesses, and kabotongo periodically throughout 1904 and 1905.[84] By early 1906, the White Fathers housed forty-some people in their *hôpital*, some with mongota, some ailing from kabotongo (a condition understood to manifest on the skin requiring separate sleeping and eating quarters), and some with other serious illnesses of unknown origin.[85] People from the mission helped nurse the sick; some caretakers were also people in earlier stages of the illness and housed on site. The mission provided food for bamongota, relying on those able to collect wood and carry water for themselves to do so.[86] As the White Fathers maintained their hospice and continued to care for the dying, the hospital buildings then became sites of further alternate treatments, as German doctors offered quinine and atoxyl there. While the White Fathers had no medicine for mongota or kabotongo, they had in recent years offered medicine and food during episodes of other illnesses.[87]

While the Anglican mission did not become a similar hub for care and treatment of bamongota, it had been a site of other sleeping sickness research in the recent past. In May–June 1904, Pastor H. T. C. Weatherhead, relying on mission networks, surveyed people from around Bugalla, Bukasa, and several other smaller islands for signs of sleeping sickness.[88] The majority of the Bugalla population he surveyed lived on church lands (some were young women and likely mission dependents); those on Bukasa were subjects of an important and sympathetic Protestant chief, titled as Kaganda.[89] Weatherhead focused particularly on documenting swollen cervical glands—Winterbottom's sign—and rough stage of illness, passing information on to the British Sleeping Sickness Commission at Entebbe. Weatherhead returned a year later, after the Anglicans had evacuated, to follow up on the health of the sixty-some people he had first surveyed—at least of those who survived, his interlocutors on the mainland pointedly noted. Again, he directed his attention to bamongota and especially the presence of particular swollen glands.[90] On this second visit, finding seriously ill people, Weatherhead used his own transport to convey some to the Bumangi mission for care. The pastor fit these surveys into the itinerations among his readers that made him a familiar presence among dispersed island communities. For communities around the Bugalla mission and village, as well as further afield on Bukasa and Kome, Weatherhead's examinations meant that the cervical gland-focused work of Koch and the German expedition was hardly new. Ssese voluntarism to undergo examination of their necks and of unusual swelling under Koch at Bugalla must be read against these other local experiences, as they constitute continuities in the kinds of attention that others—particularly Europeans—paid to one's body in the context of mongota. In the diverse therapeutic milieus they inhabited and created, people fit German researchers' practices into extant Ganda and mission healing practices and evolving understandings of illness.

The sites in which German researchers worked were laden with intimate history, as missionary spaces and relationships were co-opted to locate and facilitate new therapies. The final and wider context within which I read use of German sites on Bugala Island is that of the practices and networks connected to the Ssese Islands themselves: important historic sites of healing, fecundity, and the amelioration of misfortune. By any account, Robert Koch had no knowledge of the history of the Ssese Islands as a place where people along the lake's littoral had historically sought relief from misfortune or serious illness or actively pursued fertility and prosperity; neither was Koch aware of the dispersed networks of shrines and mediums that connected wider populations to the Sseses. The missionaries who were aware of the islands' association with lubaale veneration thought of it largely as a means to chart the success of their conversion efforts: as the balubaale and their mediums' influence fell, so might their own Christian churches rise. But while Koch and his researchers were unaware that they had established a site offering medicine and promising healing amid the historically most potent sites of healing power in the region, Ssese and Ganda people knew. The islands had historically functioned as a key nexus of regional medical, ritual, and political networks that shaped daily life in times of plenty and in times of crisis. The emplacement of German research on the Sseses would have had significance for their patients from nearby and farther afield.

Regionally, the Ssese archipelago—the "hive of deities" in kubándwa practice—functioned as a place where healing could be achieved at numerous shrines as mediums provided intercession and directed therapeutic action. Ssese shrines, as with those of balubaale around the lakeshore, offered access, via offerings and spirit mediumship, to emplaced forces that "conditioned . . . prosperity and fecundity" by maintaining ties to ancestral figures.[91] While evidence in contemporary narratives of mongota and treatment-seeking does not detail how those people who flocked to Bugala Island understood their actions in the context of traditions of lubaale consultation or supplication, their movements evince the vitality and persistence of intellectual networks that had connected people with diverse possibilities for relief in past times of crisis. Beginning in the eighteenth century, people who availed themselves of the lubaale Mukasa's powers were historically "linked laterally through a central place"—the main shrine on Bubembe Island (and later Bukasa Island)—where shrine practices fostered "connections that crossed language, region, and standing" to constitute a littoral community.[92] These historic networks of information-sharing and creativity in addressing illness and trouble, though altered in the late nineteenth century, would have helped information about events on the Sseses to circulate around the northern shores of the lake. For some adults, whose elder relatives would have had direct experience with the ameliorative and generative powers of balubaale (especially Mukasa) in their lives during the mid-nineteenth century, the location of new healing resources on the Sseses would have called upon lived memory of past successes and healing with strong resonance.

The wider resonance and implications of emplaced research on the Sseses within Buganda surface forcefully when examining Bugalla in comparison to other sites of colonial sleeping sickness work. Throughout Koch's work on Bugalla, research at the British Sleeping Sickness Commission's Entebbe laboratory continued. Quite early in its tenure, the Entebbe lab had acquired a reputation as a place associated with odd and suspicious behavior. In 1903, scientists wanted monkeys for their trypanosome research and enlisted missionaries and islanders on the Sseses to trap and send animals to the mainland. This generated speculation about scientists' "indelicate practices" and some Ssese spoke of an avowed determination "not to fall into their hands."[93] Animal and human experimentation were part of the commission's work, as were post-mortem dissections of human and animal corpses. Over time, the Entebbe lab became tainted by "notions that are prevalent among the natives as to the gruesome nature of the work" there, which had implications for isolation camps; policymakers advised strongly against conducting post-mortems where people would also be isolated and treated. Because such activities were known to occur at the Entebbe lab, people were convinced, Governor Hesketh-Bell wrote in 1907, "that sleeping sickness was 'manufactured' by the experts in the laboratory, and that the bodies of the dead were opened for some fell purpose of the white man!"[94] Koch conducted animal research concurrently with atoxyl treatments at Bugalla (paying particular attention to crocodiles and monkeys), but performed few, if any, formal autopsies despite a clear desire to do so. His thousands of patient-subjects, further, were a very different cohort than the selected prisoners, colonial servants, and invalided soldiers who comprised the Entebbe lab's small experimental population.[95] This may have been a key difference in allowing people to evaluate German research and situate it within their own moral and ethical frameworks. Also impactful was the provision of therapy and, certainly, the openness of Bugalla's physical site, where onlookers could view nearly every activity occurring in the open-fronted work tents, in the treatment banda, or with animal specimens. Similar rumors of the "fell purposes" of the German site did not circulate, in extant records.

Bugalla's contrast to a sister site is all the more telling. Concurrent with the latter part of Koch's experimental work at Bugalla and Bumangi were the beginnings of coordinated British efforts to establish isolation stations. (Bugalla, for instance, transitioned after Koch's departure in 1907 to become the second in a series of British colonial isolation camps.) Throughout this period, and after the Bugalla site was taken over by the British, the sleeping sickness camp on the Sseses had a reputation as a place where people continued to engage with British sleeping sickness interventions. It continued to be a site where people circulated into and out of the purview of colonial medicine willingly, an aspect that British officials welcomed, even as they noted their own skepticism about ongoing use of atoxyl. Use of Bugalla contrasted sharply with the first British isolation and treatment camp in Busiro, located 20 miles inland from the lake, east of Kampala and north of the Ssese Islands. The new British camp was on a hill at a place known as Buwanuka (shown on map 1.1). British authorities located it there to place it at some remove from areas of sleeping sickness transmission on the lakeshore, but near enough to move affected people readily into isolation.[96] It was also along a well-traveled route connecting Kampala and Hoima and readily accessible from Kampala by bicycle. But the Buwanuka camp was a troubled place. Four different British medical officers supervised the camp in its first year, one of whom was killed in a hunting accident after only a short time on site.[97] Further, "through its long death-roll," reports in 1906–7 noted, Buwanuka "gained a specially sinister reputation, and there is increasing difficulty in inducing the sick to go there. . . . The unsophisticated natives associate the place mainly with the idea of death and shun it accordingly."[98] Officials noted that the camp's poor provisioning also made it unattractive—people were isolated there, often with insufficient suitable food.[99]

But while the logic of associating a place where people went to die with death was, for British researchers, sound enough, Buwanuka had other meanings that colonial officials did not consider. Contemporary British scientists commented blithely that it was known as a place of violent thunderstorms.[100] Indeed, it was a place closely associated with Kiwanuka, the lubaale of thunder and lightning in Ganda cosmology. Buwanuka hill was his home.[101] Kiwanuka had an unpredictable, dangerous power. He was renowned for bringing violent storms and had humbled the powerful nineteenth-century kabaka Suna.[102] British officials noted a preference for a sleeping sickness camp on the Ssese Islands over one on the mainland that likely had a material basis in poor food and abundant death. But this may have also been a choice imbued with knowledge and experience of different kinds of situated power in Ganda cosmologies: between a place strongly associated with Mukasa, the most powerful lubaale, who healed and granted fertility and productivity, and one of Kiwanuka, who could endanger both individuals and collective prosperity with unpredictable shocks of violent weather.[103] Bugalla and the Sseses stood in contrast to other colonial sleeping sickness–focused places that were tainted by rumors of strange activity, experiences of death, and imminent danger.

For many living around the lake, sharing information about treatments on the Sseses amid the worsening epidemic of mongota fit into the daily activities of work, travel, and trade on the inland sea, with word-of-mouth traveling between ports and landings in large and small canoes. Researchers' and missionaries' comments about the origins of the people who came to Bugala Island can be read as itineraries rather than simple provenance: those who assembled at the Bugalla camp had come most recently from Bukasa Island, from nearby Buddu, and from Buvuma Island, for example, but likely traveled to these points of departure from areas further afield. These itineraries speak to circulation of people and information through areas on the lakeshore where kubándwa shrines and mediums had facilitated connections between people with common concerns and through which vigorous travel continued. News about goings-on traveled within and across littoral linguistic and cultural communities. Talk of treatments at Koch's camp may have recalled historic kubándwa practice, for some, or been congruent with treatment-seeking in diverse mission contexts around the lake—the Anglican hospital complex at Mengo, for instance, or the White Fathers' hospice at Kisubi—for others. Regardless of the nature of individual treatment-seeking, we know that some of the urgent mobilities to the Ssese Islands late in 1906 fit into long-established ways of orienting periods of work and productivity chronologically. Lunar cycles and the phases of the moon's visibility had structured times of work and rest for littoral communities historically, with the arrival of the new moon marking a brief period of mandatory rest and altered activity throughout the wider region. On Bubembe Island, at one of Mukasa's key historic shrines, the phases of the moon's visibility had structured gatherings oriented around healing and fecundity for several generations, within the lifetimes of the grandparents of Ssese islanders and Buganda mainland populations and thus within accessible lived memory.[104] Historic lunar charts indicate that increased mobility directed toward Bugala Island, noted by missionaries as "a fever" in mid-November 1906, correlated with the arrival of the new moon on November 16 that year.[105] Whether understood as an auspicious and appropriate time to seek healing in the context of past kubándwa practice, or simply as a period of altered behavior during which a person might prioritize activities other than work, the concurrence of the new moon and noticeably vigorous mobility reminds us that the thousands of people who engaged with Koch's camp on Bugala Island did so while locating his work in the rhythms and priorities of their own lives.

The surges and flows of people into Koch's purview made an enduring impression, as did particular articulations of treatment-seeking that he collected. The perception of strong enthusiasm that Koch and others gathered from mobilities oriented toward Bugala in late 1906 allowed him to understand his sleeping sickness camp as a public health intervention that could rely on voluntarism and popular willingness to displace one's life in order to receive treatment. But perceived voluntarism and urgent engagement had many possible motivations, mediated through individual experiences as well as common practices or references. Voluntarism and treatment-seeking on Bugala Island was contingent and embedded in local, specific cultural contexts. Koch's understanding of the parameters of a sleeping sickness camp depended on the specific littoral world where he located his research.

THE FAILURE OF ATOXYL AND THE FUTURE OF GERMAN COLONIAL SLEEPING SICKNESS PREVENTION

By late 1906 and early 1907, around two thousand people had come into the German treatment regime at the Bugalla camp and the hospice at Bumangi. Many others continued to assemble in anticipation of receiving treatment. By April 1907, however, only several hundred remained; some had died, but many withdrew again from Bugalla, returning to their homes on the islands or the mainland. Several factors led to the swift withdrawal of so many people, but chief among them, and with most significant implications for colonial public health agendas in East Africa, was atoxyl treatment itself.

Given the numbers of people who appeared at the Bugalla and Bumangi sites initially, Koch's supply of atoxyl was insufficient to dose everyone, and so both camp intakes and atoxyl treatments were meted out carefully and in a limited manner to only the most characteristic cases.[106] Processes of screening and triage left some of the crowds at Bugalla without access to camp-based therapies. Koch reportedly wanted to only admit local Ssese people, initially, though he shifted to screening patients by stage of infection rather than provenance.[107] But screening out less-characteristic cases left scores waiting around the margins of the camp. Some left, then, because they never received treatment. Many people appear to have waited, keeping the numbers of prospective patients high for several months and adding to the bustle around Bugalla as well as the pressure on local food supplies.[108] Neither people inside the camp nor those waiting around it were given food by the expedition, but rather had to see to their own provisions; some were housed in new camp buildings, but many lived in villages nearby.[109] Those who traveled from further afield had to acquire food locally, and some traveled with money to do so but became immiserated when their cowries ran out.[110] Patients' capacity to remain at Bugalla dwindled as they expended resources, particularly for those already at some distance from family and kin support and after a series of lean years.[111] Conditions at Bumangi were somewhat better, as priests provided meat in addition to regular meals and organized collection of water and firewood for the sick. Funds provided by the German expedition for use of their buildings assisted with this, but still did not allow missionaries to provide for the sick consistently or to accept all who came.[112]

Despite the uncertainties and precariousness of life around the camp, however, people stayed for weeks and months. But when expectations of therapy and healing were not met, trouble brewed. An early sign of the nature of those expectations came in late November or early December 1906, when the expedition suddenly stopped accepting and treating patients. Injections were by that time central to the experience of treatment at Bugalla and Bumangi, which involved, first, being treated in particular spaces by particular people, and, second, a suite of internal and external treatments. When stocks of atoxyl ran too low to treat all patients, the camp stopped dosing people for fifteen days.[113] The abrupt interruption in treatment apparently caused a great commotion among the sick, many of whom by that time had been receiving atoxyl for several weeks. Following the pause and with the arrival of more atoxyl powder, injections resumed and routine returned. But more significant disruptions lay just ahead, as the multi-week course of atoxyl was coming to its end for some patients. The completion of a course of injections, or perhaps simply the extensive period of regular treatment, invited evaluation on the efficacy of treatment. The expedition's missionary collaborators gleaned significant discontent: the ranking priest told Koch in January that "the rush for treatment has finally fallen off." The missionary further observed that "the sick do not heal, many die, others are tired of having to follow this regimen of the thermometer, quinine, and injections, several leave to return no more and others only want to follow. . . . Beyond the improvements of the first days there is nothing new to report, other than wasting in the sick, fatigue; no one can be said to be cured or close to being so; it is an overall moral and physical decline."[114] This dire assessment at a moment perceived by priests as pivotal—at the end of treatment, when Koch and his team should have seen definitive results—belies a continuous and deeply felt skepticism held by the White Fathers about atoxyl. The turn away from the camp that they reported was grounded in patients' dissatisfaction, however, not their own. Camp-based medical practice was extensively demanding, and after an initial perceived benefit for some, the camp regimen did not continue to heal a person's symptoms; illness remained. People might have felt less weak or sick after a few weeks of treatment, but trypanosomes would have remained visible to researchers for months. This required atoxyl treatment beyond initial improvement, an aspect of atoxyl regimens that was persistently challenging for researchers as they dealt with patients who had their own ideas about healing and curing. A laboratory-defined "cure" and an experientially understood improvement occurred on asynchronous timelines and, further, evidence does not indicate that either priests or researchers discussed the ambivalence of "curing" with atoxyl with patients at all. Koch reported obliquely in April 1907 that an extended treatment involving subcutaneous injections was "difficult" with African patients.[115] Scattered notes from researchers also indicate that the painfulness of atoxyl injections—which White Fathers also remarked upon—proved a deterrent for some patients and had led to less regular treatment than the designated every-ten-days schedule.[116] The attraction of a new therapy ultimately fell short, as against the pain it caused and the time it required.

Other, more serious changes also affected withdrawal from the Bugalla camp. Roughly simultaneously with when priests noted that the "rush for treatment" began to abate in early 1907, Koch and his team had changed dosage regimens to test the effect of different exposures to atoxyl on trypanosomes, offering the drug first in daily, high-dose injections, and then orally. Analyzing observations of patients' engagement with treatment (what missionaries saw from outside the camp) alongside German researchers' reports on their work (what occurred inside the camp) suggests strong correlations between these changing regimens and altered Ssese engagement. Missionary reports of people leaving treatment and others desiring to do so in January 1907 correlate with the end of the first months-long courses of treatment for some patients and the beginning of high dose injections. Koch first attempted a shorter, more intense atoxyl course—daily injections of 1 gram of atoxyl, double the highest dose and 166 times higher than the initial dose in the previous, progressive regimen. This resulted in acute dizziness, nausea, and pain in patients; despite such adverse symptoms, Koch persisted because these effects seemed to be transient.[117] Then one of the patients on the high-dose regimen became blind in both eyes, quickly and without reversal.[118] (It is very likely, given atoxyl's toxicity and capacity to damage the optic nerve, that other cases of visual impairment and total blindness among camp patients also arose and went unreported or undetected.) At this case of blindness, Koch ceased the daily high-dose regimen, changing to a still-high 500 mg injection, which was met with patient resistance. Outside the camp and unaware of these shifts, priests at Bumangi commented in February simply that "[we] note the departure of many patients, tired of treatment."[119] Koch then pivoted again in the following weeks to an oral administration of atoxyl in 500 mg doses.[120] One hundred and fifty patients received the medicine by mouth, a notable enough change in the camp's work that it was relayed to nearby priests by the kweba in mid-March.[121] But this oral therapy was ineffective, by measure of the trypanosomes that survived, and Koch ultimately reverted back to injections at his original intervals of ten days for the severely ill and twenty days for the less sick. Koch had also been alerted to the possibility of atoxyl resistance in trypanosome research in animals, so experimentation with different chemicals picked up speed as differing atoxyl regimes proved ineffective.[122] These changes in the medicines offered would certainly have registered with people accustomed to a daily regimen, notwithstanding the accompanying, perceivable impacts of different doses' side effects. Further, beginning in April 1907, the expedition dosed people in the Bugalla camp with injections of Afridol blue (*Trypanblau* or trypan blue) and *Trypanrot* (trypan red), both dyestuffs and benzidine derivates that were, like atoxyl, supplied by Paul Ehrlich's laboratory at the Institute for Experimental Therapy in Frankfurt.[123] People did not tolerate either substance at all, as noted by doctors, nor did dyestuffs have a noticeable effect on trypanosomes in patients' blood or lymph.[124] After months of regular, evenly spaced doses of atoxyl, expedition researchers' therapeutic experimentation had taken a frenetic turn. While dyes were a quickly evident failure, how well patients truly "tolerated" atoxyl is also doubtful. Although blindness purportedly did not arise until later in the treatment regime, patients also periodically developed abscesses at injection sites, muscle pain, and nausea from the regular atoxyl regime. Koch tried to adapt atoxyl regimens in order to maintain his pool of patients, but the strategy backfired spectacularly.

The same social processes that drew people to the camp also shaped how they would leave. Talk about the camp's therapies circulated in and around Bugalla, percolating up to the highest local Ganda political authority, as the kweba heard news of changing treatments. A tangible shift in interactions between patients and doctors resulted after rapid changes to drug regimens. Koch's April report from Bugalla is peppered with references to patient resistance, refusal, and concerns about "tolerance" that imply widespread, rather than individual, actions that forced Koch to scale back changes to atoxyl regimens.[125] The camp's fortunes suffered as the cures offered there were not durable and severe, detrimental side effects from atoxyl began to appear. People circulating around and through the Bugalla and Bumangi sites saw not only illness and debility, but also death. Many who died in the camp would likely have been buried nearby, perhaps en masse, creating indelible spaces connecting the camp with dying.[126] Here, as would emerge later elsewhere around Lake Victoria in sleeping sickness isolation camps, rejection of camp medical regimens registered in the outflow of patients. The sharp increase in patients and potential patients around Bugalla between August and November 1906 is mirrored by an equally precipitous decline in cases in February and March 1907. The cumulative impact of altered Ssese and Ganda engagement with the camp was stark: by May 1907, only six hundred people remained of the nearly two thousand total (plus others awaiting treatment) who circulated around Bumangi and Bugalla the previous December.[127] These last cases were, by all accounts, either severely ill or people who lived locally; many of those who had departed were returning to other islands or the mainland.[128]

Researchers operated just at the edge of atoxyl's potential toxicity, experimenting with doses for which the long-term impact remained unclear, but the short-term impact was also under debate. They had also created expectations of consistent kinds of activity through their creation of and dependence on a routinized but multifaceted clinical encounter. Ultimately, clumsy calibrations to patient resistance and tolerance while escalating experimental work with atoxyl were untenable. These experimental decisions and their intellectual underpinnings have durably and convincingly animated critical scholarship on Koch's ethics and his legacy. Powerful critiques by Wolfgang Eckart and Christoph Gradmann document the racialized deployment of atoxyl and other experimental drugs under Koch's supervision in Europe and Africa, proving that African patients were subjected to more toxic doses than Europeans.[129] While Koch's work clearly fit into other colonial experiments with atoxyl at the time—his later publications from the Sseses were closely watched by Belgian doctors also experimenting with atoxyl and by the British who used atoxyl on the Buganda mainland—the rush to prove a conclusive cure alongside his careless approach to African patients was dangerous.[130] In the rapidly changing regimens of early 1907, Koch played fast and loose with experimental protocols, varying doses and medicines widely. Thresholds of clinical use he himself had established were tossed aside in the urgency to produce conclusive results and a cure for sleeping sickness. This capricious, anxious experimentality caused irreversible harm to patients at Bugalla and was profoundly disruptive for the Bugalla camp.

People came to Bugalla for healing and therapy, understood and interpreted through multiple but potentially enmeshed intellectual and therapeutic frameworks. The gradual outflow of people from Bumangi and Bugalla likewise involved assessments of healing, the pull of familial or seasonal responsibilities, and perhaps the recognition of continuing decline and a desire to seek other treatment or die at home.[131] Those who withdrew from Bugalla individually, or who returned irregularly for treatment, undermined the consistency of longitudinal data and were a frustration for Koch's team, but were taken on board as anticipated losses. More remarkable and more foreboding were the departures of people in scores, successively and then more rapidly, in correlation first with the end of initial courses of treatment and second with particular treatments and drug effects in March and April 1907. The former circumstance suggests a reassessment of the efficacy of German treatments and hoped-for cures and a turn toward home; the latter moves constitute an additional rebuke of those treatments and decision to remove oneself (or one's family) from increasingly unpredictable, painful, or harmful therapeutic regimes. People from around the lake received new treatments and new approaches to illness from German doctors with cautious acceptance and persistence, but decisively abandoned them, along with withdrawing from the people administering them, if their utility was not apparent.

The shifting use of the Bugalla camp and the yield of nearly a year of research produced a different set of conclusions than Koch had hoped. When it began to emerge that people treated with atoxyl still relapsed into sleeping sickness, but could have few trypanosomes detected in their bloodstream for several months, Koch departed from his earlier claims that atoxyl could cure the disease. Instead, he argued that atoxyl interfered with disease transmission by reducing parasites in the peripheral blood and lymph, such that tsetse flies could not become infected and transmit parasites further—people could thus be rendered "harmless." Atoxyl here operated similarly to quinine, his persistent reference point, but without its curative potential.[132] In an environment where complete fly eradication seemed impracticable, colonial control on mobility was not meaningful, and people traveled frequently in and out of areas with sleeping sickness, atoxyl was, "though not an infallible remedy, still such a powerful weapon in the battle against sleeping sickness, that one must at present utilize it as such as much as possible."[133] The use of atoxyl would continue to structure German strategies for controlling sleeping sickness around German East Africa. With no firm plan for how long treatment with atoxyl should extend, Koch envisioned a system of camps to allow prolonged treatment with atoxyl and to isolate the sick from both disease-carrying flies and from healthy people nearby. More broadly, atoxyl remained in use across European colonies in Africa, to varying degrees, despite the known acute and cumulative negative side effects.[134]

Koch's work on Bugala Island shaped the subsequent German sleeping sickness campaign, defining its practical organization, its theoretical framework, and the scientific and clinical methods to which staff would adhere. For a focal disease like sleeping sickness, the work of surveillance, research, and treatment would be structured geographically and spatially. The sleeping sickness camp—called variously an isolation camp, collection camp, and concentration camp—was not, in its essence, a great innovation.[135] Isolation hospitals and quarantine stations had been trusted methods of dealing with epidemics for centuries prior.[136] From the first sign of potential success with atoxyl in November 1906, Koch foresaw the development of "stations" modeled after his camp at Bugalla as a key element in further efforts to prevent sleeping sickness.[137] In the German anti–sleeping sickness campaign, camps were to be located among particular populations and rooted in place, with all activity circulating around them, constituting permanent sites of treatment and observation. Each camp would effectively function as a new hub for medical treatment, one focused on sleeping sickness but also addressing its attendant secondary infections (and the "complications" from side effects that developed as a result of drug treatments). The camps also enabled an ongoing research program, allowing doctors to follow cases under their supervision and contribute to a growing body of statistics on the progress of sleeping sickness cases. The camp concept, as an intervention rooted in a particular place and reaching the population in its catchment, implicitly rested on voluntarism and the engagement of the sick and their families. But these elements were, on the Sseses, the product of complex and specific local dynamics.

Ongoing German colonial energies were aimed at two sites near the Uganda border to begin camp-focused sleeping sickness control—near Shirati and in the Mori and Mara River catchments on the eastern side of the lake, and near Bukoba on the western side, in the kingdoms of Kiziba and Bugabu.[138] These sites were particularly relevant for ideas connecting African mobility with the epidemic's spread, which meant intense scrutiny of travel for people going to and from Uganda and who made their lives on the lake. A person who spent time regularly on the lakeshore or islands—rowing, fishing, collecting rubber, getting water—was understood to have greater likelihood of exposure to the fly vector and thus the parasite. Life and livelihood itself around the lake were pathologized in these ways of understanding of the epidemic. On the German colonial mainland, sleeping sickness did not appear to be as widespread and its geographic scope was not so naturally bounded as on the Ssese Islands. Camps there, as a consequence, would have to be places of stricter isolation and control, drawing from a wider population, keeping the sick away from the well, and feeding, sheltering, and monitoring the sick. The camps would have to absorb any and all people with trypanosomes identified in their bodies in order to prevent the wider spread of the epidemic. Though Koch did not demonstrate any awareness of it, the intention to intern people with sleeping sickness would mean coming to terms with their long-term care and provisioning as well as attracting a proportion of a local population away from villages and family homes, as the sick and their caretakers were effectively relocated.[139] The German camp system relied on a distinctive prophylactic vision for atoxyl, although the drug was in widespread use by other colonial medical administrations. Ultimately, as Koch's influence on practices in East Africa diminished after 1908, some of the daily practices of sleeping sickness camps would shift to accommodate changing circumstances, but his ideas about their parameters and purpose, based on the Bugalla camp, would endure.[140]

The Bugalla camp was, first and foremost, conceived as a way to isolate people with sleeping sickness from the fly vector that carried the disease and consequently hinder the disease's spread. It was an ad hoc laboratory, research site, and clinic. As the sick were treated with a variety of drugs and experimental remedies, researchers followed the progress of the disease in patients and treatment efficacy was evaluated.[141] The camp's siting and the layout of its buildings carved out a new space for biomedical research on the islands.

Koch's camp at Bugalla served as a model for the German colonial anti–sleeping sickness campaign in East Africa. Here was a system built around isolation and drug treatment, a modernized lazaretto for the age of chemotherapy. Here, too, was an ideal colonial intervention, where African voluntarism seemed to facilitate medical modernity and a portable, sustainable program to secure public health and curb the epidemic emerged. Where violence or coercion underlay the camp's smooth functioning, Koch's reports never revealed it. Bugalla allowed Koch to argue that atoxyl could be useful to interrupt transmission of the disease by lowering the parasites available in the bloodstream for the fly vector to ingest and transmit; the streams of people seeking treatment painted a rosy picture of African enthusiasm for treatment and willingness to travel for it, often to include rearranging family, hearth, labor, and home to do so.

But the initial success of Koch's Bugala Island sites depended on the Ssese and Ganda intellectual worlds into which they were integrated. The specific local dynamics and regional resonances around Bugala Island and the Ssese archipelago played a central role in driving mobility to the camp, as well as initial, widespread engagement with German researchers' methods and treatments. Bugala Island was not just any site, but the site of two long-standing missions with a tradition of engagement and care. And the Ssese Islands were not just any place, but were the cosmological origin and ritual home of the most significant healing power to which people living around the lake might appeal. More recently, too, the islands were a place connected to the mainland by Christian religious networks, robust trade, and rowing expertise. Bugalla, the first German *Schlafkrankheitslager*, was located in a productive and complex milieu, where historic healing traditions, ongoing therapeutic diversity and creativity, and place-focused political and healing power overlapped and intertwined. The history of the Bugalla camp illuminates the resonance of emplacement and local context for the trajectory of public health and colonial research and demonstrates how Ssese intellectual worlds shaped both the unfolding of scientific research and the contours of the German colonial public health campaign to come.

PART II

The Kingdom of Kiziba, c. 1890–1914

The Kingdom of Kiziba, c. 1890

An Overview

PEOPLE LIVING on the Ssese Islands had relations with populations around the northern shores of the lake, in Buddu to the immediate west, and in the Haya kingdoms that lay below the Kagera River to the south and west.[1] People from the Haya kingdom of Kiziba, northernmost of the small states, circulated to the Ssese Islands in the late nineteenth century; Ziba mobilities would ultimately involve the kingdom's populations in new epidemics and in new disease-prevention interventions in the decades to come. This brief overview orients the reader to important aspects of Ziba society in the late nineteenth century, focusing first on Kiziba's geography and environment, then addressing social, political, and economic aspects of Ziba life. Locating Kiziba in the changing economic and political trends of the nineteenth century, it also offers an overview of the advent of European, and particularly German, contact in the region and of early colonial politics. These elements of life in Kiziba set the scene for the different modalities of Ziba responses to outbreaks of widespread illness at the turn of the century, and for the German colonial public health interventions and Ziba royal strategies deployed to meet them, which are the subjects of chapters three and four.

The central western shores of Lake Victoria and their hinterlands, south of the Kagera River, were the domain of Haya populations. *Haya* here identifies a cultural unit, of which Buhaya is the land and Kihaya/Oluhaya is the language spoken. As the region's kingdoms emerged in the seventeenth century, Haya settlements, dense but unevenly distributed, came to cover the landscape.[2] Topography in Buhaya shaped settlement, as a German doctor would later observe of Kiziba in the early twentieth century: "the country is made up of a row of ridges that tower island-like over the swamps which separate them. The ridges are fairly flat, their tops for the most part very long and wide and fully covered by banana groves. The spread of the banana groves is often hours-long, the number of huts contained within them often several hundred."[3] The hillsides and high ground of these ridges were places of society and politics in Haya culture, while surrounding dense forest and swamp, though traversed and utilized, were spaces apart from village life, offering refuge or ensuring exile.[4] Banana farming and vegeculture were well established by the late nineteenth century, fitting within wider regional patterns of bananas as the staple crop alongside cultivation of endemic *robusta* coffee trees, pulses, and vegetables. Cattle kept alongside banana groves provided manure for cultivation as well as milk and butter, and their exchange also constituted a key means of cementing patron-client relations and securing wider social and political bonds.[5] Haya households were organized around the requirements of subsistence production of plantain staple crops as well as ritually important coffee in gendered labor regimes. A family's *kibanja* (farm) surrounded the home and could be subdivided among male heirs. Social and political institutions, chiefly patrilineal clans and patrilineal land tenure (both of which also implicated agnatic relationships and maternal lines), allowed consistent claims to productive land that could be maintained within families and kinship groups. Alongside agricultural production, Haya communities also exploited iron deposits and developed traditions of smelting and tool production, as well as skilled craft production of pottery and barkcloth. Cultivation and other household tasks were traditionally gendered, as women, for example, cared for the daily maintenance of fields, worked common fields outside of villages, and tended vegetable crops, while men cultivated bananas, cleared new fields, and forged metal tools or made barkcloth.[6]

Economic activity in Ziba villages near the lake was diverse, relying on widespread pockets of fertile land for subsistence farming, and also fit into lacustrine trade. Particularly in villages nearest the lakeshore or river, people fished for household consumption and marketed their catch locally and regionally.[7] Travelers and migrants from Kiziba circulated within vigorous local and regional trading networks, both in and around the Great Lakes region and connecting to the Indian Ocean coast. Around Lake Victoria and its hinterlands, Ziba people historically marketed fish, barkcloth, raffia, and iron hoes in trade for salt, goats, or ivory.[8] They were also later deeply involved in porterage and the caravan trade, connected regionally within the wider Karagwe caravan network and linked to trade with

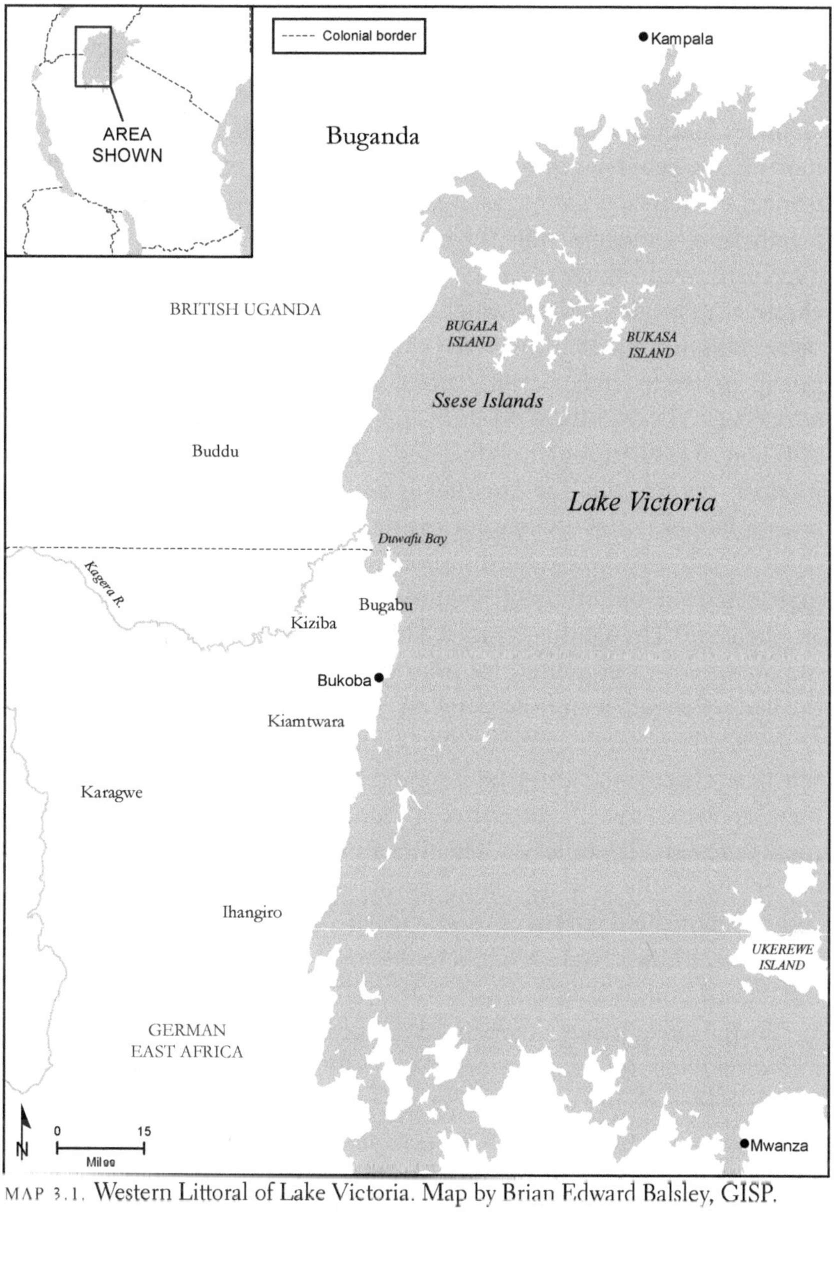

MAP 3.1. Western Littoral of Lake Victoria. Map by Brian Edward Balsley, GISP.

Buganda, along African infrastructure that connected the western littoral with east-west caravan routes to the south.[9] A broad main road punctuated by way stations ran along the western side of Lake Victoria, connecting those east-west caravan routes with Usui (Biharamulo) and points north, including Buhaya.[10]

Related politically to the kingdoms of Bunyoro and Buganda, the Haya population was by the mid-nineteenth century divided into several kingdoms. The Haya kingdoms, which shifted in size and power amid alliances and wars between their ruling *bakama* (kings, sing. *mukama*), marked out generally discrete areas of the varied terrain of the region, divided by rivers or swamps, plateaus, and forests. Villages were governed locally by a hereditary chief, or *mukungu*, but at the district level by an appointed royal minister, a *mukungu mukuru* or *mwami*, as well as local elders.[11] Atop this political structure was the mukama, an institution of sacred kingship where royal authority was embodied in a hereditary, patrilineal monarchy linked with cosmologically and politically significant ancestors.[12] Historically, Ziba monarchs were deeply involved in the acquisition of cattle through raiding and warfare, with both cattle and their products used to link the king to his people and to consolidate control among familial rivals. Cattle, for example, could be used to reward regiments or villages (through the distribution of meat) and punish subordinates or defeated enemies (through requirements of gifting for compensation of an offense).[13] Though the transfer and possession of cattle constituted an important aspect of politics for prior generations, access to land formed the basis of clientage in Buhaya following the decimating cattle epidemics of the late nineteenth century.[14] The mukama was a "receiver of milk," a title that reflected the historical importance of the circulation of cattle and their products, as well as the labor of a kingdom's subjects, to maintaining royal power and legitimacy.[15] The title also connoted the king's extractive power and ability to make claims on his subjects' labor, goods, and harvest.[16] Tribute, in the form of "first fruits of a farm, such as bananas, grain, livestock, or hides," were brought to the royal court, and the mukama also influenced ironworking and barkcloth production.[17]

By the late nineteenth century, control of land had been consolidated into royal hands for several generations, within a hierarchical politics that blended roles of clients, administrators, and landholders.[18] Large, consolidated farms might be held by clan or family heads and also granted to subchiefs by favor of the king within the system of *nyarubanja* (a large kibanja, denoting an estate or clan/family landholding); these were "important sources of patronage, labor, and tribute for their noble estate holders" as well as for the mukama.[19] Ongoing processes of the alienation and granting of land by royal fiat connected clients to the king. In addition to the distribution of land and patronage through nyarubanja, predicated on all farms being the king's possessions, the open land that lay between villages also belonged to the king and could be allocated by him, according to late nineteenth-century observers.[20] The mukama's prerogative to distribute unoccupied land or land without heirs, an aspect of his position at the apex of political and social hierarchies, fit into more intimate relations of clientage between elder and younger family members or within clan relations.

The northernmost of the major Haya kingdoms, Kiziba had a population estimated roughly to be between twenty-five and forty thousand at the turn of the century.[21] Kiziba historically spanned the territory immediately south of the Kagera River and along the deep cut of Duwafu Bay in Lake Victoria, bounded by a southward-running river valley and seasonal swamp to the east, which separated it from the smaller kingdom of Bugabu. On the west, the Kagera bound Kiziba once again as it turned southwest, and the Ngono River and the adjacent territories of the kingdoms of Kiamtwara and Kianja defined its southern boundary. While these other kingdoms of Buhaya play a lesser role in our story, they speak to the political complexity that shaped life in the western lake region. Rivalries, both internecine and between kingdoms, occasionally erupted into open war in Buhaya, but belie a sense of common lineage and culture among the closely set polities. Ziba monarchs, for example, married members of the ruling elite of the larger kingdom of Karagwe to the west and articulated lineage links to the powerful Nyoro dynasty to the north.[22] The Haya kingdoms also shared the predicament of managing relations with a Ganda state with aggressive regional ambitions in the nineteenth century. Much like the Ssese Islands, the Ziba kingdom experienced the predations of the Ganda state; unlike the Ssese islanders, however, Ziba were not skilled rowers nor custodians of a historically important shrine. The tributary relationship between the powerful Ganda state and Kiziba was one of sharply defined subordination. Warfare and resulting disruption of life—as food supplies were consumed or lost, agricultural labor diminished, or security decreased—had wide impacts in Kiziba. Important, too, was the potential of captivity or enslavement that might follow military defeat, a process with changing implications as the region became more deeply involved in regional trade in the nineteenth century.[23]

Involvement in regional caravan trade and the wider machinations of the Ganda, Arab-Swahili traders, and Christian missionaries around the lake in the 1870s drew European attention to the Haya kingdoms by the late 1880s. European concern was inconsistent and generally oriented around competition between German and British diplomatic interests; insofar as wider plans for territorial acquisition were a part of British and German calculations, these plans did not bear fruit in any organized or deliberate fashion at Lake Victoria.[24] The colonial station at Bukoba, an attractive anchorage on the western lakeshore with a long, wide beach backed by low, rolling terrain, was founded in 1890. The Bukoba *boma* (Kiswahili, fort) was inconsistently staffed for much of its first few years; once the Anglo-German treaty of 1890 had secured territorial boundaries, German practice in the area combined "divide and rule" manipulations of the individual Haya kings with a spendthrift ethos of indirect rule.[25] The region was by any stretch remote from the primary foci of German energy and resources in East Africa: the Swahili coastal trading ports of Tanga, Bagamoyo, and Dar es Salaam, and the northeastern agricultural settler highlands in the Usambara Mountains. At the turn of the century, the German East African shores of Lake Victoria were thinly staffed with colonial officials at three widely dispersed stations: Shirati, Mwanza, and Bukoba. Bukoba town and its environs were otherwise home to a few German civilians farming coffee, Indian traders, and polyglot European missionaries.[26] While Bukoba's thriving local market town and populous hinterland were a potential source of tax revenue for the colonial administration, they were not a source of profit for the colony.[27] Furthermore, the colonial government had scant knowledge of local languages, little to no infrastructure, and inconsistent—if not hostile—relations with Haya authorities.

The processes of European intervention and colonial incursion specific to western Lake Victoria involved a particularly contentious era for Kiziba specifically. Mukama Mutatembwa, who had ruled Kiziba for nearly two decades by the time scattered German expeditions arrived, is remembered as a dogged resistor of colonial interference; this manifests in historical scholarship in narratives of near-consistent, low-level conflict with the German boma founded at Bukoba in 1890, characterized by punitive German forays into the kingdom and manipulation of local rivalries to Mutatembwa's disadvantage.[28] Difficult relations with White Fathers missionaries, who arrived in the area in 1892, further complicated interactions between Ziba authorities and German officials throughout the 1890s. The German colonial government in the region transitioned from military to civilian rule in 1906, at which point the Haya kingdoms became part of the Bukoba Residency, its borders roughly congruent to their boundaries.[29] The neighboring kingdoms of Rwanda and Urundi became their own residencies, the three administrative entities remaining distinct within the German East African protectorate. The residencies were underpinned by the idea that strong, well-articulated, monarchic political structures, when subjugated after military conquest, could be used to rule the densely populated territories they covered.[30] The residents themselves, typically former military men, had significant leeway in shaping local policy and managing the administrative budget, and African rulers maintained their traditional political authority while also serving administrative functions for the colonial state.[31] Ultimately, German civilian rule in the Bukoba Residency brought an increased interest in developing raw materials and exploiting Haya trade and production for the emerging colonial economy.[32] The impetus for changing African agriculture and trade was hardly unidirectional, however, and Kiziba provides an excellent example of the give-and-take through which African farming was connected to a global market in the early twentieth century as political and agricultural traditions met with changing economic imperatives.[33]

In the latter half of the nineteenth century, then, key challenges to Ziba social and political prosperity came in the form of poor rainfall and thus poor harvests, as well as in the threat to security and free movement posed by Ganda raiding parties or warfare between Haya kingdoms.[34] Alongside local or regional conflict and slave raiding, but also occasionally exacerbated by them, Ziba populations also contended with diverse maladies that challenged individual and collective prosperity; they managed sickness and death within durable social and political frameworks. German colonial incursion, and eventual colonial public health interventions, added further complexity to efforts to ensure prosperity and health undertaken by both Ziba elites and the kingdom's wider population.

3 The Prince and the Plague

Politics, Public Health, and Rubunga *in Kiziba*

I ARGUED in the preceding two chapters for the importance of the social, political, and intellectual worlds of Ssese islanders in shaping the parameters of Robert Koch's sleeping sickness research, and for the centrality of Ssese and other treatment-seekers' engagement to his conceptualization of the sleeping sickness camp as a widely transferrable colonial public health intervention. Before delving into the complex history of the flagship German sleeping sickness camp at Lake Victoria, located in northern Kiziba, I here examine political and spatial dynamics in Kiziba in decades prior that were crucial to the management of outbreaks of illness and to the wider maintenance of health, prosperity, and political legitimacy in the kingdom. This history demonstrates, at a fundamental level, how experiences of illness deeply colored Ziba and German political relations. My exploration of an illness called *rubunga* provides an important counterpoint to the history of sleeping sickness prevention that follows in chapter 4.[1] Efforts to mitigate the impact of rubunga offer insight into Haya strategies of managing illness and death amid shifting political dynamics in the late nineteenth century.

Several overlapping dynamics with deep historical roots came into play in the late nineteenth and early twentieth centuries, when new challenges to the kingdom's health arose: the juxtaposition between home/village and savannah; the reorientation of daily life in Haya social geography in times of crisis; and the responsibility of the *mukama* to mitigate illness and disaster for his subjects. People in Kiziba maintained health and prosperity through a combination of daily activities and observances—ways of being in the world—that encompassed social relationships and royal politics. Populations structured and maintained both interpersonal and extended social relations to ensure security and prosperity; such efforts also included the spatial relations through which people located themselves in the landscape. In Kiziba, as we will see, practices of establishing the location of home and community within familiar spaces and landscapes changed when rubunga spread; these changes allowed communities a measure of resiliency in both social and economic production. In such times of crisis, the mukama's central political-ritual role meant that his actions could have broad implications for the kingdom's health, and, likewise, that illness and misfortune could impact his political legitimacy. This chapter explores evidence of historic experiences with widespread illness and the extant practices to mitigate and resolve such illness in order to understand the therapeutic, political, and ritual resources that Kiziba's populace could call upon in the late nineteenth century. It situates these efforts within the social landscapes of Haya life that connected to both royal authority and traditions of land use. Rubunga would also be a practical model for how health and politics could intertwine in the early colonial era, serving as a foundational experience for the implementation of both Ziba and German understandings of disease prevention during a time of significant change in diverse aspects of Ziba life. It was, for Ziba and German actors, the nearest and most compelling precedent for sleeping sickness.

Outbreaks of rubunga occurred amid succession struggles for the throne of Kiziba during a period when German colonial incursion accelerated. The interweaving of medical and political interventions to address the illness and the strategic prioritization of controlling epidemics had broad repercussions for relations between colonizers, missionaries, and Haya elites. Rubunga catalyzed increasing colonial medical and public health attention to Kiziba, as anxieties about controlling the spread of *Pest* (German, plague) intensified. I analyze Ziba measures to manage rubunga alongside colonial plague research and public health policies to understand how ideas about political legitimacy, prerogatives of land use, and ensuring health intertwined. Established Ziba measures to isolate the sick, as well as to establish physical distance between those who were well and a place where sickness had broken out, provided the precedents upon which colonial mandates could be layered. Examining how Ziba princes coped with rubunga, I demonstrate that plague-focused research, colonial health work, and Ziba strategies to weather rubunga had wider political implications and relevance and would prove important when sleeping sickness prevention efforts arrived, in due time, in the same field.

SECURING HEALTH AND MITIGATING MISFORTUNE IN A CHANGING POLITICAL LANDSCAPE

Ziba society and the Ziba kingdom's politics were profoundly impacted by its unique terrain and environment. Fundamental to human mobility and variable population density was Kiziba's terrain, defined by Lake Victoria, the Kagera River, and the hills and swamps that rolled away from the lakeshore. *Kagera*, in the Ziba dialect of Oluhaya, historically meant simply "river," and this great river was centrally important in shaping the kingdom's local social geography.[2] The meandering course, fast-moving waters, and silty mouth of the great Kagera River defined the landscape of the northwestern shore of Lake Victoria to the south of the Ssese Islands.[3] One early European visitor to the region described it as resembling "a giant blackish octopus that projects its tentacles in all directions . . . no mountain valley left untouched by its reach."[4] The Kagera's role in shaping local mobility was significant, but also variable. The seasonal floods of the river, its tributaries, and low-lying areas in its watershed defined patterns of overland travel for people moving through the region, such as Arab, Swahili, and other African traders, caravan porters, and parishioners moving among the area's missions. Social and political life centered on the hillsides and ridges of land to the west of the inland sea, making travel between villages on adjacent high ground difficult when the swamps between them flooded.[5]

In Kiziba, as in Buhaya more generally, villages were loose groups of homes and farms dotted across the hilly landscape. The typical Ziba home was a traditional circular, domed *msonge*, each surrounded by a farm, a *kibanja* (pl. *bibanja*), planted with banana and coffee trees.[6] Away from main roads, countless small paths branched off to connect homes and farms set back among banana and coffee trees. In the 1890s, Carl Hermann, the German administrator at Bukoba, remarked on the difficulty of navigating paths through dense banana groves to find villages, and of finding the way from house to house, writing "the houses are scattered about, connected by a tangle of criss-crossing, looping paths, which are bordered by high hedges. Only people of the village themselves, rarely a neighbor, not to mention a foreigner, can find their way and get out again."[7] Fr. Edmond Cesard, a White Fathers missionary living at Kashozi, wrote in the 1930s that a typical Haya village still "offers the view of a banana forest. Homes are hidden in the bibanja. The odd footpaths connect the huts one to the other in such a way that at night, it is easy to lose one's way."[8] Cesard's diagram (figure 3.1), though stylized in a 1930s aesthetic, illustrates the dispersed plan of a Haya (and therefore Ziba) village along

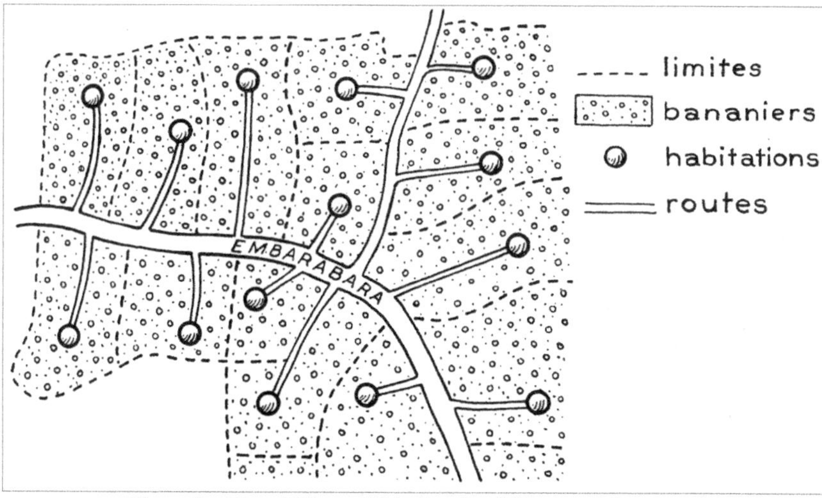

FIGURE 3.1. **Plan of a Haya Village.** This sketch of Haya social geography accompanied White Fathers priest Edmond Cesard's anthropological publications on Haya life in 1937. Haya homes are depicted set deep within banana farms and off of a main road (*embarabara*). Source: Cesard, "Le Muhaya," 29, after Rehse, *Kiziba*, 12–13.

a main road. The "*limites*" shown here were likely planted with hedges or trees to separate the farms and may have delineated the footpaths between homes that so confounded outsiders who ventured off the main roads.

Haya populations structured village life around clan and kin relations and also fit it into wider royal political geographies of power. What seemed to Europeans a confusing tangle was a complex social geography of densely populated and intensively farmed pockets of settlement, organized and governed within a centralized, hierarchical political and social order. Demarcated by hedges and footpaths, Haya households and farms sat on adjacent parcels of land such that "the village as a whole is a continuous group of households on perennially cultivated land."[9] Homes and farms situated thusly were surrounded by open savannah grasslands—what early German observers referred to as "steppe"—that generally lay between one village and the next. Forests and streambeds or swamps were also features of non-village land near Lake Victoria, their resources occasionally used and their spaces traversed, but not regularly inhabited or brought into cultivation with tree crops.[10] The distinction between villages (*ebyaro*), defined by the presence of perennial tree crops and households, and open savannah grasslands (*orweya*) was historically a significant one. Ethnographies conducted over the long twentieth century point to the durability of this distinction and the immutability of ebyaro and orweya lands, which had differing fertility for tree crops and therefore different value to sustain productive society.[11] Small grassland orweya plots adjacent to a farm might be brought into cultivation by people (and particularly women) of a household annually and also served as pastureland for cattle, sheep, and goats.[12] But these marginal grassland plots were not consistently cultivated and maintained as were the farms themselves; they were, until later in the twentieth century, generally outside of traditional clan or familial land tenure claims.[13] Orweya was not where people made their homes in normal circumstances. By contrast, established, productive farms were highly valued and their control mediated through family and clan relationships. Individually controlled and patrilineally inherited, larger parcels might be subdivided among male children or grandchildren; if a person died without an heir, however, land reverted to the mukama, who could redistribute land to his allies or subchiefs.[14] Wider regional political dynamics could also shape domestic landscapes. In a political context where warfare, historically, frequently involved raiding for captives and slaves, non-nucleated villages where homes were difficult to find afforded a measure of protection against outsiders. Indeed, we might read the spaces created by Ziba (and Haya) society—the dense banana groves surrounding homes, the concentration of homes on ridges—as offering protection in times of insecurity in multiple ways: nourishing bodies, connecting generations through custodianship of land, and providing shelter or refuge.

Historically, Haya populations structured and maintained both interpersonal and extended social relations to ensure security and prosperity; such efforts also included the spatial relations through which people located themselves in the landscape. On a more intimate scale, actions of the members of a family or household could also have wide-ranging impacts on the health, fertility, and prosperity of their relations, for example through the observation of gendered culinary taboos that affected fertility and virility, or through delegation of certain kinds of labor.[15] Haya spaces were also fundamentally constitutive of social and political relations. Fathers and sons, or husbands and wives, understood and articulated their relationships with one another through the embeddedness of these relationships in specific spaces—the hearth, the home, the farm, the savannah, or a gravesite, for example—and the ways that people in different social roles or positions oriented to them.[16] These social and spatial dynamics would prove crucial to the management of outbreaks of illness.

Atop the political and social hierarchy and responsible for the kingdom was the mukama, whose action or inaction could affect harvests, food supplies, widespread illness, and vulnerability to attack and warfare.[17]

While the mukama's title spoke to the symbolic and practical power of cattle possession and redistribution, he was more broadly responsible for maintaining the well-being of the land and the people, ensuring productivity and health and preventing misfortune through spiritual and ritual practices.[18] Honoring and interceding with the powerful deity Mugasha (Mukasa), whose home was on the Ssese Islands, was a mukama's particular responsibility; the palace alone maintained a shrine and *embandwa* medium dedicated to Mugasha, within which shrine objects symbolizing the kingdom's prosperity (cowrie shells, bananas, a miniature oar or dried fish) were placed. The "close connection" of the royal family to Mugasha allowed the palace to influence the regularity of rains, the safety and productivity of fishermen and others on the lake, and the prosperity of banana groves in Kiziba.[19] Apart from the mukama, only fishermen could appeal directly to Mugasha, a relationship that mirrored observance of rituals and prayers to Mukasa in Buganda. This political-ritual nexus gave people a kingdom-wide political context and a cosmology within which to situate experiences of illness and misfortune, alongside their experiences in their own households and their sense of connection with ancestral spirits and the natural world. The mukama's potential to affect the livelihood of the wider kingdom, and his responsibility to act to ensure health and prosperity, redounded to the subchiefs and clients through which political power was distributed and whose actions also affected perceptions of his legitimacy. Other important deities in Ziba spiritual practice, such as Wamara, associated with the spirits of the dead, or Irungu, associated with the earth, were not subject to such restricted, royally mediated access; rather, intercession was mediated historically by *cwezi-kubándwa* mediums and, in the early twentieth century, via shrines in many households.[20] People in Kiziba maintained health and prosperity through a combination of daily activities and observances—ways of being in the world—that encompassed social relationships and royal politics.

Ziba patients in the late nineteenth century utilized diverse therapeutic resources, facilitated by practitioners whose powers derived from different kinds of knowledge and authority. Determining remedies and mediating their use was the realm of particular people skilled and initiated in divining the cause of illness or pain from diverse sources, a reality that colonial observers recognized, even as they dismissed mediumship or divination as sorcery or witchcraft. Both men and women served as healers, and often these people were elders in a given community.[21] Healing and spirit mediumship could be deeply intertwined, in Kiziba as elsewhere in the region, even as healers' relations with those around them and with political authority might shift in variable political or social winds. Embandwa, diviner-healers affiliated with cwezi or kubándwa spirit possession cults, were also part of the therapeutic landscape at the time, and offered alternative sources of healing and intercession alongside knowledgeable family elders and clan members.[22] Embandwa might be called upon to identify the cause of an illness of long duration or seriousness, or address multiple illnesses within a household—both signs of deeper imbalance or disturbance in relations between people and their ancestors or other cosmological forces.[23] Cwezi-kubándwa healers still practiced locally through the early twentieth century, and, indeed, seem to still have been attracting initiates: missionaries at Kashozi were spectators at a "party in honor of *muchwezi*" in 1907 and a German doctor in northern Kiziba would report that a "school for *mbandwa*" was active in nearby Bugabu in 1908 amid the sleeping sickness epidemic.[24] As well, the mukama maintained visible ritual connections with regional deities (such as Mugasha/Mukasa), linking chiefly authority to powerful forces affecting health and illness. Not all healers were mediums, however—some were consulted as knowledgeable men and women whose experience and learning had led to accumulated knowledge of botanical remedies, apart from the powers of spiritual intercession.[25]

The Ziba armamentarium included remedies for common ailments such as stomachache, headache, fever, or diarrhea as well as treatments for ailments that European observers linked to particular biomedically defined diseases such as gonorrhea, leprosy, smallpox, or tuberculosis.[26] Several of the healers' remedies cataloged by Rehse in the early twentieth century included botanical preparations or mixtures intended to be ingested or rubbed into the skin to ameliorate pain, lesions on the skin, or places of injury. Numerous remedies for broken bones and treatments for different kinds of wounds signal, for the modern reader, the important role that healers also regularly played in the aftermath of violent conflict or the daily enterprise of precarious, difficult physical work. The healers that Rehse encountered in the late nineteenth and early twentieth centuries dealt with ailments resolved through use of botanical medicines as well as those that necessitated propitiation of spirits.[27] Diviner-healers (embandwa) also provided preventive remedies, making medicines that people might wear on the body as amulets, as well as medicines to use under particular or precarious circumstances—a drink to be taken by a father during his child's birth, for instance, to secure the child's health.[28]

Traces of more grave illnesses, distinctive for their ability to touch wider proportions of the population, also surface in the historical record. Such episodes of illness indicate how the mukama's leadership could impact the kingdom's health and prosperity. Observing scarring that they equated with survival of smallpox, Europeans concluded that the disease had been known for generations, but was of a milder type and perhaps had resulted in only scattered infections—never "devastating epidemics."[29] Pox-causing illnesses were, by the turn of the century, known widely enough that people used the juice of a particular plant, *kyabakiriao*, as a salve rubbed on the inner leg if a case of pox appeared locally.[30] Ziba histories offer a more complex narrative of pox, however. Pox-causing illnesses known as *burundu* were known historically in Kiziba, associated with upheaval, disruption, mobility, and war, as illustrated by an example from the late 1850s. Lwamgira recounts epidemics of smallpox (burundu) following the Ganda army's arrival in the kingdom on the way to attack Kiziba's rival kingdom Kiamtwara. Ganda armies, which arrived late in the time of *kabaka* Suna and Ziba mukama Ruhangerezi II, were so numerous and so hungry that foodstuffs were exhausted for both the Ganda and Ziba: "because they ate all the food available and finished it and they resorted to eating the roots of banana trees and the country became a semi-desert as in the plains it was where the Baganda stayed."[31] In Kiamtwara, Ganda armies encountered illness, sickening Suna and triggering a return to Buganda once again via Kiziba. Raided by now unruly Ganda forces and threatened by neighboring kingdoms, Kiziba was also besieged by pox. Ziba responses brought further social dislocation alongside illness and death. In southern Kiziba, people took extraordinary measures, gathering the sick in "camps," Lwamgira's history recounts, because the sick "could not stay in villages with others *according to Kiziba*."[32] Ruhangarezi's authority and legitimacy were undermined alongside the strength of his armies and people, and local armies laid waste to the land—the mukama fled to the north and the surviving people migrated elsewhere due to famine and disease.[33]

The crisis during Ruhangarezi's reign in roughly the 1860s serves to underscore the connections between political legitimacy and population health—alongside insecurity from war and famine—in Kiziba. The heroic and elite-focused narrative collected and recounted by Lwamgira links illness with hunger and conflict as an exceptional phenomenon that occurred amid widespread upheaval. The narrative also indicates that practices existed in the kingdom to respond to widespread or spreading illness that involved local displacements and the reorganization of village life. Such practices—asserted here as "according to Kiziba"—coordinated local as well as palace political authority. In Kiziba, the return of prosperity was achieved through resolution of royal enmities with Kiamtwara, sealed with a cattle gift. A son of Ruhangarezi II would ascend as mukama soon thereafter in the 1870s and take the name Mutatembwa; he would reign during the arrival of German colonizers and, along with his sons, meet the arrival of rubunga and then the early warnings of a deadly wasting illness called *isimagira.*[34]

RUBUNGA: POLITICS, POWER, AND ILLNESS

Although populations knew burundu and other serious maladies, looming largest in the historical experiences of illness and wellness in nineteenth-century Kiziba was rubunga. Multiple sources discuss the novelty of rubunga in Kiziba and the Haya kingdoms, dating its arrival to sometime in the latter part of the nineteenth century. Lwamgira writes, "Rubunga never made a person wait for two days. He would fall sick and soon after die. All those that went to mourn at Mugenge's [sic] died and nobody survived. It then spread to other villages and killed people; and there had never been such a disease which killed people in that way. . . . People could not move because of fear of dying on the way or in the bush. . . . The disease stayed in Kiziba for two generations and all that time people continually died."[35] Lwamgira's narrative reminds us that trade around the lake connected populations along well-traveled routes, not necessarily contingent on the demands of Ganda, Swahili/Arab, or European proto-imperial trading networks that emerged in the nineteenth century. Indeed, the Oluhaya word *rubunga* is etymologically related to the verb *ku-bunga,* historically associated consistently (across nineteenth-century dictionaries to the present) and across space (in comparison with other Great Lakes Bantu languages, including Kinyarwanda, Kirundi, Lunyoro) with wandering travel, via the root verb *-bungo.*[36] Variants have taken on values of trade, filiality, or indigence.[37] Rubunga in its name speaks more to how illness might travel than the nature of the illness in an individual or its etiology. The quality of rubunga as a thing that traveled (ku-bunga) is underscored in the story of a returning traveler who had circulated in distant kingdoms. The traveler who brought plague into the kingdom, in this narrative, had gone far to the north, perhaps near Lake Edward in Ankole, and returned on a different route via Ishanje, likely moving along long-standing salt trading routes.[38] The traveler Mugengere rejected warnings in Ishanje of the deadly rubunga there, asserting that "Rubunga had never killed the Baziba."[39] Implied here is deep familiarity with the disease but also the potential of variable virulence known to those who had experienced it. But this time, kill rubunga did: two days after the traveler's return to his home in Kanyigo in northern Kiziba, his two sons died, and mourners who came to his home died as well. This event marked the arrival of rubunga in the kingdom's nineteenth-century history, according to royal narratives and subsequent colonial surveys of the Ziba population.

Rubunga's qualities included, foremost, a swift and visible decline in health and a fast death. German- and French-speaking observers readily equated rubunga with bubonic plague (as *Bubonenpest* or *peste bubonique*, also using the more general *Pest/peste*). Some of what made Pest distinctive to German observers also likely made rubunga alarming for Ziba observers—overheated, weakened bodies with unusual and dramatic swelling in consistent, specific places (the armpits, the groin) and suppurating, blackening skin at sites of swelling. Death usually followed these signs. Ziba narratives do not offer an etiology for rubunga in the same way that Ganda sources addressed the cosmological origins of kaumpuli's appearance in populations. We should not assume a direct equivalence between rubunga, kaumpuli, plague, and Bubonenpest in Kiziba, despite German desires to establish this. But in analyzing the ramifications of rubunga, I operate with the understanding that Ziba and German narrators described congruent visible impacts on the body when some said "rubunga" and others said "Bubonenpest," irrespective of diverging nosologies or etiologies.

Though German colonial reports do not document personal observations of Bubonenpest until 1898, combining Ziba, missionary, and German narratives allows us to approximate when rubunga began to cause widespread illness and death in northern Buhaya generally and Kiziba specifically in the nineteenth century. Lwamgira's narrative of Mugengere returning from the north with rubunga echoed in German surveys of the Ziba population; the story of a traveler returning from across the Kagera and bringing rubunga/Pest apparently was widely known and remembered as occurring around the mid-1870s.[40] The perceived advent of rubunga in the late nineteenth century offered an intuitive connection, for European observers, between the arrival of plague and the expansion of regional and subcontinental trade networks; this fit with their own understandings, from European history, of how a disease like plague might travel.[41] (In wider geographical and chronological perspective, however, Ziba chronologies of rubunga notably situate it well before the advent of the third pandemic of bubonic plague in 1892, and therefore indicate that global histories of plague must account for African outbreaks on a different timeline.)[42] By the 1890s—in range of the "two generations" during which rubunga troubled Kiziba, according to Lwamgira—German colonial officials had become acutely concerned about Pest/plague in the coastal areas of the Bukoba Residency.[43] At this point, sources identify the illness present variously as rubunga, Pest, *Beulenpest*, plague, or even the Luganda kaumpuli, signaling that a set of potentially overlapping and entangled contemporary nosologies were in use.[44] Reports of the serious and deadly illness become regular elements in colonial and missionary texts from the mid-1890s. In June 1896, Pest cropped up in a colonial report alongside notes that, in Kiziba, the year was dry enough that the bananas were suffering badly, implying that the spring rains had failed and correlating Pest with hunger.[45] White Fathers missionaries recounted in November 1897 that colonial medical officer Dr. Maximilian Zupitza's "study of kaumpuli/lubunga" focused on the southern part of Kiziba, specifically the farms around Bugandika controlled by Rweshabula, a powerful chief and son of mukama Mutatembwa.[46] By July 1898, missionaries at Kashozi began to note that "lubunga . . . increased more and more" in the northern areas of the kingdom as well.[47] Rubunga thus emerged as a wider problem in Kiziba, causing sickness and death and entangled with political turmoil. Its spread also signaled, to Ziba populations, profound trouble for authorities.

While Ziba and Haya sources reveal relatively little about the precise signs of rubunga that people may have recognized, accounts from the 1890s reveal the adaptability and range of Ziba measures to contain the disease, at the level of individual households and villages as well as within the purview of the mukama's court.[48] We gather from Lwamgira's narrative that people mourned victims of rubunga in keeping with established ways, by coming together at the home of the dead. But the illness's severity and widening impact led people to change their usual behavior—to fear travel, to stay put—because of the risk of falling ill and dying away from one's home and kin. Because rubunga was an illness that seemed to spread, too, Ziba responses had multiple impacts on mobility within the kingdom. On the one hand, people wanted to avoid taking long journeys—a significant change, given the importance of regional trade in local economies and Ziba engagement in it. On the other hand, people apparently altered how and where they lived during rubunga, moving away from their homes and banana groves and out into the "steppe," that is, onto uncultivated land away from homes and banana groves.[49] There, people built other temporary homes, continued to travel back to their banana groves to farm, and waited out rubunga.[50] They also, in some cases, set sick people apart from those unaffected in those new homes and allocated someone to care for the sick.[51] By 1902, the colonial narrative of Ziba beliefs and practices in response to plague reflected an accumulated consensus, as summarized by colonial medical officer Dr. Oskar Feldmann: "In the observations of the natives, every epidemic of plague is preceded by a die-off of rats. In an outbreak of plague, all of the inhabitants leave their homes (including those sick with plague, as much as they still can) and install themselves anew outside of the affected farm. The sick live strictly isolated, and a relative, who voluntarily takes over care, shares the seclusion."[52] Time spent living in these steppe homes, apart from one's permanent home and one's farm, could range into many months, until illness subsided and return was judged safe. People dispersed (and, colonial officers hastened to note, were harder to find) but generally remained nearby their homes and farms.[53]

Building steppe homes in proximity to established residences and farms indicates that continuity of farming and land use was an abiding concern, fitting with systems of land tenure and agriculture centered around banana cultivation and the maintenance of patrilineal and clan claims.[54] Ziba practices of establishing the location of home and community within familiar spaces and landscapes changed while rubunga spread; these changes allowed communities a measure of resiliency in both social and economic production. Some people turned to nearby missionaries as well. The Catholic mission in southern Kiziba at "Marienberg" (now Kashozi) provided another source of assistance amid the turmoil of rubunga. The Kashozi compound was situated centrally in Kiziba, along main roads to Bukoba and Kiamtwara and near the Ziba court; travel between kingdom, mission, and German boma flowed overland on the same routes traveled by caravans. White Fathers priests stationed at Kashozi wrote in October 1899 that "people flock to our villages. Several, fleeing the plague [*la peste*] raging in neighboring countries, take refuge with us."[55] The Kashozi mission and its environs would have provided people with a place to live, or perhaps peripheral space to build a temporary home, as well as food amid the turmoil of relocating during rubunga.

German colonial rule in Buhaya expanded concurrently with rubunga, as colonial knowledge of populations increased as a result of efforts to monitor the illness and widespread efforts to avoid its toll. Scholarship evaluating the relationship between colonial rule and public health campaigns typically addresses these phenomena, when explicitly linked, as two sides of the same coin: European imperial and colonial efforts to shape African culture, economy, and politics were achieved through efforts to combat epidemic disease or change healing practice, the efficacy of which in turn was used to underpin assertions of European superiority.[56] However, the example of rubunga in Kiziba at the turn of the century indicates that colonial public health interventions could dovetail with existing African strategies for mitigating the impact of illness on broader populations. Available sources suggest that certain preventive measures met with positive reception, not (to paraphrase German appraisals) because of their inherent rationality, superiority, or effectiveness, but because they fit with established Ziba ways of navigating widespread illness. In other words, German anti-plague/anti-rubunga interventions gained traction not because African subjects were gradually being brought over to European colonial, biomedical rationales, but because they contained elements that accorded with Ziba intellectual worlds. For elites, many measures also had the added benefit of political expedience. Two preventive measures—burning the corpse and possessions of the dead, and responding to the die-off of rats—therefore illuminate the multiple, overlapping frameworks within which early colonial disease prevention efforts had meaning for affected Ziba people.

Measures to isolate the sick, as well as practices to establish physical distance between those who were well and a place where sickness had broken out, had precedents in Kiziba upon which colonial mandates could be layered. Colonial officials mentioned Ziba elites' enthusiastic adoption of preventive measures such as burning the corpse, possessions, and home of a person who died of plague. Dr. Eggel, recently arrived in the region, noted in a January 1900 report, "I have made it compulsory for all of the sultans and katikiros to incinerate every plague victim's corpse (*Pestleiche*) in their hut with all possessions. All have agreed and readily complied with this measure, because it is in the sultans' interest to reduce cases of sickness and death as much as possible."[57] Eggel's aggressive measures and his acknowledgment that minimizing illness and death in the Haya kingdoms was "in the sultans' interest" underscore the importance that disease control had for the German colonial administration, and therefore the administration's appraisal of and support for the Haya kings who cooperated with German directives. In the context of his total report, however, this comment reads as a flash of recognition regarding the political importance of population health for African authorities and the heightened stakes of rubunga. Controlling misfortune and enabling prosperity within a polity were responsibilities inherent to royal and chiefly legitimacy. Intentionally burning bodies, possessions, and homes may not have had extensive precedent in Haya life proximate to illness; it appears to have more commonly been a consequence of accident or warfare.[58] Diverse sources suggest, however, that incineration of corpses would have strongly contradicted traditional, ordinary burial practices, in which bodies were interred and graves maintained on family or clan-held farms. Burned bodies or homes required particular rituals to mitigate endangerment to the living.[59] The distinctive illness and death that rubunga caused, however, opened up room for such departures from normal practice to occur. But the serious implications of such departures for social relations and connections between the living and dead would have required management by ritual-political authorities—chiefs and the mukama—and perhaps also coordination with embandwa mediums.

Colonial reports from the same era are peppered with notes of regularly discovering *Aschenplätze* (ash-places), the burned-out spots that doctors took as telltales of a recent plague case. In correlation with reports of plague mortality at the time, these remnants indicated that communities consistently burned down the homes of rubunga victims, and indicated ongoing societal coordination and mediation by Ziba political authorities. By 1901, anti-plague/rubunga practices had regularized such that Dr. Robert Ahlbory (who would later work in the anti–sleeping sickness campaign) could note, "Prevention measures were the usual and those familiar for a long time among the Negroes: isolation of the sick and those with contact with them, above all those who lived in the same huts. Burning down of huts. Destruction of rats. Closing of traffic and quarantine measures."[60] These typical measures incorporated practices with local historical precedent (isolation of the sick and their contacts, refraining from travel) as well as imported practices such as formal quarantine, in this case the observation of a ten-day isolation period upon evidence of plague. But these measures were also extraordinary departures from the everyday, reactions to a traumatic event that necessitated such drastic activities as burning the bodies of relatives and destroying one's home and valuable household possessions.

But what to make of the "destruction of rats," a regular aspect of colonial reports from 1901 onward?[61] Doctors' accounts of populations in northern Kiziba eagerly introduced rat-plague connections, with one observer asserting that people looked out for sickening and death of rats in the banana groves, which was "seen as a harbinger of oncoming plague [*Pest*]."[62] Elsewhere, a doctor noted that evidence of burned-down homes was absent, because villagers had fled into the "steppes" as soon as rats began to die, thus avoiding infection.[63] German reports were eager to emphasize a historic awareness among Ziba communities of rat illness and death as a harbinger of disease in human populations; this established parallels with historic European sensibilities regarding plague and spoke to some scientists' desire to hew to a broader consensus about plague in the age of empire.[64] Despite colonial doctors' grasping at rat die-offs as events precipitating subsequent avoidance of "infected" houses or people, however, such a direct association by Ziba populations between rat and human deaths is not evident in Ziba or wider Haya sources. But this does not preclude an association between animal and human death within Haya rubrics of wider misfortune or illness. Another 1901 report mentioned that dogs, chickens, and snakes could be struck by rubunga as well as rats, according to Ziba informants—a point that clearly flummoxed the German doctor increasingly committed to the rat-flea-human cycle of plague transmission. Ziba understandings of rubunga incorporated its effects on humans and animals in the domestic sphere (i.e., chickens and dogs) as well as the realm of ritually symbolic animals (i.e., snakes).[65] Its impact was wide.

The question of human-animal associations in plague/rubunga between German and Haya interlocutors also speaks, fundamentally, to processes of knowledge production. Addressing the potential connections between human illness and animal illness offers further opportunity to consider how extant and imported disease prevention measures meshed, but also drives home the instability of European etiologies of plague in a time of aggressively expanding public health work. Consistent German attention to Ziba associations of rats with plague/rubunga seem, rhetorically, to signal German doctors' awareness of rat-flea-human cycles of plague transmission posited in the late 1890s. Their insistence that the roots of Ziba beliefs and activities lay in German persuasion, rather than extant Ziba practices, reinforced their own conversions to a now-scientifically verified, standardized understanding of plague etiology evident, for instance, in their increasingly regular discussions of microscopic verification of the presence of the plague bacillus.[66] We also know that German doctors circulated more frequently through Ziba and other Haya communities and gathered information about illness and death with more regularity in the years when plague presented an acute colonial concern. They apparently carried into the field particular (if not always stable) early ideas about rat-flea-human transmission of plague, which structured the kinds of inquiries doctors made about individual, household, or village experience with rubunga/Pest, and therefore possibly cued their informants' particular responses or prioritizations of events in a sequence. Effectively a sampling bias, this possibility once again underscores the extent to which Germans and Ziba may not have been speaking about the same problem when discussing rubunga.[67] We might also read colonial discussions of Ziba practices to represent tentative efforts to find commonalities between European and African "medical" systems, though never absent assertions of the superiority of European biomedicine over everything it might meet. Ultimately, evidence offers an enduring perspective that efforts to understand and prevent plague/Pest/rubunga involved both overlaps in practice and congruence in the aim of preventing further deaths. Such overlaps or congruencies would enable collaboration or compliance, regardless of whether all participants—Ziba, Ganda, or German—acted based on the same logic or within the same nosological or etiological frameworks.

COLONIAL PLAGUE RESEARCH AND ZIBA POLITICAL CRITIQUE

Ziba responses to rubunga call attention to another way that the 1890s were a strange and challenging time in the kingdom. Episodes of rubunga were concurrent with turmoil within the Ziba palace, as long-standing enmity between two powerful rival princes intensified as their father, mukama Mutatembwa, failed in old age.[68] Plague-focused research, colonial health work, and Ziba strategies to weather rubunga had wider political implications and relevance. Politics in Kiziba in the 1890s oriented around triangular relations between the Ziba royal court and its district chiefs, Catholic White Fathers missionaries, and the German colonial boma at Bukoba town. Coping with la peste/rubunga/Bubonenpest brought missionaries, colonial officials, princes, and ministers into closer engagement with one another, but their priorities sharply diverged. Initial inquiries into the sources of death and illness in southern Kiziba in 1896–97 generated tension between boma, mission, and palace; the ensuing conflict would play out over the next seven years. The impact of rubunga and its management would redound in the longer term to the sleeping sickness epidemic.

First, a brief synopsis of the core events.[69] The aging mukama Mutatembwa's two most prominent and powerful sons, Rweshabula and Karutasigwa, were half-brothers. Each served the palace as a district chief responsible for a large territory of the kingdom, Rweshabula in the south and Karutasigwa in the north. Poor relations between the two sons flowered into outright conflict as they jockeyed for the mukama's favor and control of the kingdom upon his death. As early reports of sickness and death began to reach the boma in 1896–97, colonial attention fell on the villages and farms around Bugandika, in the area of southern Kiziba under control of Rweshabula (see map 4.1). Dr. Maximilian Zupitza chose Bugandika as a base to conduct his research on an outbreak of illness suspected to be bubonic plague in late 1897.[70] As Zupitza's research proceeded, wider regional turmoil intervened as the Bugandan kabaka-in-exile, Mwanga, escaped from captivity in the Bukoba fort and found refuge among allies in Kiziba (including Karutasigwa). This drew the ire of the German authorities, who feared looking foolish to their British counterparts to the north, for whom Mwanga was a constant political frustration. So the boma kept a wary eye on both princes, if for different reasons.

Zupitza's desire for research material—blood and corpses—to examine in his field laboratory at Bugandika soon made a bacteriological inquiry into a political mess. His research required blood samples from the sick and, ideally, conducting autopsies on the dead. From the perspective of late nineteenth-century bacteriological research, microscopic identification of a specific pathogen was central to establishing the cause of an epidemic; further exploration of the pathogen's impact on specific organ systems in a cadaver confirmed the presence of a specific disease causing death.[71] But procuring bodies, blood, or lymph was and still remains a fraught process, all the more so in circumstances absent consensus between researcher and subject on such matters as the relationships between the living and the dead, the systems governing the treatment of the sick and the dead, and the motives and meanings behind how sick people or corpses were treated. Kiziba in 1897 was precisely such a situation where researchers and subjects came into conflict, as Fr. Van Thiel of the White Fathers recounted: "Telling people . . . 'Bring me your plague-stricken, sick or dead, so that I may study them, cut them, and slice them open': it's a bit unsettling. Hence the absolute refusal to bring [Zupitza] the sick. Likewise the dead, who they inter in secret and, counter to their customs, without mournful singing. Hence also mocking remarks they aim at the Doctor: 'This White, is he a sorcerer? Does he want to kill, flay our brothers? . . . eat them? Cut the hearts of our dead and bring them to him: never!'"[72] Local recalcitrance to provide Zupitza with bodies for autopsies led him to formally demand corpses from mukama Mutatembwa, on the premise that Mutatembwa could command all his subjects to comply.

Zupitza's demand for plague-stricken bodies, alive and dead, constituted a significant departure from common practice, even if we consider that rubunga likely already disrupted many aspects of Ziba life. To retain, expose, and cut open a dead person's body interfered significantly with proscribed processes of burial, into which were also folded transitions of familial leadership, management of the care and livelihood of surviving spouses or children, and distribution of the possessions of the dead. The potential for wider disruption in an individual's life or in one's relationships as a consequence of meddling with the bodies of the dead underscored Haya practices around death and burial. Funerals also provided opportunity to reinforce social ties through sharing of plantain beer and gatherings around mourning and burial, with burial generally occurring within a day of death.[73] The burning of homes and corpses of rubunga victims within them was already an extraordinary measure. Here, German research sought to alter long-standing funerary practices by changing social participation in the transition between life and death. The claiming and taking of bodies for the European doctor's use not only flouted Ziba funerary traditions, but also implied malevolent activity: cutting the bodies of the dead was rumored to be part of sorcerers' work, involved in the preparation of deadly medicines used to poison unsuspecting people.[74] The taken parts of rubunga victims' bodies could thus endanger anyone, threatening peace and health.

Mutatembwa refused to provide Zupitza with corpses of the plague-stricken. In response, Zupitza demanded the dispatch of a Ziba katikiro (a court minister) to Bugandika, which sat in Rweshabula's districts. Zupitza or accompanying *askari* (African soldiers in colonial service) then hung the katikiro upon his arrival at the doctor's camp. The killing of the minister set Mutatembwa and Karutasigwa against Rweshabula, whose cooperation with Zupitza was presumed to be part of wider intrigues. Palace reprisals against Rweshabula included burning homes in his villages, which then triggered a German pursuit of Mutatembwa and Karutasigwa, ostensibly in defense of their ally. Mutatembwa evaded the Germans for months, ultimately leading the Germans to cease outright war, nominally depose Mutatembwa, and divide Kiziba between the control of Karutasigwa and Rweshabula, with the latter serving as a sort of prince regent. Here, the preferences of the Catholic mission and the German boma aligned, rooted in Rweshabula's generally sympathetic reception of Catholic missionizing.

The wider political repercussions of plague research continued to be significant. While Zupitza's appeal to the mukama certainly fits within German colonial models of indirect rule, the hanging of the katikiro was an unexpected aggression, even acknowledging that colonial judicial processes and determinations of punishment tended toward escalating violence. To kill a king's emissary was foremost an attack on the king; further, because Zupitza's work with sick and dead bodies already hinted at dangerous, malevolent, supernatural activities, the katikiro's murder likely had even higher stakes for the perceived safety of the mukama and the kingdom. Zupitza appears to have touched off the war between Rweshabula and the allied Mutatembwa and Karutasigwa, with widespread and years-long implications for life and livelihoods in Kiziba.[75] The events of late 1897–98 seem to have been a conflict that none of the actors involved would have chosen, where plague research provided a spark to the tinderbox of familial and political rivalry. For a few years following the disturbances in the divided kingdom, Rweshabula enjoyed the favor of the colonial administration and ongoing good relations with the White Fathers, the missionaries also helped to mediate between the Germans and Karutasigwa and remained engaged with both Karutasigwa and Mutatembwa. The aging mukama preferred, by most accounts, that Karutasigwa succeed him, and the two appear to have been allied during Rweshabula's regency. The division of the kingdom kept an uneasy peace for a time.

Multiple tellings of the controversy in Bugandika illustrate the diverse ways in which plague prevention and politics meshed. For Ziba villagers, a central concern was the threat posed by technologies and practices of colonial plague research to the ability to care for the sick and dead, as well as the wider problem of avoiding rubunga in customary ways while also contending with warring leaders and wider insecurity wrought by conflict. For Ziba leaders, though hardly unified, the implications of ongoing (and somewhat unpredictable) German interventions for the stability of kingship and political control loomed large. It would not only be in matters of politics or trade, but also in matters of health and illness, that elites would have to contend with colonial interference and the threat of punitive violence. For German officials, the perception of obedience and cooperation was paramount—a lack of control amid a looming epidemic was great cause for concern. For missionaries, conversely, the potential long-term gains of a stable, fertile context for further growth of Christianity remained central and required a degree of neutrality. While they lamented the violence and disruption, they worked consistently to stay in contact with all rival parties and continue recruiting catechists and students from throughout Kiziba.

But even as German concerns about plague introduced elements of complexity and uncertainty to Ziba politics, they could also be navigated successfully, perhaps to the wider benefit of Ziba leaders. A period of brief calm in Ziba politics and society ended when further cases of plague appeared in 1899. Thereafter, German opinions about Karutasigwa, and politics in Kiziba more generally, began to change markedly. It was Rweshabula's lack of cooperation with colonial interventions that elevated Karutasigwa's legitimacy and suitability to rule in German eyes. This outbreak of plague in 1899 had, according to Bukoba Station doctor Eggel, "purely external and indeed political origins." Eggel described Rweshabula later in 1901 as an "extraordinarily scheming and greedy person, who lived in constant strife with his brothers, [and] oppressed and heavily taxed his subjects so that peace never prevailed in the country." Rweshabula's governance created strife and instability, triggering migration into other districts, so that "plague was time and again carried anew and all of the measures taken against it proved insufficient, because peace did not prevail in the country." By contrast, Karutasigwa and another brother, Kamgumia, were "calm, beloved by their subjects and fully acquiescent to German rule," he noted, and he considered Karutasigwa additionally "very intelligent." The doctor asserted that Karutasigwa, further, thoroughly endorsed the German administration's rules of action against the epidemic, "realized their benefit," and notified authorities immediately when plague appeared.[76] As a consequence of Rweshabula's perceived inaction, German authorities punitively diminished his territory to a single village. When Rweshabula was deposed as prince regent in 1900, Karutasigwa gained effective control of his ailing father's kingdom. German officials asserted a wholesale turnaround for Kiziba with regard to plague as a result: Karutasigwa and Kamgumia's compliance with German regulations allowed the suppression of further cases of plague and "the character of plague in Kiziba seemed to gradually become more benign," with lower mortality rates.[77] Eggel's report, though still dedicated to the seriousness of the problem of plague, is shaded with a palpable sense of relief at the possibilities that "peace" and calm in Kiziba could bring for the diminished spread of plague and its more effective control. When Mutatembwa died in 1903, Karutasigwa succeeded him, taking the name Mutahangarwa, with the full support of German authorities at Bukoba. Rweshabula, for his part, remained suborned as a minor chief of a single area near Bugandika, on the southern edges of the kingdom.

Stories of out-migration from Rweshabula's districts into those of other Ziba leaders in 1899–1900 suggest that increased mobility in response to rubunga persisted, but the added detail of movement into other chiefs' territories indicates that these moves were of a different character than temporary oscillations between village and steppe. Faltering prosperity and increasing illness would have undermined Rweshabula's legitimacy as a good custodian of his province, even if he was not the heavy-handed, greedy oppressor that German reports made him out to be.[78] German perceptions of uninhibited mobility in his kingdom suggest a loss of political power: as conditions worsened, Rweshabula's ability to influence his subjects diminished. Rweshabula fell afoul of his subjects and German administrators in trying to manage the difficult quandary that rubunga presented. Karutasigwa, however, did not, despite the continuing presence of Pest/rubunga in Kiziba—cases continued to crop up.[79] The enmeshed implications of rubunga, with bearing on both Ziba palace politics and Ziba-German relations, must not have been lost on Karutasigwa. His knowledge of the retributive murder of a high palace official as a result of Rweshabula's noncompliance with Zupitza's plague research in 1897 was certain, but his awareness of the importance of a healthy, prosperous, and unharassed population in his own districts was equally so. Cooperation with German anti-plague regulations established, and then cemented, Karutasigwa's reputation as an "intelligent" ruler, an appraisal that would endure through the end of German overrule years later.

The controversy over colonial plague research in Bugandika underscores that widespread reported compliance with colonial antiplague instructions to burn the houses of the sick or ill amid rubunga masked complex processes deeply enmeshed with Ziba politics. Accounts of voluntarism further require a cautious approach, because violence, personal or collective, was always well within the realm of possible German responses to Ziba activity. Though discrete punishments for not complying with anti-plague directives in the villages and outlying districts are absent from doctors' reports on plague, fines and imprisonment were specified as consequences for avoiding or escaping quarantine at Bukoba, for instance.[80] The threat of coercion was ever-present. But we might also read the story of Rweshabula's fall and Karutasigwa's rise as resulting not from the convincingly dominant, coercive power of the incipient German colonial state, but from an astute, entrepreneurial use of the different meanings that rubunga or Pest had for the various forces affecting the Ziba kingdom. Karutasigwa ascended to the Ziba kingship because he balanced his subjects' expectations regarding chiefly responsibility for widespread security—here, minimizing and managing German interference *and* taking steps to mitigate rubunga—as well as German expectations regarding how a cooperative, "modern" chief would behave. Karutasigwa and his subchiefs, attending to illness and death in their districts, seem to have understood some German-introduced activities as effective, perhaps familiar, and valuable. His aggressive measures to limit illness—destroying old habitations, taking seriously animal and human sickness, and isolating the sick—grafted new and old practices, new and old politics, in the interest of maintaining stability, prosperity, and power. This, as would emerge during the sleeping sickness epidemic in the decade to come, proved a strength of Karutasigwa, enthroned as Mutahangarwa. Rubunga provided a practical model of how health and politics could intertwine in the early colonial era, one that shaped subsequent responses to widespread death and disease. It served as a foundational experience for the implementation of both Ziba and German understandings of disease prevention in this early colonial era, during a time of significant change in diverse aspects of Ziba life.

4 Gland-Feelers, Elusive Patients, and the Kigarama Camp

EXPERIENCES OF rubunga in the late nineteenth and early twentieth centuries connected to subsequent episodes of widespread illness and death in multiple ways. Rubunga would provide a durable reference point in Ziba society, offering strategies for reorienting domestic life to avoid illness with displacements and returns to villages. Coping with rubunga opened up possibilities for more drastic changes in social and cultural practices around care of the sick and the dead, for which mediation by royal and chiefly political-ritual authorities was crucial. Further, experiences of controlling rubunga/plague provided a model for engagement between Ziba political authorities and colonial administrators. For the wider Ziba population, experiences of rubunga set the basic parameters for what to expect of colonial public health interventions as well as their leaders' engagement with them in the early colonial era. The political and tactical status quo produced through efforts to curb rubunga would prove important in the years to come.

As rubunga abated and the Ziba kingdom quieted under Karutasigwa's regency, a new kind of illness and death began to appear. Known locally as *isimagira* and possibly also as *botongo*, this form of illness and death weakened and sickened the afflicted, changing their bodies and temperaments, and caused strange patterns of sleepiness. Many died. The epidemic disease that the Germans called *Schlafkrankheit* had come to Kiziba's shores. With sleeping sickness, after rubunga, also came innovations in both politics and health. This chapter examines one such innovation, the sleeping sickness camp near Kigarama in northern Kiziba. As the chief German sleeping sickness camp on Lake Victoria between 1907 and 1914, the Kigarama camp was a new site of treatment, research, and commerce. The camp's work to find and treat sleeping sickness cases in Kiziba, particularly colonial efforts to draw in and retain patients, affected economic and political relationships within the kingdom.

This chapter explores the establishment and expansion of the Kigarama camp by German colonial and Ziba royal authorities. I fit the camp into the political economy of land and labor in Kiziba and link the village and camp to persistent Ziba political and social institutions and ideologies, expanding on historical scholarship on sleeping sickness interventions that have focused primarily on colonial institutions. This particular sleeping sickness intervention, which had shifting Ziba royal and German colonial support throughout its tenure, altered economic and social aspects of relations between colonial authorities, the Ziba court, and the kingdom's population. The camp remained connected with traditional forms of power, but also created relationships that allowed individuals and families to negotiate for material goods, cash, or access to land. It functioned as a place to access resources and therapeutic alternatives that people initially chose to utilize and later chose to avoid, expanding and eroding both royal and colonial authority in turn.

SLEEPING SICKNESS PREVENTION AND THE POLITICS OF LAND AND LABOR IN KIZIBA

In the early twentieth century, explosive mortality triggered by epidemic sleeping sickness and widespread famine hit Buganda, the Ssese Islands, and Busoga. Coastal areas on the northern shores of Lake Victoria were depopulated as people died, fled, or were later forced to relocate according to British colonial policies. Ziba activity in farming and fishing in coastal areas on the western lakeshore changed as a consequence of these wider disasters. Ziba people continued to fish in the rich coastal waters, using fishing camps on Bugala Island, but people also moved to farm abandoned banana groves and gardens on the Sseses.[1] This increased migration also connected Ziba people with sleeping sickness, as people previously unexposed to the disease in Kiziba now farmed and fished in areas that were thick with tsetse flies. As these seasonal migrants and travelers returned to Kiziba, they brought the disease with them.

Ziba sources do not mark the initial appearance of epidemic sleeping sickness in the kingdom, but colonial medical interest first fell upon Kiziba in 1903, when the specter of sleeping sickness first arose for the German

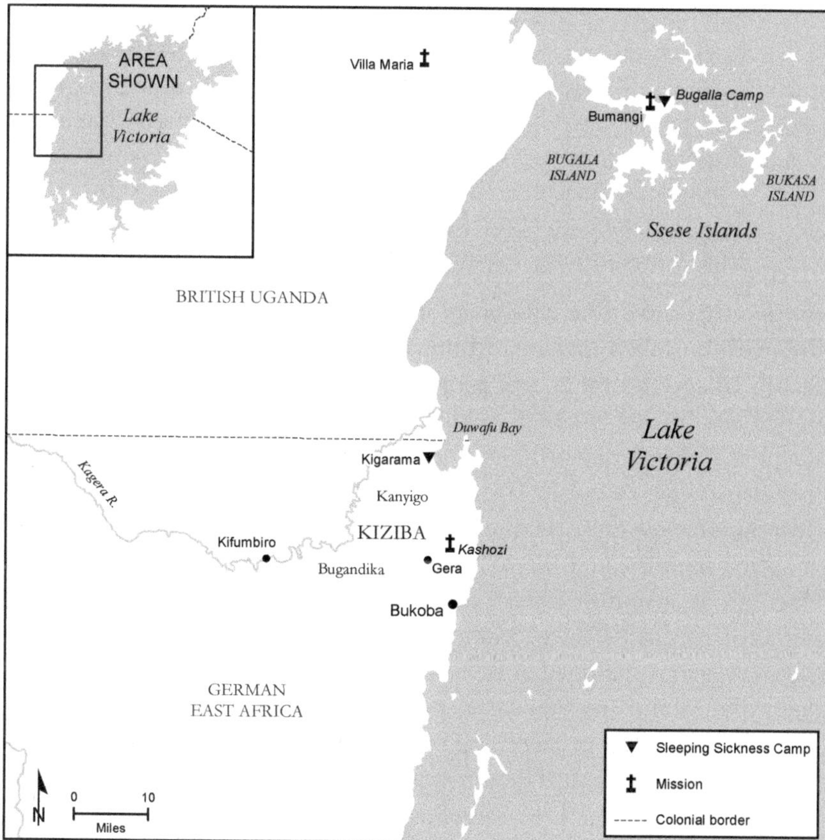

MAP 4.1. Kiziba. Map by Brian Edward Balsley, GISP.

colonial administration. Initial attention focused on the small kingdom of Bugabu, which ruled the peninsula across Duwafu Bay from Kiziba and lay due north of the German headquarters at Bukoba.[2] But the illness was more widespread. Missionaries at Kashozi, inland between Bukoba and the Bugabu kingdom, reported in correspondence in late 1903.

> Until the last year, there was never a question of this terrible illness. But several months ago, we became worried, the military station's doctor had a look and finally, we found some twenty sick people, all of whom we took to the hospital located near Bukoba, whether or not they came willingly. Some ten of the sick have died there. But it is notable that all these sick people, at least as recognized up to this point, have brought their illness from Uganda where they had gone to get rich. . . . Nevertheless, the authorities thought it necessary to take certain measures,

and for several months, we have been in quarantine, which does not bode well for the exchange of couriers who get lost or at least come to us after long delays. According to our best and brightest, the famous *Glossina palpalis*, a type of tsetse that spreads the sleeping germ, a parasite named, so it seems, Trypanosoma Castellanii, does not exist at all in our neighborhood.

Let the Good Lord keep us from this terrible fly that wreaks such havoc among our neighbors to the north.[3]

Around the same time, additional cases were discovered in Kiziba when the newly installed mukama Mutahangarwa, carrying out instructions to search villages for other sick people recently returned from Uganda, reported discovering two men hidden by their relatives.[4] People found with sleeping sickness were all believed to be "young men from the north of this district, who went to Uganda to seek work and, after they acquired sleeping sickness there, had returned to their relatives."[5]

Ziba mobility and ties to coastal Buganda kept colonial eyes on the kingdom as sleeping sickness mortality continued to spread around the northern rim of the lake. Monitoring and surveillance intensified. The German doctor stationed at Bukoba, Oskar Feldmann, examined the residents of northern areas for signs of disease, and, following practice during the outbreak of *Pest* in preceding years, ordered the homes of the sick that he found to be burned.[6] The sick were moved into isolation in the Bukoba military hospital and Feldmann instructed the mukama to continue to search among his subjects for other cases. After maintaining a watchful eye on scattered cases, colonial energies came fully to bear on Kiziba as Robert Koch's expedition on the Ssese Islands identified sick Ziba fishermen living in camps on the northern part of Bugala Island. An expedition doctor would ultimately come to Buhaya in 1907 to investigate reports of sleeping sickness in the area, focusing on Kiziba and tracing the routes of Ziba seasonal migrants and fishermen between Kiziba, the Ssese Islands, and coastal Buganda. Coming to Kiziba, German doctors sought people with sleeping sickness and a place to base their research and disease prevention efforts, modeled after Koch's camp at Bugala. Initial searches of villages nearest the coast yielded the potential patients and subjects they sought; collaboration with mukama Mutahangarwa led to the creation of the sleeping sickness camp near Kigarama.

Kigarama, in Kanyigo Province, occupied a distinctive space in the kingdom geographically and economically. Kanyigo was the only province in Kiziba with direct access to the lake and was also, in the early twentieth century, the kingdom's most populous province; Kigarama, for its part, was the largest community near the lake in the province. Alternating ridges and swamps dominate the landscape in Kanyigo district, as one moves west from Lake Victoria toward the Kagera River. Kigarama sits on a narrow plateau atop one such high ridge in the northern reaches of Kanyigo. Seasonal swampland lies below to the west and Lake Victoria sits beyond a steep slope to the east. On the ridge where Kigarama sits ran a frequently used road toward Buddu to the north and thence to Buganda. Access to the lake flowed through the province to landings on what is now called Duwafu Bay. The colonial border between German East Africa and British Uganda lay an hour and half on foot from the village by routes used in the early twentieth century.[7] Further roads and many scattered footpaths connected Kigarama to nearby villages, and, to the south, to the Ziba royal court at Gera, two Catholic missions run by the White Fathers at Mugana and Kashozi, and, ultimately, to the German colonial station at Bukoba, the area's largest town. In 1907, Kigarama was a village of several hundred homes scattered among banana farms, with population roughly estimated by colonial officials to be at least a thousand people, making it one of the larger villages in Kiziba.[8] Economically, it was important locally for its landing on the lakeshore and for its regular market. Its orientation to both the lake and hinterlands made Kigarama a place of connection between Kiziba and the lakeshore communities of Buganda, Buddu, and the Ssese Islands to the north and east. As a village near the lakeshore, it was in a zone of potential exposure and opportunity—simultaneously vulnerable to the predations of Ganda armies and well-positioned for Ziba to access both nearby caravan routes and the products of the lake.

A key factor in the organization of the German campaign in Kiziba was scientists' confidence that the tsetse fly vector was absent from areas hit hardest by the disease in Bukoba district: in this case, Kiziba and the neighboring lakeshore kingdom of Bugabu. They believed that Kiziba and Bugabu were not themselves "sources" or foci of the infection but rather "a stage on the path of the disease," and interventions there were "essential to impede further introductions [of the disease], and also to prevent the possibility of carrying the disease into other territories."[9] These two kingdoms could thus serve as buffer zones between extant epidemic foci to the north and known fly belts to the south, effectively protecting key towns and caravan routes in central German East Africa from the Uganda epidemic.[10] Efforts to control sleeping sickness in Kiziba, then, were directed toward identifying existing cases, treating those cases with drugs that would hopefully suppress trypanosomes in their bloodstream, and preventing people's further movement into areas where the fly vector was present. The Kigarama camp was a targeted intervention, designed to disrupt transmission of sleeping sickness believed to be imported from Uganda, and where clinical observations of possible anti-trypanosome therapies could also continue. Drugs administered there were either experimental or known *not* to cure trypanosomal infections. In the latter case, the primary drug used (atoxyl) was valuable chiefly for its potential to break transmission and had known harmful side effects. In order to be effective at curbing transmission, the Kigarama camp staff had to either discover or attract people already sick and carrying trypanosomes in their bodies, proffering medicines that doctors knew would likely not cure the fatal infection but incentivizing examination and treatment nonetheless.

The sleeping sickness camp, a place of confinement, isolation, and treatment, did not develop de novo in Kigarama; rather, it derived from the work of the German sleeping sickness expedition in British Uganda led by Robert Koch. Links between Kigarama and Koch's Bugalla camp were direct. Koch's assistants at the Bugalla camp became the coordinator and key officers in the campaign at both Lake Victoria and Lake Tanganyika. Dr. Robert Kudicke, a colonial medical officer detailed to work with Koch, had been dispatched to Kiziba in 1907 to set up a sleeping sickness camp and brought supplies from the now-departed expedition for his own laboratory there.[11] Like Bugalla, Kigarama was located where sick people were known to be.

The camp sat on Kigarama's eastern periphery and fit into historical patterns of land distribution under the purview of Ziba royal authority. The location and construction of the camp in 1907 also followed royal consideration that weighed social and political factors in Kiziba in balance with a developing relationship with colonial authorities. After notifying German authorities when the disease "increased" in the area in 1907, mukama Mutahangarwa directed the location and building of the camp.[12] A rudimentary camp first quickly took shape on land designated and made available to German doctors by the king and likely according to royal fiat. Mzee Heslon Lutimba, a former teacher whose father had worked for German doctors and who lived on a farm adjacent to the camp's site in 2008, recalled local memory of the camp's location and its history: the land for the *kambi* (Kiswahili, camp) was "given by the king."[13] Medical reports do not indicate whether Mutahangarwa or any other Ziba landholder was compensated for the land upon which the camp was built. This account fits both with historical information about the mukama's control over land and his prerogative over its distribution, as well as traditions of royal and chiefly responsibility to intervene during outbreaks of illness, such as dispersing a village's inhabitants away from affected homes during episodes of rubunga. Royal authorities also directed laborers in the building and later expansions of the camp.[14]

There was powerful political symbolism for Mutahangarwa in collaborating with colonial authorities on terms he could claim to set and on ground he defined. The camp's location east of Kigarama village and toward the lake situated it outside of land under cultivation, on the open savannah that typically surrounded Haya villages historically.[15] Such land, as *orweya* for seasonal cultivation for women on nearby farms, was important in Haya agriculture and economic activity. But lands outside established villages were also important resources during outbreaks of disease. This open savannah was the same kind of space that had provided a safer haven for temporary homes for Ziba people trying to evade illness in previous generations. The camp's placement there mirrored displacements from villages to evade illness, and, further in the past, in Mutahangarwa's grandfather's era, patterns of setting the sick apart from established homes during outbreaks of pox-causing illness. Both the siting of the camp on the savannah near Kigarama and the subsequent dispersal of grasslands around the camp fell under the purview of royal power but also summoned royal responsibility. Managing the location and building of Kigarama represented efforts to assert chiefly responsibility for mitigating illness and misfortune as well as to deploy chiefly prerogatives for allocation of land.[16]

The Kigarama camp lay a small distance from the lakeshore near the crest of a hill, similarly to Bugalla. The layout of the sleeping sickness camp does not survive in extant sources, but descriptions indicate that the Bugalla camp was a model. Kigarama also had open-air bandas and rooms for laboratory examinations, homes for German doctors, and barracks for the sick all within its confines; the cleared and open space of the camp looked different than the homes and farms around it. But establishing the camp was not just a matter of creating physical structures and a newly demarcated space—the sick soon came. Initially, the camp was designed to isolate confirmed cases of sleeping sickness and treat them with Koch's months-long atoxyl regimen of two consecutive daily doses, repeated every eight to ten days, for six to eight weeks. This regimen had the goal of suppressing an individual's trypanosomal infection in order to quell further transmission of sleeping sickness. Data collected at the camp also fed into wider understandings of sleeping sickness morbidity and of atoxyl's long-term impacts on trypanosomes.[17] Early movement of people to the camp was facilitated by mukama Mutahangarwa, who played an ongoing, active role in identifying suspicious cases and compelling their movement to the camp.[18] As months went on, people came to Kigarama in increasing numbers. Many arrived from Kiziba, Bugabu, and the nearby colonial borderlands. A core group of those treated—around 50 of the 336 men and women in the camp as of early August 1907—had previously been treated at Koch's Bugalla camp.[19] In late 1907 and into 1908, camp statistics report an initial population of 365 people, growing to over 570 in the next eight months—nearly half the estimated size of the village nearby.[20] People arrived with their relatives, who sometimes stayed to help to care for them, providing food and helping to shelter them.[21] An initial policy of strict internment in fenced-in camps lasted only a matter of months before being revised by Dr. Friedrich K. Kleine, Koch's right-hand man on the Ssese expedition and the head of the colonial administration's *Schlafkrankheitsbekämpfung* (sleeping sickness campaign). This shift in policy developed in response to African patients entering and leaving the camp of their own volition at Kigarama and elsewhere in German East Africa and an acknowledgement that holding people against their will to isolate them and treat them with atoxyl, as initially conceived, had been "completely futile."[22] Isolation and internment were neither feasible, given the thin German presence on the ground, nor effective, given that people avoided circumstances that would lead to their confinement.[23] Instead, an open camp system would allow people to come and go, gathering them for examination and allowing them to return for scheduled injections with atoxyl and other drugs. It also accommodated the presence of patients' kin on the periphery of the camp. Patients housed within the camp were typically those in later stages of the disease who could no longer move independently, as well as people classified as "mad" or violent cases, as a result of the trypanosome's attack on their central nervous system.[24] The campaign's new focus on ambulatory treatment increased its dependence on camps as points of intervention; it was applied consistently around Lake Victoria and at Lake Tanganyika, the campaign's other area of concern. But this policy also ensured that the most effective way to continue to reach a population—and to draw in people for examination and keep them in atoxyl regimens—remained a point of debate at the administrative level.[25] Coercion and incentivization underlay the campaign from its conception.

The Kigarama camp soon emerged as a hub of activity, its environs built and improved by the people staying there. One impact of increased activity near Kigarama was a change to the surrounding landscape and use of the land. Relatives of the sick who settled near the camp to help with their care worked, alongside those patients well enough to work, on building "huts" and clearing paths within the camp, for which they were paid a small wage.[26] Regular settlements arose on its periphery, a pattern encouraged by the doctors posted there from 1908 onward. Small dwellings and seasonal gardens tended by the relatives of the sick sprang up on land set aside for them just south of the camp, separated only by a berm or fence from the camp proper.[27] These small gardens served a dual purpose early in the campaign, allowing relatives to stay nearby when German policies focused on interning and isolating positive cases, and also deferring some of the cost of provisioning patients from the camp's limited budget. After mid-1908, as the campaign transitioned to a focus on ambulatory treatment with various drugs, colonial doctors understood that these small plots fringing the camp could encourage people to stay connected to the camp and return regularly for injections.

The camp itself also grew, enlarged gradually by a few farms through arrangement by the Bukoba resident in 1908.[28] These additions to the initial land dedicated by Mutahangarwa for the sleeping sickness camp expanded its capacity significantly by 1909.[29] Dwellings for seriously ill patients and their families now numbered 116; these dwellings, located in farms absorbed by the camp, meant a population of several hundred people was scattered around its core "as in their homes."[30] The camp itself had also grown more formal in addition to increasing in size, with the addition of a new "laboratory building" constructed between July and October 1909, which housed a treatment room, laboratory, pharmacy, and writing room in a sizable structure.[31] The rooms were bright and spacious, camp supervisor Dr. Kudicke noted, and the structure was made from fired bricks on a foundation of local stone. In addition to the multiuse building there were a new enclosed building for the confinement of people Kudicke defined as "raving mad," and living quarters for a nursing sister.[32] These brick-and-stone structures of the camp were distinctive in and around Kigarama. Departures from traditional Haya *msonge* homes, they were among only a few other stone and brick buildings locally, including the Catholic mission church at Kashozi and Mutahangarwa's residence at Gera. Architecturally and spatially distinctive, the camp's laboratory rooms and open-air treatment bandas manifested the most permanent biomedical presence people in Kiziba had yet seen locally.[33]

Colonial diagnostic and therapeutic ideals structured admission to the Kigarama camp and life inside it. Crossing into the camp first meant admission into German doctors' system of record-keeping, with each person assigned a discrete case number. A person's sex and age were recorded along with their provenance, particularly in the case of travelers or people identified as "foreigners" living in the area. A single doctor typically manned the camp, supported by a small cohort of African assistants who looked after patients and likely helped manage communication across German, Kiswahili, and Oluhaya, and, later, a nursing sister.[34] The camp doctor palpated people's bodies, testing for characteristic swelling of glands on the back of the neck (Winterbottom's sign), considered a key presentation of the disease.[35] The different doctors stationed at Kigarama each continued to examine blood and lymph to diagnose sleeping sickness, despite the vagaries of these methods; when a person appeared lethargic and severely ill, however, laboratory confirmation of a diagnosis was often left aside.[36] Once diagnosed, people were categorized as lightly ill (*Leichtkranke*), seriously ill (*Schwerkranke*), or very seriously ill (*Sehr Schwerkranke*)—subjective distinctions that differentiated between the earlier or later stages of a trypanosome infection, correlated with its impact on the nervous system as understood at the time. These distinctions allowed doctors to assign people to different regimens of treatment and also to gauge the overall severity of the disease within the local population. Atoxyl injections provided the meter for camp activities' rhythms. On a camp-wide scale, this meant the movement of small cohorts of people to be tested and injected on specific days, cycling at eight- or ten-day intervals; the severely ill would have moved only within the camp's confines, while those well enough to get around might have come from nearby farms or the small plots allocated to the sick and their relatives around the camp. In its early years, entering the camp meant leaving relatives or kin outside, an initial separation that would have been marked, if not necessarily permanent. Moving patients through the camp's different spaces—treatment bandas, examination rooms, and shelters—fell largely to a few African orderlies. On any given day, the people receiving atoxyl would have experienced the nausea and pain at the injection site on one's back that regularly accompanied injections, along with any ongoing symptoms of their trypanosomal infection. Between injection days, people would have turned to the work of daily life and household obligations, either on plots near the camp or at home on their own *kibanja*, before returning for the next round.

THE RESOURCES AND RESONANCES OF THE KIGARAMA CAMP

Work at the camp at Kigarama focused on identifying potential cases of sleeping sickness, treating confirmed cases with a variety of accepted and experimental trypanocidal drugs, and monitoring people within months-long dosage regimens. In time, staff also followed up on subsequent or recurring sleeping sickness infections. Camp doctors promoted their medicines as a means to heal the striking changes in a person's body and temperament that one's kin would recognize as isimagira (sleeping, dozing), botongo (decay, collapsing inward), or perhaps even mongota (sleeping, nodding) for people with connections to Buganda. For Ziba suspected of being sick and those with confirmed cases of sleeping sickness, dealings with German doctors at the camp would not have been isolated interactions, but rather repeated and increasingly regularized encounters for individuals and their relatives and neighbors. The camp altered aspects of daily life for Ziba people and, in its ongoing work, was fit into their intellectual frameworks.

Though visible only through its architectural footprint in the early twenty-first century, with piles of bricks where houses had been and low berms around its perimeter, the site of the former camp at Kigarama was still known locally in 2008 as the place where the "German hospital" or the "sleeping sickness hospital" had been. People living near Kigarama in recent years still recognized the site of the kambi or *hospitali* (Kiswahili, hospital), pointing out piles of bricks that had been houses for German doctors there and marking out the general boundaries of the camp by trees that had been planted adjacent and a path that ran down to the lake along one side.[37] The kambi was where people were treated for an ailment called variously *botongo* or *isimagira* in Oluhaya, *ugonjwa wa malale* (sickness of sleepers or sleep) in Kiswahili, and *Schlafkrankheit* in German.[38] People living in and around Kigarama recalled these two Oluhaya names in discussions about local memory of the Kigarama camp, linking the camp to these illnesses as well as the more common Kiswahili *ugonjwa wa malale.* The malady's contemporaneous names, *botongo* and *isimagira,* though now associated with distinct medical entities, bring forward the gravity and physicality of changes to a person's body that drew people into the newly marked space of the camp. *Botongo* could be derived from a verb root *ku-boota* meaning "to weaken," or, relatedly, "to rot, or to waste away"; linguist Henry Muzale glossed its connotations in a conversation with the author as akin to what happens to a traditional Haya home when the central pole fails when damaged by termites—a caving inward, a collapse. A White Fathers priest resident at Kashozi in the early 1900s glossed *ku-botongana,* a related verb, as sleeping in the middle of the day, as children might.[39] *Isimagira* relates to the verb *ku-simágira,* in use at the time, connoting sleep or dozing and recalling another sign of illness.[40] While *isimagira* has parallels to the Luganda *mongota,* each rooted in a person's notable sleeping or sleepiness, *botongo* may have been more capacious in describing decline and collapse. In Luganda, *kabotongo* named a condition with suppurating lesions that Ssese Island missionaries had associated with syphilis, and that, later, was used to name nonvenereal syphilis or yaws.[41] The overlapping use of the term *botongo* in the era of sleeping sickness and through the era of heightened concerns about syphilis and yaws in the interwar era presents ongoing challenges for historicizing etiologies within Haya nosological systems. This overlap suggests for historians the different possible, simultaneous registers in which illness was recognized and named—as weakness, as sleepiness, as physical infirmity—and indicates that these meanings would have been underpinned by potentially divergent, complex nosologies. More broadly, in historical epidemiological perspective, the designation of different names for similar new forms of sickness—mongota, isimagira—around Lake Victoria within a short span of time suggests simultaneous but separate processes of explanation and identification of an illness or symptoms and underscores the likely novelty of this illness in epidemic form for communities at the time.[42]

Less than a year after it was established, the Kigarama camp had secured a place in the local economy and established its footprint on the landscape of coastal Kiziba. The camp provided an alternate hub of activity from the village nearby, one that centered colonial definitions of illness and processes of treatment. Doctors also provided access to cash, durable goods, and food, as incentives to keep people in or near the camps. At admission, for instance, some new patients received barkcloth for clothing and sometimes also a sleeping mat.[43] People were also provided with food at the camp, particularly those who were too sick arrange for their own food or without relatives to do so. The sick were given either small amounts of cash (two heller, or .02 rupees, in May 1908) or food in kind. The camp also had a small store and tried to attract a regular market. Early on, some well enough to work received small payments to maintain and expand the camp.[44] After the camp transitioned to focus strategically on ambulatory treatments from mid-1908 onward, German doctors offered varying amounts of cash—one to three rupees—to people who voluntarily came to the camp, offering the same amount to relatives who brought confirmed cases of sleeping sickness.[45] By comparison, the annual hut tax in this period was four rupees; the acting resident would ultimately propose forgiveness of the four-rupee hut tax for the sick.[46] The camp also animated economic relationships in other ways. Ziba patients were not the only local people inside the camp, which depended on the work of African orderlies helping to care for the sick and guards to mind its boundaries. A number of unspecified assistants also worked in its laboratory and clinic and were employed as servants in doctors' homes. The different kinds of expertise required in various jobs suggest that doctors employed a combination of newly hired local Ziba people and people who had already been attached to Koch's expedition or were *askari* (African soldiers) in the colonial police or military corps.[47] With its offer of wages, a small store selling meat, and perhaps also a small regular market nearby, the camp constituted a new, colonially defined space for treatment that was simultaneously a Ziba space of commerce, acquisition, agricultural production, healing, and exchange.[48]

German doctors saw Kigarama and its environs as an ideal research site for its geography and its epidemiological circumstances. But the Kigarama camp held significance in Ziba social and political geographies beyond its therapeutic functions and economic meanings. The fortunes of the camp were rooted in the place's multilayered history and how it fit into Ziba economies of land and labor. Similarly to other interlacustrine societies, political power and social relations were embedded in the Ziba landscape. Kigarama fit into a complex sociopolitical geography that gave the camp's site a variety of possible meanings to those who passed through it, calling upon histories and experiences of clan affinity, royal ritual-political power, treatment-seeking, and prerogatives of land use. Focusing on Kigarama, in place and over time, offers a locally centered perspective on Ziba uses of the camp and how these changed.

For Ziba populations in the early twentieth century, Kigarama fit into knowledge of the kingdom and people's history. The village's own history connected to Ziba intellectual worlds that understood continuity of royal lineage—and the ongoing maintenance of sites that signaled this continuity—as central to political legitimacy. As a place, Kigarama connected with several clans, organizations of kinship and affinity constituted from patrilineal and agnatic networks and associated with common places and stories of origin. Historically, clans in the interlacustrine kingdoms "formed categories of people rather than closely knit corporate groups," with members linked through mutual obligation, association with or avoidance of taboo animals or objects, and the performance of specific roles in the royal court by clan members.[49] Clan associations with particular places or villages were widespread in Buhaya and related to the nature of clan affiliation—the common ancestor or founder whose history stitched clan members together was often fixed in a particular place, which then retained ongoing importance. Several clans held Kigarama as their village; all had strong links to the mukama, his palace and court, and his political lineage and dynastic history. Indeed, no other village within Kiziba was associated with as many clans as Kigarama, nor did any other villages with which multiple clans were associated have the same extraordinary concentration of clans associated with kingship and royal ritual. Kigarama was associated with the clans of the king (*babito*), of court jesters (*bakurwa*), of "priests" (*bashamula*), of former kings (*babiki*), and of first-arrivers and those responsible for demarcating the king's *kikale* or court (*bahinda*).[50] It was the site of the first Ziba mukama's palace and of two subsequent palaces, between the fifteenth and seventeenth centuries, and was the burial site for several ancient kings.[51] Thus it remained a significant place for royal tradition and for people who, through their clan, connected with historic ritual roles in the Ziba court and palace; such communities and histories would have remained durably relevant for royal rituals and thus royal legitimacy into the early twentieth century. This was despite an individual mukama's potential mobility: Ziba *bakama* traditionally established new kikale throughout their reign and kingship was, in this way, mobile throughout the kingdom and had changeable seats of royal power. Mutahangarwa himself established a residence at Gera, not far from his father's last palace at Bwanjai, in the center of the kingdom.[52] Nevertheless, the clans linked to Kigarama rooted royal authority and Mutahangarwa's lineage there. Overlaps in the geography of political power in Kiziba and the siting of the sleeping sickness camp would have resonated for Ziba people within the context of Mutahangarwa's responsibility for addressing the illness that struck people in the kingdom, because of his own lineage associations with the village and his wider role in securing health and prosperity in the kingdom. Moreover, recent history manifested these deeper values of good and legitimate kingship: Kanyigo had been Mutahangarwa's own province as a prince and chief during his father's reign. Unlike his rival half brother's provinces around Bugandika, his farms and estates were not emptied by emigration to other areas during rubunga.[53]

But Kigarama—variously spelled Kigalama—also called upon other ways of understanding and remembering places and their histories with regard to healing and power. Its name invoked the power held in oaths, paying homage, and taking actions to ensure health and prosperity. Linguistic evidence concerned with political and ritual power in Ziba history offers further context for understanding Kigarama in overlapping and sometimes competing registers, as a place of royal significance and of healing forces. Thus, *kigarama* is not just the name of a village, but also a word that associated powerful forces with the place itself (through the locative *ki*). As a site historically connected to Ziba kingship and political legitimacy, where important ancestors were buried and through which royal power circulated and legitimacy was maintained, Ziba people may have understood Kigarama in the context of the Great Lakes Bantu root

-la(a)ma, which glossed saying important or magical words, in the swearing of oaths between generations and in paying homage to kings. But in neighboring Bunyoro, a kingdom to which Kiziba's own royal lineage connected, the same root glossed engaging with *mucwezi*, mediums who were historically sources of significant healing power.[54] The overlap of these meanings recalls the interplay in the deep past between royal power and *cwezi-kubándwa* healing, as kings' political-ritual power was mediated by ritualists and mediums. Nearer to the place's modern pronunciation is the Bantu root *-làm*, which invokes healing and health with population-wide implications and which glossed being healthy and well throughout the wider region.[55] In Kiziba, the root *-làm* centered language around the long life of the king.[56] An early twentieth century dictionary from Kiziba glossed *-lama* as "being healthy," an idea that the word *kigalama* would locate in a specific place.[57] It is possible that the place had been a location of access to healing power not exclusively associated with royal history. Colonial records indicate that the Kigarama camp was in proximity to places where Ziba *embandwa* healers gathered, although the sites of shrines and the nature of healers' work are obscure in extant sources.[58] Nevertheless, tracing these linguistic roots alongside more recent historical evidence indicates that widespread understandings of the interrelation of a kingdom's health and a king's health, of political legitimacy and prosperity, and of healing powers located in particular places all circulated around Kigarama. The generative possibilities in Kigarama's name and in historically overlapping political, ritual, and therapeutic practices in Kiziba suggest that its history allowed the new colonial intervention to be situated in a space with deep associations with healing and royal power.

Dispersal of land around the camp fit with Haya traditions of managing cultivated and uncultivated land for agricultural production, but here connected directly to both the colonial public health campaign and to the king in new ways. Mutahangarwa's allocation of the land for the camp and for its subsequent enlargement fell within royal rights to land, but deployed those royal rights in a new modality—directing land to the sick and their kin.[59] The camp sat beyond the village on land likely not ideal or desired for perennial crops, given patterns of cultivation in coastal areas of Buhaya.[60] The camp's location and its immediate environs might have historically been orweya lands put under seasonal cultivation by nearby households, but there are no indications that the camp displaced active cultivation. When Kudicke arranged for small plots for families of the sick near the camp, he effectively brought new land into cultivation that had not been worked consistently before. Further, these small plots near the camp put land into new hands, as the sick were not all from the farms and homes nearest to the camp site. Many of the sick were accompanied by their spouses or young relatives. The camp therefore provided an opportunity for people without direct claim to land—certainly women but perhaps also young men—to gain access to small plots not otherwise under cultivation.[61] This likely had immediate ramifications for food security, particularly if a person's illness had diminished a household's agricultural labor or other remunerative work, or if the home and farm of a sick person were at too great of a distance from Kigarama for kin to both care for the sick and keep crops growing well simultaneously. The camp functioned as a place to access resources connected to both colonial doctors and to the mukama.

Mutahangarwa's demarcation of the Kigarama camp also signaled other new manifestations of royal power in Kiziba. Although dispersal of land was historically the mukama's prerogative and the use of orweya lands fit into household seasonal production, Mutahangarwa undertook significant changes in agricultural regimes and land use after coming to power in 1903. The establishment and expansion of the Kigarama camp also accorded with ever-greater claims made by the king and his officials on clients' use of land. Further, it fit within new trends in royal claims on land and labor that shifted toward production of coffee as a cash crop, among others, a shift that was encouraged by German colonial officials.[62] In Kiziba, varieties of *ebitoke*, plantain banana, were the staple food crop and banana trees were intercropped with *robusta* coffee trees whose fruits were chiefly property. Coffee, traditionally associated with royal political and cosmological power, had occasionally been traded for cattle before colonial incursion; cultivation began to expand in the early twentieth century into cash-crop production at the urging of both kings and colonial officials.[63] Between 1905 and 1907, Mutahangarwa asserted a stronger hand in managing the kingdom's economy, largely by encouraging coffee cultivation.[64] In 1905, the king and his chiefs overcame a popular reluctance to plant more coffee with the threat of punishment. Lwamgira notes that "whoever refused to plant coffee would be arrested and detained or beaten. So the people planted coffee by force."[65] Haya coffee production was linked to the world market, and coffee and its cultivation underwent subsequent shifts in cosmological, political, and economic meanings.[66] Coffee was no longer a crop produced for social and ritual use, grown in small amounts with its berries offered and chewed, but was now produced in tons and exported to Indian Ocean ports.[67] Intensifying Ziba engagement with global markets, brought about by pressure from the mukama and his officials to cultivate coffee, was concurrent with other royal and colonial interventions into daily life, such as efforts to control sleeping sickness. Occurring simultaneously with Mutahangarwa's interventions to expand coffee cultivation, then, the establishment of the camp echoed new ways that Mutahangarwa exercised royal privilege to redirect land use and labor allocation to fit new political and economic agendas. In all of these possible ways of fitting into Ziba politics and geographies of social life, the Kigarama camp was freighted with meanings and potentials far beyond colonial doctors' intentions to establish it as a place of medical treatment.

GLAND-FEELERS: AUXILIARIES AT WORK

Ziba auxiliaries employed at the camp embodied both changing royal politics and durable royal prerogatives expressed in the realm of public health. The anti–sleeping sickness campaign envisioned by German authorities required intimate knowledge of disease epidemiology, vector habitats, and human behavior. The campaign's success depended on knowing who the sick were, where they lived, and how severely they were affected, as well as on optimizing popular adherence to surveillance measures and drug regimens. But, early on, significant gaps opened up between medical officers' expectations, colonial knowledge, and levels of participation by Ziba communities. Local circumstances forced the doctor in charge, Robert Kudicke, to work closely with Mutahangarwa to shift the campaign's tactics and focus accordingly. A cohort of Ziba medical auxiliaries, known as *Drüsenfühlern* (gland-feelers) resulted from this interchange; these men worked in the interstices of royal politics, public health, and the local economy. The employment of these African auxiliaries began as an attempt to mend gaps in colonial knowledge about the situation in Kiziba, which included a lack of local languages and dialects, scant awareness of crossing points of rivers and swamps, and only basic information on the size, location, and distribution of the population in rural Buhaya.[68] Gland-feelers, exclusively men, were trained to find potential cases by focusing on swelling of the cervical lymph glands, understood as an early presentation of sleeping sickness. Kudicke trained them to survey a population for sleeping sickness, relying on visual recognition of the correct glands as well as physical detection of enlarged glands through palpation. In addition to identifying those suspected of being sick, auxiliaries were intended to work along with senior Ziba authorities to bring the sick to the sleeping sickness camp near Kigarama. These auxiliaries simultaneously fit into an existing niche in the political, social, and economic landscape of Kiziba, while also facilitating new relationships created by the sleeping sickness campaign, through which they mediated both royal power and colonial authority.

The history of Ziba gland-feelers provides a glimpse into early colonial dynamics of health, labor, and politics prior to the era of high colonialism. They have cropped up in histories of the East African medical profession, of German tropical medicine, and of the Haya kingdoms, where they are mentioned in passing as early auxiliaries to German colonial public health.[69] Similar intermediaries at work as field assistants, clerks, and interpreters were central to colonial governance, just as translators and guides had been indispensable to earlier exploration and trade.[70] Access to wage labor, education, and professional networks conferred new forms of social mobility and cultural capital upon these auxiliaries, who took advantage of opportunities to expand their own circles of influence in the course of their work for missionaries, researchers, and administrators. Medical auxiliaries did this and sometimes more, traversing African and European approaches to illness and wellness; some became crucial participants in colonial scientific and medical work, influencing research agendas and shaping research outcomes.[71] Active in a time when neither colonial political strategies nor public health approaches to sleeping sickness were entrenched, the Kigarama gland-feelers encourage consideration of the diversity of auxiliary labor involved in colonial medical and public health efforts and its connections to existing forms of political and ritual power. Their history highlights the precariousness of public health surveillance work in an era when royal authority, colonial power, and scientific knowledge were all in flux and no durable cure for sleeping sickness was available.

As policymakers reviewed the campaign in its first year, the camp at Kigarama was seen as "commendable" compared to others around lakes Victoria and Tanganyika. Officials attributed its success to Kudicke's "efforts to get along with the natives without compulsory measures"—not forcing people to come to the camp against their will.[72] But Kudicke's exemplary camp, unlike others in the region that relied upon searches conducted by European personnel and negotiations with individual village chiefs, used compulsion of a different kind. At Kigarama, African auxiliaries—Kudicke's gland-feelers—had begun to augment and sometimes supplant the work of German doctors in delivering people to the camp.

As the anti–sleeping sickness campaign in Kiziba gained momentum, colonial officials believed that many more cases remained hidden in the surrounding countryside, posing the threat of continued transmission into "healthy" areas and demanding vigilant surveillance. But the social and agricultural organization of Kiziba, the campaign's "field of action," made finding individual cases of sleeping sickness difficult. The hillsides and ridges of Kiziba presented colonial officials with a labyrinthine network of dispersed homes and loosely connected villages, interspersed with uncultivated fields, open savannah, forests, and swampy land.[73] This local geography allowed people to evade colonial authorities, just as they might have evaded Ganda raiding parties or a rival king's armies in previous generations. Colonial knowledge of exactly how and where the population was distributed remained inconsistent in the years just before World War I.[74] German doctors' unfamiliarity with and their inability to penetrate the variegated Ziba landscape to locate the sick presented the German sleeping sickness campaign at Lake Victoria with its primary challenge.

Determining whether a person had sleeping sickness required close contact—measuring body temperature, palpating glands, perhaps taking blood—but this process was often hindered by people fleeing at the approach of a European doctor. Kudicke addressed the problem of finding, examining, and diagnosing suspected cases of sleeping sickness by training Ziba men as auxiliaries. In October 1907, he proposed a novel solution to his superiors in the campaign: "in order to treat these [sleeping sickness cases] as soon as possible, I will attempt to have individual villages searched by natives who have been trained in the palpation of glands. The sultan Mutahangarwa has sent me 10 young people for this purpose, whose training I have already begun."[75] Kudicke gathered this cohort of men, whom he called gland-feelers, through his connections with local leaders and the Ziba king. Gland-feelers would identify suspicious cases and bring people to the sleeping sickness camp at Kigarama for further examination, allowing him to confirm a diagnosis of sleeping sickness in his laboratory. Their survey work was no great departure from what he himself might have done in visually and physically surveying a population for signs of sleeping sickness. Gland-feelers' work was to function, in many cases, as the first survey of sleeping sickness in the interior of Kiziba. Initially focused on communities nearest the Kigarama camp and the lakeshore, these auxiliaries went on to conduct searches for the sick in ever-widening swathes of territory to the west, extending into Kiziba and German Buddu. Around twenty-five gland-feelers worked for the sleeping sickness campaign at the height of their activity, covering an area that extended at least 150 square miles.

Gland-feelers operated within a system that already offered incentives for cooperation with colonial medical authorities, such as the cash and durable goods distributed directly to the sick and their relatives at Kigarama. Kudicke then attached a specific economic incentive to the discovery of proven sleeping sickness cases: gland-feelers would now earn a premium per case identified, alongside their monthly wages. Receiving the premium for each positive case depended not simply on getting suspected cases to the camp at Kigarama, but also on the confirmation of infection with trypanosomes through examining blood, lymph, or spinal fluid. Both wages and premiums shifted over the course of the campaign, varying from three to five rupees for a base monthly wage, plus the additional money for each confirmed case.[76] A reward of one rupee per positive case was provided in late 1908, increasing in 1909 to three rupees.[77] The total earnings from the base wage and premiums for just a few cases, then, were largely equal to the six-to-eight-rupee indemnity offered elsewhere by the Germans to people forced to relocate out of tsetse fly areas and abandon their farms.[78] Throughout, wages and premiums would have readily fulfilled the four-rupee yearly hut tax in place during the same period.[79] For this cohort of men, sleeping sickness surveillance was remunerative work at a time when cash demands were becoming more common.

Gland-feelers' work focused on the identification of the sick and their delivery to the camp for injections. Stories of the camp remembered to the author in 2008 recounted this work, as elderly men located stories of the camp within stories of local experiences of German colonial rule and the history of Kiziba's kings, as well as relative to World War I. In their tellings, sleeping sickness was a disease of the past that their grandparents and parents knew. One such story, passed down from grandfather to grandson, recalled the involvement of both Europeans and Africans at work at Kigarama: "The sick were helped by people to go to the hospital. The Germans send [sic] people to search for sick people. These are people who worked for the Germans and they were paid for that. The Germans did not go themselves."[80] German doctors and Ziba auxiliaries divided the labor of public health surveillance. Ziba auxiliaries allowed the campaign to survey a far greater number of people more quickly than Kudicke could alone. He was at the time serving as the sole medical officer responsible for the kingdom of Kiziba, Bugabu, German Buddu, and the fringes of Karagwe, as well as crossings on the Kagera—a large and populous territory. Between January and June 1908, in the meantime, Kudicke supervised the treatment of upwards of five hundred people at the camp each month, leaving little time for tours of the area.[81] Gland-feelers allowed Kudicke to remain in his laboratory at Kigarama, identifying trypanosomes under the microscope, monitoring long-term regimens of atoxyl, and keeping track of disease mortality and the drug's effects.

Searching for the sick became the work of young Ziba men. These auxiliaries initially produced promising results: Kudicke noted that "by far the most newly admitted [cases] have been found by Africans trained in gland palpation."[82] Within eight months, the initial group of 10 gland-feelers had expanded to 23, with 85 new cases admitted in the first quarter of 1908. The total number of people treated by late May 1908 was 581. Kudicke's report described how gland-feelers did their work: "In the field of action of the sleeping sickness camp 23 such assistants are working and with the following assignment: every hut will be searched individually. A representative of the katikiro should if possible be at each examination. The gland-feelers note the names of all people with swollen neck and armpit glands. Only people with at least bean-sized glands will be sent, with the help of the katikiro, to Kigarama."[83] Finding the right glands with the right level of swelling was not a straightforward matter, though, despite historic and consistent reliance on "Winterbottom's sign," swollen glands on the back of the neck. German doctors in the field in East Africa referred to a scale of gland size based on common items to determine when swollen glands were alarming. Apart from the "bean-sized" swelling that Kudicke instructed Kigarama gland-feelers to attend to, other doctors referred to a wide variety of items—peas, cherries, hazelnuts, pigeon's eggs—to gauge the standards by which they diagnosed the sick and explain these to their colleagues and superiors.[84] What exactly these different grades of inflammation meant for a person's health was not precisely known, but attention to swollen glands remained the most widely used method for initial screening, in tandem with watchfulness for other serious symptoms of sleeping sickness such as swelling of the hands or face, wasting, or sleepiness.

Just as the Ziba palace had been involved in the establishment of the camp at Kigarama, so, too, was Ziba political life involved in gland-feelers' activities. Kudicke had not, after all, simply plucked men out of the general Ziba population to be trained as gland-feelers: they were sent to him by the king. The cohort of gland-feelers fitted into economies of male labor focused on royal prerogatives to direct and delegate manpower throughout the kingdom. While able-bodied men of all ages were required to perform unpaid work for the mukama (or for chiefs) in the institution called *nsiku*, the dominance of youth among Kudicke's gland-feelers suggests they fit more exactly into the social and political institution known as *muteko*.[85]

Broadly, the political economy in Buhaya of labor outside of the household focused on the demands of the mukama and his district and village subchiefs. Royal claims extracted the "first fruits" given to the king and his chiefs, but also required able-bodied men to travel to the palace and "work there without payment for one month . . . cleaning plantations, erecting buildings, cutting firewood, herding cattle, etc."—the compulsory labor of nsiku.[86] Claims on male labor fit within other practices that connected women to the household and men to the palace. Domestic responsibilities differed for unmarried young men and women. Colonial ethnographer, linguist, and sometime planter Hermann Rehse observed that young women helped with cooking and keeping courtyards clear, while young men spliced rope, fished, hunted, and brought "taxes" to the royal court, while also under obligations for agricultural labor to clan and family elders.[87] Young men would have brought tribute from a family or clan farm to the court in the form of a tax in cowries and first fruits of the harvest. Though Rehse does not record that boys "carrying the taxes" spent an extended period of time there, a responsibility to travel regularly to the court also served to socialize Ziba boys and orient them in the political hierarchy.

Muteko, however, was distinctive: an age-set group composed of young men of a common generational cohort who had been selected by local political leaders to serve together at the mukama's palace.[88] Arriving at the palace between ages ten and twelve, boys learned "the arts of war" but were also responsible for pasturing cattle, cleaning out cattle enclosures, cutting grass, and other general work to maintain the palace complex.[89] Promising young men remained at the court for several years after the standard three years of service, receiving additional training in civil and ritual matters, with a successful few entering the king's service as officials at the court or in outlying districts. Some in the muteko may also have been sons of the mukama's chiefs or young blood relations from within his extended descent-group or clan.[90] Schmidt's late twentieth-century informants' discussion of their muteko reinforces that contact with the court was an important marker of generational identity and status.[91] One's muteko provided a foundational collective social reference point for young men that remained durable as youth grew into adulthood.[92] The muteko that circulated through Ziba palace functioned as a means for training in statecraft and military service; success in the palace helped define a man's trajectory of upward political mobility. In broader perspective, age-sets are a generation-based social form with wide distribution in Africa: the masculine *murran* in Maasai society, the Zulu regimental system, and the page system in Buganda are well-known examples.[93] Age-sets were an important way to bind young people together and provide a sense of social and communal obligation outside of the nuclear family; members of an age-set developed internal bonds of affinity and responsibility, while also orienting toward elders or superiors within the political structure they inhabited. In some cases, age-sets formed a ready-made fighting force for communal defense, while in others they formed nodes in a geographically dispersed political network within a kingdom—in both cases, generational identity served to connect individuals and political seniors. In Kiziba's muteko, the age-set functioned to link its members to the state and particularly to the Ziba mukama.

Muteko, more than other royal claims on all adult male labor, fits with the cohort of gland-feelers in several ways. The group of young men Kudicke received at the camp together indicates a corporate coherence and specific demographic that nsiku laborers would not have had. Likewise, the duration of their service, extending into many months, also indicates an exceptional status: these young men were removed from agricultural and economic obligations within household, clan, or village structures, their manpower directed by the mukama alone. Young men in the later stages of royal service in a muteko were still bound by royal claims on their labor.[94] The gland-feelers' demographic uniformity, strong association with the royal court, and duration of service, understood within this social and political context, suggests that a muteko was the only form of social organization relevant for organizing such a cohort of men.

The cohort of young men of common age would have been identifiable as a muteko, receiving their political and social education by serving at Mutahangarwa's court. Particular skills, objects, and clothing would also have signaled gland-feelers' specialized work and distinctive status. They were literate—able to note the sick or those suspected to be for camp records—and certainly carried paper and writing instruments to gather the names and locations of the sick elsewhere, as their surveillance work required. The skills and tools of literacy also suggest this cohort of men had special status within evolving relationships between palace, mission, and *boma*. At the time, young men would have learned to read and write at either the small government school in Bukoba or, more likely, at a nearby White Fathers mission school in Kiziba—in both cases, only with the approval of political authorities.[95] Their age in 1908–9 locates them among the earliest attendees of mission educational efforts. Along with the tools and skills of literacy, gland-feelers were also likely given other materials along with shoes, a hat, or cotton clothing.[96] Clothing that came from the Germans would have been a visible mark of distinction from common modes of dress, generally either barkcloth made locally or printed cotton cloth acquired through trade.[97] Both European-style clothing and the long *kanzu* tunics typical of the Swahili coast would have been familiar markers of association with the colonial regime. In recent years, changing modes of dress had also gained ground among Haya elites. Mutahangarwa himself had self-consciously refashioned royal attire during the period, becoming "the first King in Kiziba to dress like Europeans," as his historian noted.[98] The only surviving photograph of the mukama, taken during a German nobleman's tour of the region in 1907, depicts him in a crisp white jacket and hat, wearing polished leather shoes and dark trousers, fingers resting lightly on a cane, with an askari in uniform and men in kanzu in the background (see figure 4.1). Pictured standing on stuccoed steps, he was likely at the German boma at Bukoba. Mutahangarwa, here, signaled a new set of outward manifestations of Ziba royal authority that emerged in dialogue with colonial presentations of power—modes of dress, types of buildings—in a process that colonial officers read as a growing interest in modernization. Gland-feelers' dress, then, would have signaled their association with the colonial regime but also with the palace. Elsewhere, Mutahangarwa called these new manifestations of royal power into being on the Ziba landscape. The mukama's stone house, built upon succession at his new court at Gera, also fit into these adaptations of the visible representations of royal power. It was the first of its kind in Kiziba and comparable only to mission buildings.[99] The other stone structures in Kiziba that Mutahangarwa had a hand in building were the doctor's house and laboratory at the Kigarama camp.

As "middle figures" moving among the population, gland-feelers conducted essential epidemiological groundwork: surveying communities, documenting the incidence of cases in a population, and establishing site-specific records.[100] The success of Ziba auxiliaries merited acclaim and expansion, according to the resident: "The attempt of the leader of the sleeping sickness campaign in this district to have the sick located through [Negroes], who are trained in the palpation of glands, has evinced fine success and led to the discovery of a whole number of typical cases of sickness. This searching for typical cases must be continued, if in a still more intensive and expanded measure than previously."[101] And so, in 1908, gland-feelers continued to canvass areas to the north and west of Kigarama. Several auxiliaries trained at Kigarama also began to expand searches into southern parts of the residency, particularly the large kingdom of Ihangiro, from a base at Bukoba. Others had been posted alone or in pairs at key crossings on routes in and out of Uganda, such as the ferry on the Kagera River at Kifumbiro and on the main road south of Kigarama at Kikongoro. Soldiers—either men on the king's retainer or colonial African askari—were also posted with these outlying gland-feelers to assist with transporting suspected sleeping sickness carriers to Kigarama. These circulating auxiliaries were a newly visible form of both colonial and royal authority—young men of the palace, of a muteko, circulating to bring people under colonial medical surveillance. While their utility for Kudicke was clear, their meaning for Mutahangarwa was also significant. The

FIGURE 4.1. **Mutahangarwa, *Mukama* of Kiziba, c. 1907.** *Hans Meyer Collection, Leibniz-Institut für Länderkunde, Leipzig.* This is a rare surviving portrait of *mukama* Mutahangarwa, taken at Bukoba during the Duke of Mecklenburg's tour of eastern-central Africa in 1907–8.

circulation of people connected to the new therapeutic site demonstrated royal custodianship of the kingdom's health and the mukama's active role in responding to isimagira or botongo. Both the assertive role that Mutahangarwa played in siting the German camp and his ongoing facilitation of specialized auxiliaries circulating from the camp opened space for the mukama's traditional ritual responsibility for his people's health and prosperity to be executed in a new register. This was centrally important in the period just after his contested succession to the throne as well as the scattered but decisive violence of German colonial incursion—peace and health had both suffered. Both the camp and the gland-feelers, while associated with colonial public health, connected to royal political power and legitimacy. From a colonial perspective, royal engagement around sleeping sickness intervention was of a piece with Mutahangarwa's collaboration around rubunga outbreaks a decade prior and his recent shrewdness that they read as a desire to "modernize." But for Ziba onlookers and those examined, camp-based therapies were part of wider social, political, and ritual activities to mitigate illness and misfortune in the kingdom.

TROUBLE IN THE FIELD

Gland-feelers' position between newly imposed colonial technologies of disease prevention and the potential targets of these interventions afforded them significant discretionary power. Judgments about the people they encountered—about who might be sick and might be removed to the camp at Kigarama and who could remain at home—were at the core of gland-feelers' work. People subject to examination used a range of strategies in response, from negotiation to outright evasion. Challenges to gland-feelers' efficacy resonated within particular Ziba political and social frameworks, in addition to signaling a rejection of colonial health interventions. When gland-feelers encountered trouble in the field, it mirrored changes in the camp's fortunes and had repercussions for both the Ziba court and colonial officials.

Despite crediting gland-feelers with a great number of new cases, Kudicke acknowledged in December 1908 that examination and screening procedures were generally flawed, and noted that "individual observations indicate that gland-swelling to the extent which we consider typical develops comparatively late."[102] Evaluative criteria and scientific thresholds of the sleeping sickness campaign—even with the tools of microscopy and bacteriology available in the field—were, when probed, not standardized or reliable. Such ambiguity meant that neither German doctors nor gland-feelers had a truly reliable means of detecting the early stages of infection. Judgments and processes to separate certainties from uncertainties were contingent, individualized matters.[103] This contingency and uncertainty shaped the role that gland-feelers played in the campaign, German doctors' considerations of their success or failure, and communities' reactions to their work in the countryside.

As the camp at Kigarama moved into its second year, Kudicke made several adjustments to its work. Several auxiliaries remained on the payroll into 1909, mostly geared toward monitoring travelers at transit checkpoints and in Bukoba. But auxiliaries posted at these key transit points in and out of Kiziba received new instructions. With an eye on trade and aware of the impossibility of perfect monitoring, Kudicke instructed auxiliaries posted at Kifumbiro and Kikongoro to avoid anything that could be cumbersome to the free traffic of healthy people. This signified a shift in German policy: Kudicke now instructed gland-feelers to pay attention to people only with "significant' swelling. This less stringent approach reflected the view that the majority of sleeping sickness cases in the district had been identified and that the disease was in check locally, even if existing cases could not be cured. The cohort of gland-feelers was correspondingly reduced in 1909 and Kigarama's intake charts showed a decline in new cases. As well, Kudicke had begun to issue permits from the camp that allowed a person treated with atoxyl to travel more widely in the district, although travel to and from Uganda continued to be closely scrutinized.[104] These passes presaged those proposed for people moving between Belgian and German territories around Lake Tanganyika, and echo the "health passports" issued by doctors in other areas with epidemic sleeping sickness before the war.[105] The German experience in northwestern Tanzania indicates that doctors' day-to-day work had to balance the need to maintain disease control measures with concerns about disrupting the rhythms of trade and work and thus colonial economies.

However, German officials had also begun to question the efficacy of gland-feelers, concerned that they were missing suspected cases of sleeping sickness—either because they were being bribed not to report them or because they were not searching areas as directed. Kudicke wrote, of efforts to search neighboring Bugabu in late 1908, "Doubtless multiple sick people have been overlooked by the gland-feelers. Inspection by Europeans therefore cannot be spared."[106] Dr. Georg Ullrich, who temporarily replaced Kudicke as the supervising doctor at Kigarama in 1909, complained:

> At the beginning of March, the sanitation under-officer [*sic*] stationed here was sent out in order to check on the gland-feelers active in the district and to survey the western part of Kiziba for the presence of [people with sleeping sickness]. Sixty-two people with suspicious glands were discovered by him, which had escaped the gland-feelers. . . . It speaks in no way for an intensive activity of the gland-feelers, when so many suspected [cases] in a relatively small district have evaded [them], which are found by a European within five days, two of which are lost to the walk out and back.[107]

Of the sixty-two suspected cases that Ullrich's junior colleague found, only one showed trypanosomes in the bloodstream. Such inconsistencies between what African auxiliaries reported and what European superiors found on their own troubled doctors. German criticisms of Ziba auxiliaries' work were deeply colored by colonial racism and employed tropes of laziness and venality. The initial, official response was to increase oversight by European officers—at best a temporary solution, given limited European personnel in the area and auxiliaries' inherently mediatory role. The gland-feelers' supervisors gradually came to see the wage structure itself as an issue. Of Ziba gland-feelers at work near Bukoba, the station doctor wrote, "In general, the gland-feelers bring many fewer people to examination themselves. I have the feeling that these people, employed with a set monthly wage, now limit themselves to making a small circuit through their area from time to time to bring this or that man to Bukoba for examination, in order to not seem completely inactive."[108] The monthly wage, in other words, did not constitute an adequate incentive.[109] Thus, in mid-1909, the set monthly wage for gland-feelers was all but eliminated in the southern part of Bukoba and in Kiziba, with the exception of men stationed at transit points. Gland-feelers bringing verified cases of sleeping sickness to a camp would receive three rupees per case only, and no wage otherwise.[110] While gland-feelers' wages were drawn down, incentives more generally ramped up, in an effort to reveal still-hidden cases: sick people who voluntarily came to camp would also receive a three-rupee payment. Access to cash was broadened to include patients and their families.

At the same time, two separate but closely related problems focused on gland-feelers also emerged: allegations of extortion and incidents of flight from gland-feelers' examinations into British territory. Gland-feelers' work was predicated on a novel exchange unlike other wage work for the colonial administration. Rupees changed hands around the delivery of a person to the camp and the confirmation of trypanosomes in that person's body by the camp doctor. The reward for each positive case was explicitly intended to "spur [gland-feelers] on to eager action," but as intermediaries between king, camp, and people they were subject to competing claims and conflicting responsibilities.[111] Kudicke and subsequent camp doctors needed the gland-feelers to be broadly familiar with their territory, but also wanted them to be neutral, unbiased, and scientific, bringing in all suspected cases of sleeping sickness to the camp, where the microscope would ideally deliver a verdict of infection or health.

But familiarity could also be a liability when a variety of social relations might connect auxiliaries to the people they examined. The gland-feelers' rootedness in Kiziba, which initially made them attractive as auxiliaries, also rendered them susceptible to claims upon the use of their discretionary power: whether to leave an older relative in peace, or to delay removal of household members until planting or harvest was done, or to identify a rival as a "suspicious" case, for example. In these predominantly agricultural areas, people offered payments of food to auxiliaries in return for overlooking a household or an individual with suspicious glands.[112] Kudicke began to recognize the potential for trouble late in 1908, noting that "in order to prevent colored gland-feelers being enticed, perhaps, to corruption, it is ordered that they must change their area from time to time (in Bugabu as well as in Kiziba)."[113] Transferring gland-feelers to less familiar surroundings would ideally restore neutrality to their work and eliminate possible "corruption," promising better returns, ironically, than deploying them close to home as originally planned. But even shifting auxiliaries to a "neutral" field did not resolve the inherent tensions created by the combination of subjective identification of the "right" kind of swollen glands, the cash bounty per positive case, and the potentially serious impact of removal. On Bumbire Island, south of Bukoba, gland-feelers came to be at loggerheads with a reluctant population. The supervising doctor sent in supporting askari, but the population complained of blackmail. In that case, a Kigarama-trained gland-feeler found guilty of blackmail was punished.[114]

Both auxiliaries and the people they examined acted to protect their own interests; the gland-feelers were not a particularly venal cohort of men. They fit into the enterprising use of colonial employment for social and economic mobility that emerged with a new class of African colonial functionaries in the early twentieth century.[115] Further, to read negotiations and counter-payments as one-sided, driven by greed and abuse of power by functionaries of colonial public health, ignores gland-feelers' position in the Ziba political world and Ziba society. These young men were not strangers to the entire kingdom, its population linked by ties of clan and familial affinity. While new power dynamics that focused on signs of sleeping sickness did not privilege Ziba households, neither did they inherently disadvantage them as without response or recourse to colonial intervention. Interactions around examination, cast by colonial officials as corruption or blackmail, reveal complex layers of negotiation between gland-feelers and households generated by the circulation of wages, bounties, and bodies from Kigarama. Negotiations and counter-payments suggest that people attempted to regain some control of whether or when they or their kin were removed to the camp. A household's offerings resisted and complicated the simple commodification of the sick for gland-feelers' cash premiums. Counter-payments in agricultural goods may also suggest an effort to recast the movement of young men of the court between royal and domestic spaces within historical forms of exchange, such as the transport of "first fruits" by young men to the palace. These moves inflected gland-feelers' new demands oriented toward public health with practices within which people acknowledged the mukama's extractive rights and reaffirmed ties between king, people, and prosperity.

Counter-payments and evasion also signal other ways that gland-feelers' work was disruptive. Finding suspected cases of sleeping sickness required touch—getting hold of someone—that was consistent with the goals of colonial public health campaigns but at odds with Haya healing practices and ways of ordering social interactions. Both colonial and African narratives of sleeping sickness investigations in the early twentieth century emphasize the importance of close proximity in identifying sickness, a process that relied upon physical contact and close scrutiny of particular parts of the body. Such examinations could be coercive, uncomfortable interactions, as Lwamgira recalled:

> The Medical Officer sent young men all over the villages to go and press the necks of the people to see those who had signs of-Botongo disease. If they found any, they would send the person to Kigarama for treatment. . . . They captured a lot of people, but when some saw how the Doctor pierced their shoulders injecting them, they feared and migrated to other areas. The doctor continued capturing people and taking them to Kigarama and those who did not have it, were not affected.[116]

Young men who circulated "all over the villages" to "press the necks of the people" describes precisely the gland-feelers' stock in trade. Their work was here cast as "capture," but what made this interaction fraught extends beyond the consequences of the interaction to involve who was touching whom and in what context. These young, elite men demanded brief, specific physical contact with people across age and gender boundaries—anyone could be sick. This universal mandate and focus on physical contact ran against early twentieth-century Ziba social conventions that directed relations between members of different clans, between elites and commoners, and, more generally, between the old and young and men and women. Ziba social relations remained, in the early twentieth century, strongly influenced by hierarchies of status.[117] The norms that shaped social interactions were constitutive of historically grounded ways of ordering marriage, agriculture, land use, and domestic labor. Transgressing these ways of ordering society could have dire impacts for one's respectability, one's relationships, and even one's health or the health of one's family members.[118] Gland-feelers' association with royal or colonial efforts to attend to population health would not have insulated auxiliaries from such complexity.

Coercion muddies the waters of therapeutic diversity. Ziba people sought treatment from embandwa, submitting to practices of diagnosis and healing that, at the time, likely involved palpation, piercing the skin, or introducing substances into the body.[119] Injections at the camp presented another option for those who might have also sought healing in the homes of experienced relatives, at ancestral shrines, with kubándwa mediums, in missionary clinics, or from other experts.[120] At Kigarama, people opened up further modalities of contending with sickness and death, as German biomedical therapies were fitted within a range of treatments and cures available to the sick and their families. But auxiliaries working for the German camp, wearing European- or Swahili-style clothes, sometimes accompanied by colonial police or soldiers, did not operate within the realm of familiar forms of diagnosis or healing. The examinations initiated by gland-feelers were different than other interactions centered on healing, which sought to resolve individual suffering, collective misfortune, or layered combinations of both, and which employed medicines and practices within particular social and moral frameworks.[121] These young men occupied a new position, not claiming to heal but compelling a person's presence for a particular kind of therapy.

Allegations of blackmail were a vexing problem, but the departure of entire families or villages after an auxiliary's circuit through the area was an entirely different matter. Such reactions to gland-feelers differed from early, unpredictable mobility in and out of the camp that drove the decision to rely on an ambulatory system rather than one of internment camps. People's movements appeared in monthly statistics and in narratives of the camp and were taken on board by camp doctors, albeit with some apparent anxiety about reporting the issue to their superiors. Movement into and out of the camp, categorized by doctors as "flight" or "withdrawal," indicates that people in Kiziba took a flexible and pragmatic approach to using German biomedical goods. People occasionally left the camp area before their course of atoxyl had ended, or evaded the required return. These mobilities, unpredictable to colonial record-keepers, remind us that utilization of the Kigarama camp was sometimes dictated by concerns other than healing, in parallel to or overlapping with desires for therapies. The need to take advantage of the camp's provisions, such as food, tobacco, small plots of land, or sleeping mats, influenced Ziba calculations regarding entering and leaving the camp. Analysis of camp statistics alongside Ziba agricultural calendars also suggests that the ebb and flow of flight sometimes correlated roughly with local dry and rainy seasons, and therefore periods of recurring hunger or want.[122] Thus, amelioration of illness fit into other household concerns in times of dearth or want when calling upon multiple resources for security would have been necessary. But in later years, as we will see, when improvements no longer seemed durable, or treatments were seen to harm rather than heal, or the calculus in favor of utilizing colonial medicine otherwise shifted, people left the camp, avoided examination and diagnosis, or refused to return as scheduled.

Uncontrolled mobility associated with gland-feelers presented an administrative problem for the campaign and portended difficulties for the mukama as well. Early in 1909, gland-feelers traveled through Kiziba, followed by a junior German officer's tour through the area; many people identified for further examination did not turn up at the camp in subsequent weeks. Of those identified as suspicious cases, nearly half "emigrated with their families and all their possessions to British territory, according to the katikiros of the relevant districts."[123] In July 1909, Dr. Ullrich reported of areas northwest of Kiziba that the "constant presence of a gland-feeler had led to agitation in the population, with the consequence of a migration in part of entire families . . . over the British border."[124] Lwamgira's broader description of auxiliaries' work in the general efforts against sleeping sickness likewise connected fear of surveillance and camp treatments to migration "to other areas."[125] Flight from auxiliaries and avoidance of sleeping sickness surveillance was a popular response to the camp's work, read through the lens of chiefly politics in the interlacustrine kingdoms, but it was also a form of political critique directed at the mukama and his chiefs. Emigration could be a check on a king's tendency toward "despotism," as Hyden argues, when "the only way a commoner could show withdrawal of support for his superior was by migrating to another area."[126] While the extent or permanence of outward migration in Buhaya is not well documented, these departures were a key means of expressing displeasure with political leaders at the time.[127] The resettlement of households or villages elsewhere, which colonial doctors saw as an evasion of surveillance and a troublesome administrative problem, also constituted a political response to gland-feelers as emissaries of the king and to the camp as an institution associated with the Ziba court. Flight out of Kiziba, or to a district further away from the palace, was a rejection of royal authority and a critique of Mutahangarwa's cooperation with colonial authorities.

It also reflected a lack of faith in the treatments offered at Kigarama and anxiety about the health of the kingdom more broadly. What people in the countryside may have known or heard about events at Kigarama affected how they received gland-feelers who circulated among their villages and farms, and, in turn, drove people to actions extraordinarily disruptive of family and home life.

Auxiliaries' trouble in finding new cases was also connected to changing views of the Kigarama camp. German officials were quick to mention, however obliquely, the power of superstition and rumor in the Ziba population to animate avoidance of the camp. But they did not recognize Ziba mobility and talk as critiques of the efficacy of the treatment offered. Ziba people were well aware of what happened after entering the camp: namely, blood and lymph extraction, subjection to several months of injections with atoxyl and other drugs, and ongoing follow-up surveillance.[128] Some of the drugs used were experimental; none were tolerated well in long-term use.[129] People died during and after injection regimens. The extended "duration of treatment" and the treatments themselves undermined the camp's effectiveness in curbing sleeping sickness. When Kudicke returned in late 1910 after an eighteen-month furlough to find the campaign in disarray and the trust of the population "gravely unsettled," it was no wonder.

> The grounds for this mistrust are of many different varieties: the people have seen that a great portion of the sick, who they themselves did not recognize as sick, deteriorated and died despite treatment. They have seen that almost all of the people who considered themselves sick—patients in the third stage—still could not be saved from death. They have lastly observed in many cases that sick people, who at the end of treatment found themselves in good condition, in the course of observation declined and in many cases perished shortly after being admitted again [to the camp]. It is hardly astonishing that these observations in many cases were interpreted in the sense that people first got sick in the camp and that upon people's death the treatment or taking of blood was to blame, not the sickness.[130]

Sleeping sickness was fatal, meaning that people treated in the camp did not ultimately improve after a stint there, and atoxyl, the promised cure, proved ultimately detrimental. Multiple or recurring treatments over the longer term were especially problematic, and Kudicke reported that relapsed patients he sought to treat "simply ran away with the whole kit and caboodle."[131] His own assessments bolstered these rejections of atoxyl. "In the majority of cases," he wrote, "treatment of relapses with atoxyl is hopeless, and in some cases is even harmful."[132] Atoxyl, particularly after repeated use, caused damage to the optic nerve that disturbed a person's vision or blinded them completely. Further, in the interim period during Kudicke's furlough, other German staff had injected several highly toxic experimental drugs at the camp.[133] Both cumulative and recent experiences with injections compounded Ziba resistance to ongoing treatment and strengthened people's counterclaims asserting the futility or danger of colonial therapies. Examination and taking blood and lymph continued at Kigarama. In Buhaya, using parts of bodies and bodily fluids—blood among them—could still animate powerful, harmful medicines; ongoing blood and lymph extraction combined with ill effects of injected drugs undercut German claims of the camp's benefits.[134] Some Ziba people may have retained deep skepticism about the benefits of colonial medical practices and harbored concerns about their harmful effects on individuals and the kingdom more broadly, despite royal inducements to submit. The trend toward evasion or avoidance represented hundreds of individual and collective decisions, occurring for differing reasons over time, but its decisive acceleration related to the deployment of gland-feelers portended difficulty for the camp and the German campaign.

Trouble for the gland-feelers and the camp meant trouble for the mukama and, in turn, eroded his support for the sleeping sickness campaign.[135] By late 1910, Kudicke commented, "the measures of the sleeping sickness campaign do not find the same support from the side of Sultan Mutahangarwa and his katikiros as before."[136] The camp's failures had political consequences for Mutahangarwa. His decision to pull away from prior support of the sleeping sickness campaign involved political calculations that weighed the importance of cooperation with the sleeping sickness campaign for his relationship with German colonial authorities against the importance of the stability and legitimacy of royal power among his people in Kiziba. Kudicke discussed the matter with Ziba authorities and learned that people with sleeping sickness living in the villages had "emancipated themselves from the authority of the katikiros more than was good."[137] The people who sought to "emancipate" themselves from traditional authority were not untreated cases of sleeping sickness, but rather people who had been diagnosed and treated at the Kigarama camp. Their relationship with Ziba leaders may have begun to change because the camp had provided an alternative source of material goods, cash, land, and also medicine. People who increasingly challenged royal authority had personal experience of the camp and knowledge of failed and damaging treatments: atoxyl and the other drugs used harmed more than healed and Kigarama became associated with wasting and death rather than healing. Trouble mounted as more and more people left the camp's purview or refused to come for prearranged "injection days." People invoked the camp to challenge chiefly authority and to offer substantive challenges to Mutahangarwa's legitimacy. Withdrawing support from the camp allowed Mutahangarwa and his subordinate chiefs to counter such arguments, as the mukama dealt with discontent created by the sleeping sickness campaign.

By late 1910, Kudicke judged gland-feelers' searches in Kiziba to be "fruitless" and stopped deploying them. This fit into a wider shift in tactics at Kigarama. His decision to discontinue the use of Ziba auxiliaries was made with the stability and future of the colonial medical work in mind, while also reflecting the decreasing number of cases of sleeping sickness in Kiziba proper. Kudicke stopped treating relapsed cases with atoxyl and, soon after, suggested that Kigarama transition to be a polyclinic where he would continue to monitor disease incidence among the Ziba population while also treating syphilis and some minor ailments.[138] Few new cases of sleeping sickness came to the camp, and treatment with existing drugs proved too problematic. More broadly, the camp's original goal of arresting the spread of the disease around Lake Victoria seemed to have been met, as the epidemic did not expand on the western lakeshore as feared. The new brief of the camp also signaled a change in German energies to direct increasing attention to the ongoing epidemic at Lake Tanganyika in German Urundi.

Although no longer deployed to search independently for the sick in Kiziba and elsewhere in the Bukoba Residency, gland-feelers remained posted at the main ferry on the Kagera River and at key crossings of marshes and main roads until the outbreak of World War I. A few of the men trained at Kigarama by Kudicke went on to work for the sleeping sickness campaign on the coastal lowlands around Lake Tanganyika after 1912.[139] Colonial officers there sought to use their experience in identifying potential cases of sleeping sickness as they extended the campaign's reach into German Urundi. Now known primarily as "Bukoba boys," these select gland-feelers trained at Kigarama became colonial functionaries of a different sort, their conduct apparently reliable and their skills valuable to European doctors in a setting far from home.[140]

For colonial public health officials, camps such as Kigarama served as bulwarks against the anticipated wider spread of the epidemic around Lake

Victoria. The Kigarama camp was singular within the German campaign because of the working relationship between German colonial doctors and the Ziba king Mutahangarwa that facilitated the camp's establishment and expansion. As a space of overlapping colonial and royal authority, Kigarama offered Ziba people—whether admitted as patients or for relatives on its periphery—not only the possibility of healing, but also the acquisition of land and material resources. Understanding how Kigarama fit into the political economy of land distribution in Kiziba casts royal support for the camp in a different light, not only with regard to small plots of land on the camp's periphery, but also with regard to activities within the camp itself. German colonial officers created incentives to draw people into the Kigarama camp for treatment, offering goods that would have been scarce for some, as illness limited a family's capacity to farm and trade. With Mutahangarwa's mandate and assistance linking royal authority to the camp, Ziba patients drew upon both German colonial and Ziba royal resources when using the camp as a space of healing or a source for material goods. But as Ziba turned away from the camp, they also effectively turned away from its assertion of overlapping royal and colonial power.

While extant sources do not reveal what Mutahangarwa thought about the drugs and injections offered at Kigarama, it was in his interest to embrace the use of biomedical technologies against sleeping sickness in accordance with his responsibility for the health of the kingdom. For a time, the king maintained a balance between this role, on the one hand, and the demands of the German colonial administration on the other. But as sleeping sickness progressed in Ziba bodies, as increasingly troublesome auxiliaries fanned out across the kingdom, and as people returned to their homes from Kigarama unhealed, that equilibrium began to break down. The quickly changing fortunes of the sleeping sickness camp at Kigarama demonstrate the precariousness of both African royal and European colonial power in the early colonial period amid life-changing epidemic disease. Further, the history of Ziba medical auxiliaries reveals the complexity of decisions made by people contending with new assertions of colonial power and biomedical claims to heal and cure. Relationships that appear in the colonial archive as straightforward collaborations and transactions were in fact subject to continual negotiation, driven by political expediency and evaluations of personal or familial gain and loss. Social structures such as the muteko age-sets transcended but also shaped the nature of both the cohort of gland-feelers and the work that they did. These auxiliaries also embodied Ziba royal power at a time of transition, as interest in "modernization" intersected with "traditional" bases of legitimacy.

Gland-feelers tested Ziba communities' tolerance for colonial public health interventions and introduced new forms of economic exchange, centered on the discovery and delivery of people infected with sleeping sickness. As intermediaries pioneered the exchange of bodies for rupees, the population of Kiziba responded with both negotiation and evasion, forcing a change in the strategy of both colonial doctors and the king and demonstrating the continuing relevance of ideas about royal power, reciprocal obligation, and social stability. The presence of gland-feelers had a direct impact on relations between the people and their mukama, pushing the Ziba king to reevaluate his engagement with colonial authority and his support for public health interventions. Gland-feelers' work also triggered a reassessment of the efficacy of those interventions by colonial medical officers. Representative of early experimentation in colonial public health practice, Ziba auxiliaries nevertheless remained men with their own agendas, acting independently in the realm of politics and health in ways that neither the colonial authorities nor the Ziba court anticipated.

As Kigarama's story demonstrates, colonial health interventions such as sleeping sickness camps fit into local political and social histories and became entangled in extant efforts to govern, heal, and prosper. Such situated histories and experiences affected how colonial medical officers' target populations received, used, and sometimes avoided those interventions. Elsewhere within the German anti–sleeping sickness campaign, the situated histories and local experiences with illness and insecurity had different impacts. At Lake Tanganyika, the campaign's other primary field of action, dynamics of local mobility and cross-lake migration shaped daily life and were fundamental to ideas about disease transmission as well as to how and where colonial public health officers engaged with targeted populations. Histories of mobility and migration between the eastern and western lakeshores defined the horizons of sleeping sickness camps in the southern Imbo both intellectually and practically. Ultimately, these pasts and presents of mobility would embroil the campaign in local politics in unanticipated ways.

PART III

The Southern Imbo, c. 1890–1914

The Southern Imbo, c. 1890

An Overview

ANTI–SLEEPING SICKNESS efforts on the northeastern shore of Lake Tanganyika began roughly concurrently with the campaign at Lake Victoria, but on a significantly larger scale. These efforts focused primarily on the Imbo region in colonial Burundi. I here focus on orienting readers to the terrain and populations of the northern rim of the lake and to the political and economic dynamics of the late nineteenth century in the region that would impact how the German sleeping sickness campaign unfolded. At the time, the Urundi kingdom was a territory of roughly 10,800 square miles in the highlands and plains north and east of Lake Tanganyika. Its southwestern edge was the Imbo plain, defined by the narrow swath of the rift valley running along Lake Tanganyika's eastern shore and northward up into the Rusizi River valley.[1] The lake's shoreline there is irregular, with scattered rocky outcroppings, narrow beaches, sand banks, and marshes, and is also interrupted by the mouths of many small rivers and streams that flow down from the highlands. Climate and ecology made Imbo distinctive within the Urundi kingdom and the northern littoral, with implications for agriculture, politics, and health. Newbury argues that, from the mid-nineteenth century, "the people of this lakeside zone developed productive rhythms, economic relationships, and social contacts with the larger Lake Tanganyika cultural community, distinct from highland areas."[2] The southern Imbo, where the next chapter focuses, was wholly within that distinctive lakeside zone.

This stretch of lakeshore, which I refer to as the southern Imbo, runs between the modern Burundian towns of Nyanza-Lac and Rumonge and is bookended by two capes (Magala to the north and Bangwe to the south) that butted into the lake. Distinct to this area was the passage between the Ubwari peninsula on the western shore and Rumonge on the eastern shore—one of the narrowest on Lake Tanganyika. The Ubwari peninsula juts out into Lake Tanganyika like a blade aimed sharply upward at the northeastern shore; at their closest point, the two shores are barely fifteen miles apart. Ubwari's main feature is a high, central ridge of mountains, with narrow lowlands and rocky beaches around its perimeter; the peninsula joins the lakeshore with a swampy plain bordered to the north by Burton Bay and the south by Kibanga Bay. Connections between Imbo and Ubwari and its environs fit into the wider dynamics of interaction between communities around Lake Tanganyika and were centrally important for life and livelihood in each place.

Agricultural complexity, particularly the reliance on multiple crops such as sweet potato, cassava, plantain banana, and beans, was a defense against food insecurity and an exploitation of the lowlands' unique climate. Lower elevations along the lake meant warmer temperatures, slightly less rainfall, and different ecological niches and different possibilities for agricultural production, both of which had implications for nutrition and health. Imbo had a single rainy season beginning in October and ending in May; this concentrated agricultural activity in the driest months of June through September.[3] In the mid-nineteenth century, people in the Imbo grew cassava, bananas, and yams as staple crops; by contrast, people in the highlands farmed sorghum, maize and bananas.[4] Such differentiation was the result of centuries of change and agricultural innovation, some of which was particular to the Imbo region's biome. At the time, cassava was cultivated only on the coastal lowlands, as were oil palms. Imbo's vast stands of oil palms, a tree species originating in western Africa, durably made cultivation along Lake Tanganyika distinct within Burundi.[5] Ecology also had political implications. The different resource base in the plains of Imbo likewise "meant that the royal dynasty of Burundi had trouble asserting any consistent control in much of [Imbo]," in part because of the differing symbolic and practical value of crops important in Rundi royal rituals.[6] Animal husbandry—in particular the keeping of sheep, goats, and cattle—broadly contributed to family and village prosperity and also drew people into complex social relationships.[7] Cattle clientage stitched together individuals and households or compounds within nineteenth-century Burundi, as well as across different climate zones, between which transhumant cattle and sheep passed seasonally.[8]

Despite the distance between lakeland Rundi and their upland countrymen, a common political structure and language prevailed, linking communities throughout the kingdom of Urundi to one another and to the *mwami.* Similarly to Buganda, Kiziba, and other interlacustrine kingdoms, the mwami traditionally held sway as an absolute ruler, controlling distribution of land and cattle and thereby influencing power and prosperity from his court at Gitega in the highlands. Social and political structure in Rundi society functioned within a hierarchy of power linked to clan, lineage, and land. Power also rested with "great chiefs," typically close relatives of the king who might also be clan elders, whose estates were dispersed parcels of land scattered around the territory.[9] Paralleling these clan elders were chiefs governing regions of the kingdom which were then divided into chiefdoms ruled by *watwale*, subchiefs (sing. *mtwale*). Politically, distant areas such as the Imbo lowlands functioned as "semi-independent chiefdoms" on the eve of colonization.[10] Political moves by mwami Ntare Ruhatsi in the eighteenth century that had transformed and restructured political space in the south and east of Burundi, for example, had not included Imbo, where populations had strong links with other societies around the lakeshore and "retained considerable autonomy."[11] Chiefs who ruled Imbo communities maintained a loose affiliation with the Rundi monarchy in the central highlands, but this connection showed signs of weakening in the late nineteenth century due to the combined effects of power struggles late in mwami Mwezi Gisabo's rule (all the more so after his death) and social and political disruption borne of the ecological crises triggered by cattle diseases.[12] Regional autonomy in the Imbo lowlands had implications at the local level, as Rundi populations engaged independently at a village-by-village level with people who arrived from the opposite shore of the lake.

By contrast, societies across the lake on the western shore existed in a very different political landscape. There was no powerful, centralizing kingdom to directly impact life on the western lakeshore. By the later nineteenth century, when missionaries began to record their impressions of populations on the perimeter of Ubwari and the nearby lowlands, accounts suggest the independence, if not atomization, of local chiefly politics. Modern informants at Uvira described historical arrangement of villages on the northwestern littoral around lineage groups, separated from cultivated fields, unlike arrangements in Burundi or Rwanda.[13] Two key chiefdoms locally were those of Pori (on Kibanga Bay) and the Mtemi of Ubwari, both characterized by dispersed, non nucleated villages with patrilineal kinship groups.[14] Pori and the Bwari Mtemi frequently

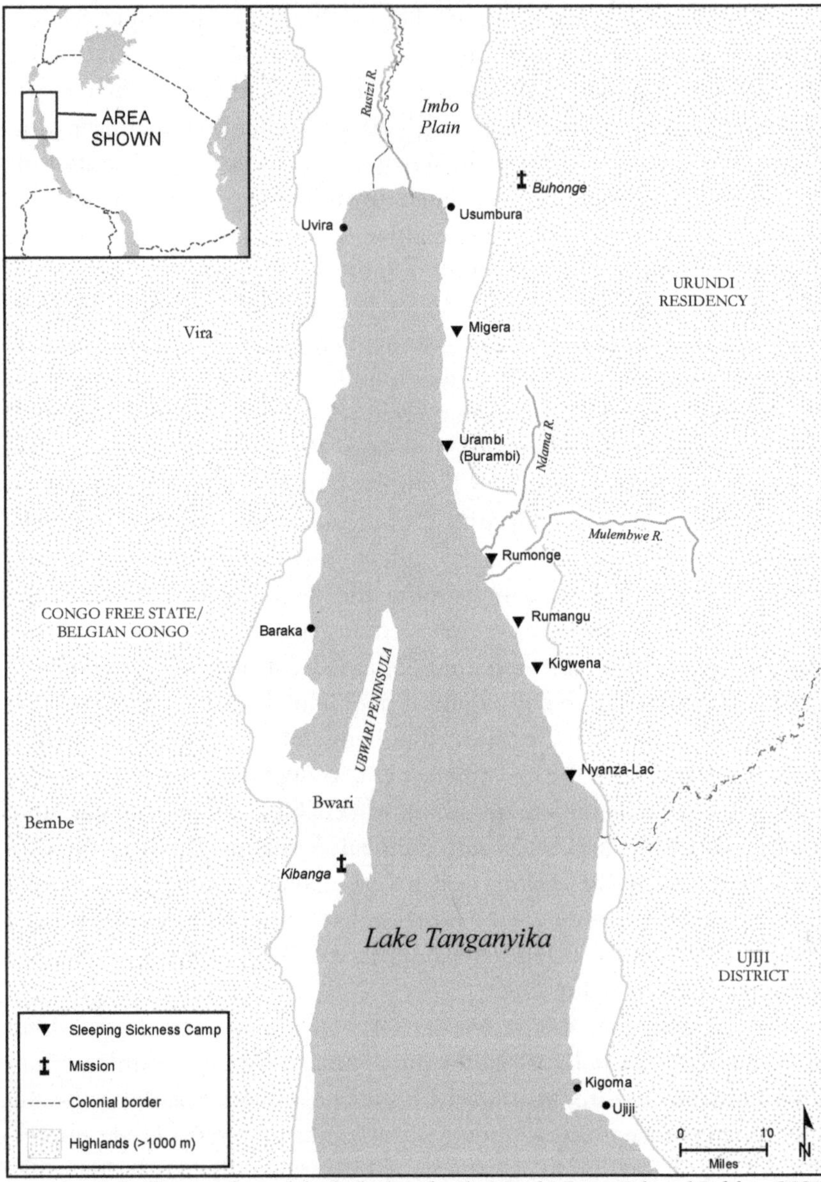

MAP 5.1. Lake Tanganyika and the Imbo Lowlands. Map by Brian Edward Balsley, GISP.

skirmished, triggering temporary displacements and local instability; in the later nineteenth century, violence between these polities connected with the presence of raiding parties associated with Swahili power at Uvira and Ujiji, seeking slaves and ivory. People of Masanse (on the lowlands north of Ubwari) and the Ubwari peninsula's perimeter also had historical connections to the better-known Bembe cultural group to the north and west, but this did not insulate their communities from the disruptions of the nineteenth century.[15]

On both the eastern and western shores, populations' use of the lake provided particular opportunities for both production and trade in the nineteenth century. Lowland Rundi use of the lake, similar to fishing in the Ssese Islands or coastal Buhaya, involved sophisticated technical and labor inputs. Using hooks, traps, and nets of various sizes allowed people to diversify their catch, while the production of fishing materials for specific aims drew in diverse members of a household: children made hooks from iron or copper wire, for instance, while women, men, and children made cord and rope for fishing nets.[16] People living on the Imbo plains deployed a variety of sophisticated fishing techniques and strategies for diverse types of fish, intended both for local consumption and preservation to trade further afield. Fishermen navigated the lakeshore with canoes punted close to shore, as well as canoes to venture out into the lake, moving around ports and villages stretching from Ujiji around the lakeshore to Uvira and the Bwari peninsula.[17] Production of *dagaa* or *akahuzo* (a small, minnow-like fish), which, in its dried form, had especially wide regional circulation as a trade good from at least the 1860s and served as a crucial dry-season food, developed out of specialized knowledge in Imbo.[18] Vigorous trade of dried and fresh fish from the lake connected people on the Imbo lowlands with local and regional commerce, but regionally unique crops such as cassava, hemp, and palm oil formed the basis of trade as Imbo and Bwari communities engaged with one another and with other lacustrine travelers.[19] People living in Ubwari and Masanse saw local trade in salt, fresh and dried fish, copper from the south, iron from the north, and palm oil from the Imbo lowlands. Sheep and goats also fit into the local economy as trade goods and provided meat and hides.[20]

New powers in the Great Lakes region, which increased just as mwami Mwezi Gisabo came to power at mid-century, grew more aggressive in their pressure on peripheral areas of the Rundi kingdom by the 1870s. Regional trade and connectivity were well established: extant African trade was highly specialized, moving salt, copper, and iron goods from their sources and differentiated from new Arab-Swahili trade in the vessels used (in dugout canoes versus dhows under sail, for instance).[21] But the invigoration of long-distance caravans and the pursuit of slaves and ivory in the region changed power dynamics for littoral communities. Ujiji became an increasingly bustling center of regional commerce from the 1840s, connecting the hinterlands of the Congo basin, Lake Tanganyika, and Lake

Kivu with the Indian Ocean coast.[22] Under Mwinyi Heri's period of control in the late 1800s, Swahili power began to touch Imbo populations as far north as Usumbura.[23] Launched from Ujiji, but with a foothold at the new hub of Uvira, Zanzibari trader Rumaliza (Mohamed ben Khelfan al Barwani) sought to expand his trading networks and influence into the western and northwestern periphery of the lake.[24] Rumaliza (whose name derived from the Kiswahili verb meaning "to exterminate, to finish") developed Uvira into a key post for forays into the eastern Congo and Rwanda, while his mercenaries pushed into nearby communities in search of slaves and ivory from minor posts between Uvira and Ubwari.[25] Raids and warfare had a deep destabilizing impact on vulnerable communities in the region, with few exceptions; Mwezi Gisabo's armies repelled Rumaliza's attempts to advance into the Urundi highlands in the 1890s, minimizing the influence of coastal powers in the Rundi highlands.[26]

European interest, and later incursion, followed established trading routes into the region. German forces gradually moved inland to found military posts at Ujiji in 1896 and Usumbura in 1897.[27] Uvira would a become focal point for Congo Free State energies in the Tanganyika-Kivu region, as interest in securing Belgian control over trade and displacing Swahili power increased in the late 1880s and early 1890s.[28] In Urundi, the Germans engaged with the aging mwami Mwezi Gisabo amid deep political turmoil and the revolt of powerful chiefs in the kingdom's periphery. Through alliances with his rivals and military defeat, German authorities forced Mwezi Gisabo into an agreement that, David Newbury argues, "traded Rundi sovereignty for German support in consolidating the power of the central court throughout Burundi."[29] In Urundi, the colonial political structure was organized around the system of the residency and based at Usumbura, which also served as the chief trading center of the northern end of Lake Tanganyika. As in Bukoba, the Urundi Residency was premised on an early idea of indirect rule and modeled after perceived British successes in Uganda, but also informed by a push for colonial reform in Germany that followed the Maji Maji rebellion after 1907.[30] Rather than installing chiefs or overlords to attempt to collect taxes, direct labor, dictate agricultural production, and mandate punishments, the Urundi Residency instead sought to graft colonial power onto the Rundi political system. Ruling the territory through Rundi chiefs forced into this relationship by military defeat, the German administration aimed at civilian rule that would govern the kingdom more cheaply and less disruptively than a continued military presence. The residency generally was the most influential in the central highlands, where the mwami's traditional power remained the strongest, but also sought to impose requirements for forced labor and porterage, to collect taxes, and to force the cultivation of particular crops elsewhere.[31] In the last decades of the nineteenth century, these political upheavals were preceded with and accompanied by relentless generalized crises, in the form of famine, disease, and crop failure, related to ecological and epidemiological change.[32] A collective narrative dictated to a White Fathers missionary sometime between 1896 and 1903, for example, describes a series of disasters that would have been especially severe for Imbo's agro-pastoralists: civil unrest, locusts, famine, diseases, and drought.[33] Similar misfortunes impacted the western shore. The historical experiences of Rundi and Congolese populations informed the range of strategies for addressing successive challenges to health and prosperity for Imbo communities, while also impacting the environments within which widespread illness from sleeping sickness would later occur. These generalized disasters and efforts to survive them invigorated mobilities around the lake and formed a key context for the outbreaks of sleeping sickness that animated German energies in Imbo.

5 Mobility, Illness, and Colonial Public Health on the Tanganyika Littoral

SIMULTANEOUSLY WITH the Lake Victoria campaigns, German public health energies also focused on a parallel campaign at Lake Tanganyika. There, chiefly along the eastern shore of the lake and in the Rusizi and Malagarasi River valleys, German doctors used the model of the sleeping sickness camp set out by Koch at Lake Victoria to coordinate sleeping sickness control. Attention focused overwhelmingly on the narrow coastal lowlands of Lake Tanganyika and the plains of the lower Rusizi valley, in an area known as Imbo. An archipelago of eight stations stretched along 150 miles of coastline, running between the bustling market towns of Ujiji and Usumbura (modern Bujumbura). The campaign in the Imbo lowlands undertook two intertwined projects. The first was an ambitious and comprehensive plan to destroy tsetse fly vector habitats along the busiest sections of the lakeshore, relying on locally mustered workers and chiefly collaboration. The second was an effort to maintain extensive atoxyl regimens among the populations surrounding each camp. These projects were a scaled-up version of the local, interconnected endeavors at Kigarama and Kishanje in the Bukoba Residency, or the parallel efforts in the Mori and Mara River valleys on the eastern lakeshore. As at Lake Victoria, the character and focus of anti–sleeping sickness work along Lake Tanganyika was shaped by the field on which it took place and the social worlds of the people who became its desired subjects. In the Imbo region, geography, mobility, and local economic dynamics were centrally important to the campaign, particularly in the southern Imbo along the narrow passage across the lake to the eastern Congo Free State (later the Belgian Congo). Mobility and migration were relevant in multiple fields: the Imbo lakeshore where camps were sited, the Bwari peninsula opposite (from which some Imbo residents had migrated), and the lake and lakeshores as the well-traveled space between them. Existing patterns of mobility and migration linked, bodily and imaginatively, the wider epidemiology of sleeping sickness on the Congolese and Urundian sides of the lake's deep valley. Central to the campaign's trajectory were the engagement and labor of Imbo populations, who inhabited these political, economic, and environmental borderlands. Littoral populations' mobilities shaped German and Belgian ideas about sleeping sickness, the epidemiology of the disease itself, and the location of sleeping sickness camps.

The archipelago of camps allowed the German anti–sleeping sickness campaign to target a local population by maintaining surveillance, detecting cases, treating people with atoxyl, and, especially at Lake Tanganyika, destroying fly habitats. The circumstances that shaped camps' design and planning at Bugalla and their elaboration at Kigarama (and, to a lesser extent, Shirati) on Lake Victoria, however, did not mesh with the circumstances on the Tanganyika lakeshore. Here, tsetse flies abounded and the potential geographical scope of sleeping sickness was immense. Likewise, the matter of who the campaign could reach from each camp was more complicated, given multiple hubs of cross-lake mobility in the southern Imbo. German officials oriented camps along the lakeshore according to their interpretation of the historic migration and settlement patterns within which African mobilities connected particular economies and sites. This foregrounding of both site and movement dovetailed with contemporary epidemiological approaches to sleeping sickness, which centered on foci created when fly habitats and human disease overlapped in time and space. Moving affected people through the camp's atoxyl regimens would ideally reduce possibilities of transmission and was an important aspect of the anti–sleeping sickness campaign. But addressing sites of transmission was another matter, as the campaign's main work evolved to focus on bush and lakeshore clearing programs targeting the fly vector's habitats. This intensive work required a consistent labor supply and meant that Imbo populations were crucial to sleeping sickness control not only as potential disease carriers, but also on the front lines of fly control work. Political and social complexity in the southern Imbo, both related to regional mobilities, therefore posed a fundamental challenge to sleeping sickness control. Initially, these local circumstances in the southern Imbo posed an intellectual and practical challenge for understanding the Tanganyika epidemic. But the issue of littoral mobilities ultimately evolved into a labor problem that undermined the campaign's conditions of possibility and embroiled the camps in broader politics in unanticipated ways.

The circulation of people, goods, and pathogens around Lake Tanganyika had historically been robust, but migration in the previous generation was particularly so; many migrants from the western shore had settled in the southern Imbo. In this chapter, I first explore the development of ideas about sleeping sickness and its transmission around Lake Tanganyika that informed the German administration's particular focus on Bwari and other "Congolese" communities living alongside Rundi communities in the southern Imbo. In the campaign on the Imbo lowlands, the scattered camps connected to one another as personnel circulated between them and Rundi and other populations were caught in each camp's shifting gaze. Practical challenges to operationalizing concerns about mobility on the Tanganyika littoral soon arose, and the campaign faced difficulties in translating these into interventions on the ground amid political, social, and economic complexity. Tension around camp labor requirements and differing perceived economic and political statuses of Imbo populations provides a closing case study for how political realities and public health priorities came to be in tension there, particularly around questions of status, labor, and mutual obligation.

LACUSTRINE CONNECTIONS AND LIFE IN THE SOUTHERN IMBO

The Imbo region's position astride pathways of local and regional mobility meant that daily life involved multiple possible orientations and connections—eastward to the highlands, westward to Ubwari and the Congo watershed beyond, along the shore to Ujiji or Usumbura, and even to the southwest to distant shores of the lake.[1] As the nineteenth century had unfolded on the Imbo plains, economic activity around the lake flourished, connecting Imbo populations with others in the region in durable, important ways. A key aspect of life in the southern Imbo was connectivity to populations around and across the lake, both to market towns like Ujiji and Usumbura, on either end of the Imbo shore, and to communities in the western littoral. Links between populations on the eastern and western shores of the lake at its narrowest point had deep social and economic relevance. Lakeside connections between Imbo and the western shore opposite—Ubwari and lowland areas of Masanse—focused on the movement of goods and people. These movements were patterned by seasonal rhythms and also punctuated by unpredictable necessities and demands. Cross-lake and regional mobilities, with particular focal points, routes, and rhythms, affected local demography and led to particular concentrations of populations from the western lakeshore in the southern Imbo. The history of Ubwari, particularly its economy and its populations' experience with widespread illness, was joined with that of the southern Imbo, as Bwari people traveled to the southern Imbo and sometimes settled there more permanently. The nature of local and regional mobility and related patterns of farming and settlement would become central to the location of German sleeping sickness camps, which were targeted toward particular concentrations of population and hubs of mobility. A tight focus on the territories on either side of the lake's narrow passage between Ubwari and the southern Imbo allows us to understand the meanings of the camps' locations and the later ramifications of their particular purview.

Historic connections between the Ubwari peninsula, the nearby mainland, and the Imbo lakeshore opposite had developed in the eighteenth and nineteenth centuries, and were well established by the mid-nineteenth century. Mobility out of Ubwari and the western lakeshore typically aimed eastward, rather than westward, due to regional geography: overland movement was more restricted than travel on the lake.[2] Bwari, Sanse, and Bembe populations on the northwestern edges of the lake (south of Uvira, the ultimate center of the northern Tanganyika slave and ivory trade) were bounded by mountains to the south below Ubwari and to the north above Uvira. On a local scale, the populations in the lowlands beyond what is now Burton Bay were all the more boxed in by the region's topography, occupying an isthmus between the mountains at the base of the Bwari peninsula and the mountains of Ubembe that rose to the southwest. And so, similarly to the Rundi inhabitants of Imbo, who were backed by mountains rising to the east, Bwari and nearby people maintained robust connections along the lakeshore and across the lake, less so with highland populations. Trade and travel across the lake encouraged longer-term migration. Near Ujiji, the bustling trading port on the eastern shore that fronted some of the lake's richest fishing waters, for example, small groups of Bwari and Bembe people settled alongside others in the late eighteenth and early nineteenth centuries.[3] The increasing insecurity of life in coastal areas during nineteenth-century intensification in the slave and ivory trade in central Africa also drove Bwari or Sanse mobility eastward. Some Bwari participated in the slave and ivory trades, profiting from integration into coastal markets and adopting Swahili status markers, while other Bwari, Sanse, and Bembe populations were made more vulnerable by these economic changes and fled from insecurity.[4]

Connections between the populations on either side of Lake Tanganyika resulted from experiences of illness and crisis and mobilities they precipitated. Flight to avoid widespread diseases also catalyzed eastward travel and Bwari mobility and migration related to particular crises or opportunities. In one example, pox-causing illness had circulated mercilessly among and around the Kibanga mission between 1884 and 1886. In the villages, people took measures to isolate the sick in huts removed from other homes; the sick took only basic provisions (sleeping mat, cooking pot) with them, and those that recovered could reenter the village only after no signs of sickness remained.[5] Death from this 1884–86 outbreak of the disease was so widespread that the crisis drove migration out of Masanse and Ubwari to the southern Imbo, with the Kibanga mission diarist noting that "the terrible scourge's ravages have been so numerous that, as best we can discern, half the population has succumbed to its ailments . . . the frightened people have fled to the opposite shore of the lake."[6] Outbreaks of pox-causing illness, famine, and Rumaliza's raiding parties seem to have accelerated cross-lake travel—in the late nineteenth century, people fled Ubwari and the coastal regions between the peninsula and Uvira town and many came to the southern Imbo. Some possibly returned once sickness had ebbed, but some likely stayed, given early twentieth-century narratives of the tenure of Bwari communities there. Regular traffic across the lake that connected the Kibanga mission and Imbo populations spread illness outside of significant epidemics, such that pox-causing diseases provided a precedent for later illness arriving with travelers from the west side of the lake for Imbo societies.[7] In a similar vein, as the deadly illness they called *malali* worsened on the Congolese shore twenty years later, populations there again fled to the relative security and peace of familiar areas of the southern Imbo.

Because of their function as hubs of cross-lake mobility between Ubwari and Imbo, two spaces in the southern Imbo were of particular relevance for colonial conceptualizations of sleeping sickness epidemiology and their strategies of targeted intervention: Rumonge and Kigwena. While people moved freely along and across the lake, connections between the southern Imbo and areas outside of the region were especially robust at Rumonge, which emerged as a particular hub of mobility and trade in the nineteenth century.[8] Rumonge sat between the lake and a wide, undulating "prairie" at the foot of the mountains, situated in the center of a small semicircular bay. It was a collection of Rundi villages among pastureland and cultivated fields, set back a few hundred meters from a long, sandy beach on the lake. The Mulembwe River flowed into the southern edge of its bay and the larger Ndama River flowed in on its northern side, each winding their way through a flat plain after dropping from the mountains further inland.[9] Although the terrain rose slightly as one went east toward the mountains from the lakeshore, and a few low hills punctuated the landscape, the area around Rumonge was broad and flat, varying between marsh, grassland, and cultivated fields. These lowlands were adjacent to the territory of powerful Imbo regional chief Ndugu in the highlands above.[10] Rumonge generally marks the northern edge of a small, oblong plain of lower elevation along the lake that is singular along the Urundian lakeshore. In the 1890s, the sheltered beach and low, open terrain beyond would have been a contrast to other possible landings on the lake to the north, where thin beaches, marshes, or rocky outcroppings dominated and mountains ran much closer to the lakeshore.

Rumonge's unique topography and its position between different ecological zones made it a flourishing hub for trade: it was a departure point for transit to the coast, via boats to Ujiji, as well as north to Uvira or into the Rusizi valley. In the 1880s, it was home to a renowned market. Sitting astride a plain that lay locally between lake and mountains and regionally between the trading centers of Ujiji and Usumbura, Rumonge's market flourished through the economic productivity and diversity of lakeshore communities. Fr. Francois Coulbois, a White Fathers missionary brought there by a local guide in the 1880s, described paths down from the mountains that followed the natural gorges cutting down toward the plain and the lake. Rumonge was a point of convergence, and several hundred people were gathered at the market when Coulbois and his party visited. Its rainy-season market featured perennial food crops, such as plantain bananas and oil palm fruit, and diverse seasonal crops as well as the products of specialized producers—skilled potters and barkcloth makers, palm oil pressers, and palm wine and banana beer brewers.[11] Other accounts of Imbo markets highlight the bustling activity connecting and lake that characterized Rumonge, with trade in "[palm] oil, mats, fish, salt, goats, honey, and all kinds of wares."[12] Jars of palm oil, alongside goats and occasionally grain, were a key exchange good moving out of Imbo and into Ubwari and Ubembe by the late nineteenth century.[13] By the end of the nineteenth century, Rumonge's traders and producers held a regular market and it served as a departure point for goods from Urundi to the markets at Ujiji.[14] It held an enduring reputation as an entry point for Imbo trade and as a hub for longer-distance commerce seeking Rundi products. More than Usumbura to the north or Nyanza to the south, Rumonge was a hub for mobility and trade in Imbo where Rundi populations in the nearby foothills, those living on the lowlands, and people from the western side of the lake came together.

Another important settlement in the region was Kigwena, a place that shared its name with a nearby river. Kigwena sat opposite Ubwari but further south than Rumonge, connected to the open terrain around Rumonge by a narrow stretch of low, marshy land along the lake. If Rumonge was defined by its low terrain and natural beach and known for its market, Kigwena's reputation lay in its stands of oil palms. Boxed in by hills that rose sharply to the east and on a narrower strip of low land, Kigwena was by all accounts not a general trading center, but rather a key site of palm oil production. Homes were scattered throughout banana farms and oil palm groves. Kigwena maintained its own connections with wider trading networks, too, by the early twentieth century. Overland caravans from Ujiji came directly to Kigwena for palm oil and a representative of an Indian trading house (likely based at Ujiji) maintained a presence there.[15]

Rumonge and Kigwena were key points of connection between Imbo Rundi populations and travelers from Ubwari and the lake's northwestern shores. Colonial observers would later note that the southern Imbo in 1905 was a zone of bustling commerce and activity on the lakeshore, with "a great number of fishermen and traders who travel the entire coastal area between Usumbura and Ujiji copiously with their boats, have no set domicile and stay for a time sometimes here, sometimes there in order to advance their commerce."[16] The perception that populations had "no set domicile" underscores how poorly German officials understood the lives of people in Urundi and the connections that people maintained between homes, farms, and fishing and trade on the lake. But the vigorous mobilities they recounted were significant and also involved populations from the eastern Congo Free State. German officials perceived discrete Bwari settlements near Rundi villages when they traversed the Imbo lowlands in the early twentieth century and argued that the area's "Congolese" population was neither homogenous nor uniformly distributed. Sanse, Vira, Bembe, and Bwari people lived around Rumonge, for instance, and the non-Rundi population around Kigwena was seen as predominantly Bwari by 1906.[17] Such colonial observations generally fit within rigid ideas about clearly defined ethno-linguistic identities and sought to map different and distinct African populations on the terrain. Nevertheless, some trends emerge. By the late nineteenth or early twentieth century, Bwari presence in Imbo had taken on particular spatial dynamics, fitting alongside Rundi agricultural life. Bwari settlement concentrated in particular areas in proximity to either oil palm groves or established markets and on the shore of the narrowest stretch of the lake opposite Ubwari, especially around Kigwena and Rumonge. Seasonally focused trade—when *dagaa* were harvested, palm oil produced, cassava flour prepared—set the rhythm for regular circulation. Distribution of Bwari populations near existing oil palm forests around Kigwena suggests that Bwari communities established themselves in close proximity to Rundi palm oil producers; Bwari people settled very near the lake and not further afield, and they came to function as key intermediaries in palm oil marketing locally at Rumonge and Kigwena and further south at Ujiji.[18] Bwari settlements in Imbo would have provided a foothold for participation in trade around the eastern side of the lake and, at Rumonge and Kigwena, placed Bwari interlocutors strategically in range of Rundi production and local markets, as well as the regional market at Ujiji; some grew wealthy through commerce around the lake. But living near Rumonge and Kigwena also kept Bwari populations in the closest proximity possible to the peninsula and allowed ongoing connections to the western shore. Similar distributions of "Congolese" populations do not appear to have been characteristic of other oil palm groves in coastal Urundi, for instance, such as in the Mtara forest located further to the north and in the hinterlands of Usumbura.

How did Bwari migrants fit into Rundi society in the southern Imbo? Extant sources do not address when and how initial trading contacts were negotiated by Bwari travelers and their Rundi interlocutors, nor how Rundi people received Bwari people who periodically arrived in the late nineteenth century, settling near familiar points of trade or fleeing from insecurity or disease. Bwari groups arriving on the Imbo shore, in smaller dugouts or in larger canoes, would have built upon existing contacts with Imbo Rundi that developed out of trade, treatment-seeking, or historical connections between communities around current and former Catholic missions, for example, as they sought to make their homes in a new place. Trading relationships regionally at the time relied upon and underscored mutual obligation and deep social ties, and would have linked new and established migrants, as well as migrants and Rundi, in clientage relations.[19] Land would have been the primary imperative, to build a home and establish a garden or farm, even if the move to Imbo was not understood as a permanent change of circumstances. Rundi structures of land distribution, executed by village leaders or perhaps a more senior or powerful *mtwale* (subchief), might have afforded Bwari (or other) arrivals a discrete portion of land nearby to, but not in competition with, established Rundi farms. This also would have established ties of obligation between Bwari arrivals and Rundi watwale. Bwari and others from the northwestern littoral, particularly if migrating in family or lineage groups, appear to have sought to recreate historic social geographies of lineage-oriented villages separated from fields, an arrangement distinctive from Rundi traditions.[20] The micro-geography of Imbo communities on the low plain around Rumonge, Rumangu, Kigwena, and Nyanza-Lac was thus a patchwork in progress, with existing Rundi homes and their farms of banana, cassava, and vegetable crops, stands of oil palms, and Bwari homes and villages. A German officer's early comment differentiated "the villages and fields of the foreigners emigrated from the Congolese shore and the banana groves of the Urundi population," capturing an emerging colonial shorthand for the two populations and their different milieux: Rundi secluded among banana farms and Bwari newcomers' discrete settlements at the margins of Rundi farms.[21] German observers consistently distinguished Bwari "settlements" (characterized as *Ansiedlungen* or *Niederlassungen*) from Rundi villages or farms, implying a Bwari mode of land use or building that was different to a perceived Rundi norm—perhaps more concentrated spatially or arranged distinctly. This served, too, to mark Bwari foreignness as implied "settlers" rather than established, "native" Rundi populations.[22] Here, German officials again built on a rigid sense of lacustrine African ethnicity established by European observers a generation earlier, premised on social cohesion but also separation: Bwari living with their countrymen, Bembe with theirs, and so on. Ideas about the distinctiveness of "Warundi" and "Wabwari" people living in Imbo drew upon racialized colonial concepts of indigeneity and "civilization" as well as erroneous presumptions about African mobility.

German ideas about cultural difference would have ramifications for both epidemic sleeping sickness and Imbo politics. By 1906, when the first surveys of sleeping sickness started, German officials referred to the Bwari as *waungwana*—free-born people, a Kiswahili term—due to their association with historic regional trade, implicitly with slaves and ivory. Bwari people, among others from the northwestern lakeshore, had played an intermediary role in Arab-Swahili trade around the lake that had circulated from Uvira and Ujiji and moved slaves and ivory into regional and Indian Ocean trade in the nineteenth century. Here, however, distinguishing Bwari populations as waungwana connoted their foreignness; German doctors used it synonymously with the German *Fremdling* (foreigner).[23] Dr. Oskar Feldmann, traveling through the southern Imbo between

Nyanza, Kigwena, and Rumonge, described populations of Rundi people among banana farms and "foreigners who emigrated from the Congolese shore."[24] Foreignness was here marked by perceived economic activity as well as a distinct provenance; Bwari populations were frequently characterized primarily as *Händler*, merchants or traders, and seen to be wealthier than their Rundi neighbors. German colonial officials emphasized the Congolese mobilities focused on palm oil and commerce that they saw in the early twentieth century, borrowing from an older vocabulary of coastal culture and status by identifying Bwari populations as waungwana. The adoption of Swahili status markers to differentiate between Bwari and other Rundi communities gave officials a shorthand with wide resonance for both European and African interlocutors. Arab-Swahili activity on the lake provided a key context within which German officials understood Bwari migration, allowing them to connect it to factors that drove mobility around the lake and in central Africa regionally in the nineteenth century. German use of waungwana mirrors that of the White Fathers, who described non-local African populations active in making war, slave raiding, or in the slave and ivory trades around Lake Tanganyika as waungwana. Priests associated such waungwana with Arab-Swahili power emanating from trading centers at Ujiji and Uvira, as mercenaries and emissaries of powerful men in the slave and ivory trade like Mohammad ben Khelfan (Rumaliza). Certainly by the time German doctors began to survey the Imbo shore for sleeping sickness in 1905, some populations around the lake had been drawn into the slave or ivory trade as intermediaries, Bwari populations among them.[25] But all Bwari migrants who settled on the eastern shore of the lake could not be characterized as waungwana with involvement in the slave trade.[26] Generalized use of waungwana for "Congolese" migrants elided complex social and economic behavior and significant change in past generations, and also simplified people's own identification within categories rooted in Swahili coastal history: waungwana, the free-born, were of higher status than *washenzi*, the slave-born.[27] Despite their clumsy use, these distinctions between "waungwana" Bwari and "local" Rundi people would be important as the German sleeping sickness campaign attempted to penetrate Imbo communities.

MOBILITY UNDER THE MICROSCOPE: MIGRATION AND SLEEPING SICKNESS

Particular environments, traditions of land use, and patterns of cross-lake mobility were central to the epidemiology of sleeping sickness. In the eco systems of Imbo, especially in areas along the Rusizi River and Tanganyika littoral, tsetse flies flourished.[28] The wider spread of the disease underscores the extent to which matters of everyday life led to frequent and repeated exposure to fly habitats. Communities' connections to the lake required them to traverse and linger in tsetse habitats. The fishing that was so central to coastal prosperity and which connected Imbo populations to other littoral communities and regional trade also exposed people to fly bites; likewise did harvesting palm kernels to produce oil and wine, and everyday activities such as collecting water. But the spread of disease also indicates the potential health impacts of increasing vulnerability, social dislocation, and mobility for populations on both sides of the lake.[29] Rundi communities lived alongside people who had emigrated from the Bwari peninsula and areas further west in the Belgian Congo. Around the turn of the century, Rundi narrators would recount, some of these migrants began to arrive showing signs of wasting, weakness, and unpredictable sleepiness. This was particularly true near Rumonge and Kigwena.[30] Through colonial officials' eyes, these circumstances provided the basis for arguments that particular populations and kinds of mobility were inherently dangerous to wider health in Imbo. In the early years of investigating sleeping sickness, German and Belgian observers put together arguments about the origins, nature, and impact of sleeping sickness on the northern arc of Lake Tanganyika's shore. For colonial medical officials, Bwari populations were the linchpin to understanding the epidemic's dissemination around the lake.

German officers had begun to travel through Imbo between Ujiji and Usumbura by 1901. Reports by European observers of outbreaks of sleeping sickness in Urundi first appeared in colonial and missionary documents in 1903–4.[31] As sleeping sickness attracted their attention, colonial officials relied on Imbo watwale for information about the disease and its origins. By the time the disease and related mortality attracted colonial notice, people in Imbo already related it to migration from Ubwari and Masanse in the Belgian Congo to areas near Rumonge and Kigwena in the southern Imbo.[32] In late 1905, Dr. Leupolt reported that Bjuko, mtwale of the territory straddling the German administrative boundary between Ujiji and the Urundi Residency, had heard of a disease that he called *ugonjwa wa usingizi*. Leupolt noted as well that the "Wabwari, that is, the Waungwana of the German coast of Tanganyika," called the illness malali and feared it.[33] Leupolt's report, the first on sleeping sickness in Imbo, established a rough sketch of the epidemiology of malali and linked it to regional mobility. The name *malali*, he reported, came from "Manyema" (a place name broadly understood to apply to interior territories of the Belgian Congo west of Lake Tanganyika); the first name German officials recorded for the illness was *malali* or the Kiswahili *ugonjwa wa malale*—"malale disease."[34] Some people in Imbo reportedly linked *malale* to the Kiswahili word for sleep, *kulala*, while other informants offered the interpretation that *malali* simply meant "sick" in languages spoken around Ubwari.[35] Rundi informants' use of a name associated with Kiswahili or languages of the western lakeshore underscores its close association with mobile populations and origins outside of areas where Kirundi was spoken. But the wider use of the Swahilized *ugonjwa wa malale* also signals the synchronicity of its spread and a particular moment in recent memory as mobility and migration accelerated amid changing regional economic and political dynamics.[36] While scattered rumors from northeastern lakeshore from the 1890s indicate that a sleeping, nodding illness was not new to the region, its wide scope and scale were remarkable for narrators and colonial informants on the littoral after 1900.[37] Many in the region held onto a sense that the illness in widespread or epidemic form had arrived from somewhere else, or had spread generally from west to east.[38]

Though German officials along the Imbo coast were aware of settlements of migrants from the western lakeshore, concerns about sleeping sickness and its importation from the Congo Free State heightened colonial interest in these populations. People from the Ubwari peninsula drew the scrutiny of colonial officials because of their perceived mobility. German doctors stationed on the lake researched details about Bwari mobility and settlement, such as the size of the local population and its distribution, while seeking to establish the origins of sleeping sickness in the area. German personnel were spread thinly—one stationed at either end of the Imbo shore—but focused on surveying African informants as they circulated.[39] Attribution of the disease's introduction to travelers from Ubwari became an early and regular refrain for German doctors in Imbo. Belief in the Bwari population's role in spreading the disease to populations in Urundi reiterated ideas about disease epidemiology and labor mobility first developed in observations at Lake Victoria by Koch and fundamental to German ideas about sleeping sickness control: healthy-seeming carriers could spread the disease and it did not have a German territorial origin. Dr. Oskar Feldmann raised an early warning to his colonial superiors in 1905 that "the German coast of Lake Tanganyika is also threatened. Statements from the natives claim that trypanosomiasis predominates on the Belgian side of Tanganyika, especially on the Bwari peninsula."[40] The first formal survey of the German lakeshore, conducted by Dr. Karl Neubert, involved conversations (likely in Kiswahili or through a translator) about where the epidemic originated and how long it had been in Imbo. "The Natives," Neubert wrote, "state overwhelmingly that they first heard of the epidemic 1–1.5–2 years earlier; at that time, it emerged on the opposite lakeshore, chiefly the Ubwari peninsula.—Certainly it reached Tanganyika from the west coast of Africa via the Congo and its tributaries.—From Ubwari it then spread the length of the coast."[41] Thus, a consensus emerged that traced the path of sleeping sickness west to east, from Manyema to Ubwari or Masanse, and thence to Rumonge and Kigwena, where it spread further on the Imbo plains.[42] The Bwari population in Imbo was a part of colonial concerns well before the formal German anti–sleeping sickness campaign started, and, further, Rumonge was considered the epicenter of the epidemic on the German coast.

In parallel, Belgian reports from Uvira and its hinterlands also focused on mortality and mobility on the northwestern lakeshore. As Belgian officials began to survey populations on the opposite shore of the lake specifically for sleeping sickness, they found that the Bwari peninsula and the coastal area south of the Belgian station Baraka were particularly hard-hit. An officer wrote in September 1904 that he found that "the village of Sima Home [*sic*] is quasi-deserted—only a half-dozen women remain. Sleeping sickness has made a great number of victims. Unaffected natives fled to German East Africa. The chief Kingoma (of Kabingala village) provided food for me and told me that chief Poret (Ubwari) and many of his men sought refuge on the German coast to avoid sleeping sickness."[43] His Uvira-based supervisor commented that such flight would be impossible to oppose and advocated that the station close off the region completely to avoid the disease spreading from the trading center.[44] The epidemic triggered a significant eastward migration into German colonial territory, noted by Belgian officials traveling through Ubwari and Masanse in the same era as Neubert canvassed the eastern shore. By 1907, the Belgian doctor in Uvira Zone, Dr. Derche, noted that sleeping sickness was common around Baraka and on the Ubwari peninsula.[45] A Belgian report of a tour of the Bwari peninsula and Kibanga Bay in late 1908 offered an even bleaker picture. Here, the impact of mobility (whether emigration across the lake or elsewhere) as well as mortality from recent decades was stark. Where travelers in the 1870s had remarked on a coastline dotted with farms in a country "magnificent and the populations abundant," Belgian lieutenant Drôs found scattered and scantly populated areas. The eastern coast of the peninsula, facing Rumonge, he found "totally deserted."[46] One village on the protected bay to the south, he noted, now relied solely on fishing for food production because wild boars troubled its farms, a problem suggesting diminished capacity in households and villages to tend and protect crops.[47] As colonial anti–sleeping sickness measures began in earnest, then, there was broad consensus among Belgian and German officials that population movement from Ubwari to Imbo was key to the spread of sleeping sickness and, along with mortality, was emptying the Congolese shore of its population.

Particular understandings of Bwari mobility and the persistent construction of the Bwari population in German territories as foreign, separate, or transient had ramifications for the pursuit of sleeping sickness control. They bolstered German narratives of an epidemic distinctive in two ways. First, officials understood the outbreak to be new: European observers contended that the high mortality at Lake Victoria was absolutely without precedent and interpreted this as the impact of a new parasite striking a previously unexposed population. A consensus within tropical medicine and colonial policy circles supported this contention, despite it being based on a relatively brief snapshot of European observation limited to the previous two decades at a maximum. This sense of the epidemic's novelty also applied to Lake Tanganyika. Second, the epidemic was distinctive because it was imported into German territories from the Congo. This idea was bolstered by wider British discussions in Uganda of the disease's arrival from the west, but supported by Koch and others' arguments about how the disease could be carried by healthy-seeming travelers. This belief in an imported epidemic implied that, if environmental conditions and tsetse vectors could be brought under control, all was not lost with the colony's health. It also meant that German concerns about sleeping sickness understood the mobility of people living on the lakeshore as inherently dangerous for wider health. Mobility was cast as nomadism, the Bwari population as unmoored and unpredictable—despite evidence of extensive, robust, and enduring ties to particular places on both sides of the lake—and now infected with a deadly disease.

Emphasizing cultural difference between perceived groups and linking to this to health ensured that provenance—coded through native language, birthplace, or self-identification—was central to the ongoing production of colonial knowledge about both the epidemic and African populations. This was contemporary epidemiological practice, but it also reflected colonial desires to associate individuals (and larger collectives) with a single, fixed ethnic or cultural identity, and, further, to set those identities on a spectrum or hierarchy.[48] It also allowed the doctors manning the sleeping sickness camps to strategically address Rundi and "Congolese" populations differently. Rundi and Bwari populations could be contrasted, if not set against each other, based on evaluations of their inherent savagery or civility, on their compliance with colonial requirements, and on their ways of living. An accumulated narrative of Bwari distinctiveness—understood both positively and negatively—also allowed the Germans to envision less stringent control measures for them, as opposed to the Rundi population. A distinctive, visible, and geographically concentrated Congolese minority required different strategies of rule than the Rundi majority, vast, dispersed, and difficult to reach.

Sleeping sickness interventions—designed by Robert Koch at Lake Victoria, elaborated by his colleague Friedrich Kleine, and executed by junior colonial officers—therefore took shape as officials mapped sleeping sickness epidemiology onto the bodies and communities of migrants from the Belgian Congo and, to a lesser extent, Rundi fishermen. The intensive use of the lake by the Congolese and lowland Rundi populations in Imbo directed the sleeping sickness campaign to specific places in the region—busy markets, oil palm groves, or narrow spaces where the lake could be crossed at Rumonge, Kigwena, and the southern Imbo lakeshore. These important foci of migration, fishing, and commerce would, as a consequence, come to serve as hubs for German sleeping sickness control interventions. Here, ideally, the campaign would manage the spread of disease by screening, diagnosing, and dosing people with atoxyl. Population mobility and the ability to reach people emerged as a fundamental challenge to these goals: the camps and German staff could readily be evaded, the affected areas were too extensive to control tightly, and mass administration of atoxyl was a weak foundation upon which to build an energetic and sustainable campaign in the long term. Environmental interventions became a key area of action, alongside drug treatments. Locations chosen for their likelihood to allow for efficiently managing the epidemic in human bodies—places where the sick were concentrated or areas through which they moved—now also became the key arena for managing the epidemic's tsetse vector. Campaign officials chose locations such as Kigwena, Rumonge, and Nyanza because extant mobilities and their routes and hubs had made them important sites for Imbo life. But spaces selected to spearhead case-finding and the use of atoxyl, as well as surveillance, now had to be approached from a different perspective: as miles and miles of shoreline and streambed to be "sanitized," made clear of tsetse flies by razing the foliage that was their habitat.

MOBILITY AS A PUBLIC HEALTH PROBLEM AT LAKE TANGANYIKA

The German sleeping sickness campaign's territory in Urundi was roughly congruent with the boundaries of the Imbo region. The overall design sought to cover the lakeshore through roughly equidistant posts between Ujiji and Usumbura. In 1909, the Urundi campaign had begun to evolve two distinctive foci: the Rusizi River and its tributaries, where it was coordinated by Friedrich Breuer from Usumbura, and the Urundi coast, where it was led by Oskar Feldmann from Nyanza and included camps at Kigwena, Rumonge, and Urambi.[49] These southern Imbo sites were foci of high rates of infection and concomitant mortality: late in 1908, officers estimated that 95 percent of Kigwena's inhabitants were infected and argued that the toll of the sick and dead in certain areas on the Tanganyika lakeshore was on par with mortality on the Ssese Islands and Buganda.[50] The Urundi coast's six posts were connected by boat travel and footpaths and staff circulated frequently between them. An examination of doctors' photographs and narratives, maps confiscated early in the First World War by Belgian officials, and German maps and reports allow us to piece together the general contours of a typical camp and the particular placement and footprint of a few specific sites. Camps were generally cleared of bushes and tall grass, with large trees left standing and grass cut short—modeling the ideal "cleared" environment where tsetse would not flourish. The camp was ideally in a flat, open space, removed from the lakeshore but near enough to allow for easy travel. Similar to the Bugalla camp on the Ssese Islands and the Kigarama camp in Kiziba, Imbo camps were constituted by spaces where structures designated for research and treatment comprised the camp's core. Radiating from these workspaces were domestic, agricultural, and commercial spaces.

Each camp formed a hub for the coordination of three main tasks: surveying the population to identify the sick, administering six-to-eight-week regimens of atoxyl injections, and clearing vegetation to destroy tsetse fly habitats in areas of high traffic or dense population.[51] Atoxyl injections and clearing work were of a piece, strategically. Use of atoxyl sought to lower the concentration of parasites in a person's bloodstream, ideally reducing opportunities for disease transmission, a matter of paramount importance in an area with dense pockets of population that overlapped with tsetse fly habitats.[52] Bush clearing would ideally reduce tsetse populations and decrease transmission of the parasite. Vigorous mobility in Imbo meant that monitoring local travel and migration remained an epidemiological problem and therefore a persistent concern for the sleeping sickness campaign.

But the realities of littoral mobilities, combined with the practical business of the campaign, daunted policymakers and doctors alike. Both understood travel as key to the disease's ongoing spread, but acknowledged that they could not monitor the entire 150-mile coastline, nor interfere with all individual travelers, nor meaningfully restrict all travel.[53] Traveling the lake for trade was crucial for African commerce and fishing was an important source of food and livelihood, both of which contributed to colonial economies' bottom lines. As a consistent counterpoint to proposals to limit traffic or fishing, then, there were arguments that more strict colonial regulation would drive smuggling (thereby depriving the regime of taxes) or encourage use of unmonitored areas (thereby potentially spreading illness even further).[54] Further, even if human mobility could be limited, there was still the matter of the fly's movements: officials also worried flies could hitchhike on long-distance vessels moving around the lake.[55]

Over time and in discrete areas, however, sleeping sickness interventions focused on African mobility gained limited traction: surveillance of travelers in caravans and on large dhows, health pass regimes along the Imbo coast, border controls at crossings of the Rusizi River, and sporadic attention to daytime fishing on smaller craft. But despite obvious and clear concerns about sleeping sickness traveling into and out of the southern Imbo and lakeside exposure to tsetse fly bites, the practical matter of intervening in human mobilities to control the disease ran into significant constraints, whether due to conflict with economic priorities or to the real feasibility of consistent enforcement. Translating understandings of sleeping sickness epidemiology into public health interventions in the southern Imbo and on the northern lakeshore more broadly required manpower, intercolonial coordination, and cooperation from local populations that neither German nor Belgian officials could ultimately muster.

Monitoring caravans for infectious diseases had precedent in colonial anti-plague measures.[56] Sleeping sickness control regulations on both caravans and dhows attempted to limit their arrival and departure to specific locations and their movements to specific routes, focusing on large groups of people and capital-intensive caravans and ships. Such measures focused on ports and caravan-route trading centers that served as bottlenecks where large groups of people could be inspected, fitting colonial public health into the business of regulating trade at major markets. Guidelines required African dhow and caravan personnel to pass a medical inspection at the major ports of Ujiji, Usumbura, and Bismarckburg and along caravan routes between Lake Tanganyika, Lake Victoria, and the Indian Ocean coast; inspection stations in addition to the three Tanganyika ports were Bukoba and Mwanza on Lake Victoria and Tabora, Kilimatinde, Langenberg, and Kassulu on the caravan routes.[57] After 1911, some people who passed inspection for sleeping sickness received a red, white, and black band, sealed and stamped with the name of the issuing station, fastened around their left wrist.[58] But the threat of inspection was also potentially deeply disruptive: at Ujiji, as one administrator protested to the campaign director, the system of *plombieren* (stamping), caused hundreds of Ha porters bound for Bukoba to drop their loads and vanish.[59] Inspection for sleeping sickness and uniform use of passes for caravan porters thus remained an ongoing point of tension between residency administrators and campaign staff. In the years just prior to World War I, too, the desire to assign semipermanent markers of health status guided German desires for reciprocal Belgian surveillance and inspection, which would ideally target travelers moving through major transit points between German East Africa and the Belgian Congo.

But concerns about ongoing, everyday, and smaller-scale mobility across the lake were another side of implementation and continued to keep connections between Imbo Rundi and western littoral populations at the fore. Rather than homing in on hundreds of people traveling with a caravan, these concerns focused consistently on areas perceived as focal points of vigorous mobility, as determined by German officers' initial surveys of the region and their attentiveness to Bwari communities. In practical terms, such checks on everyday mobility were limited by the reach of the German campaign—what could be managed from each camp. Episodes at Rumonge illustrate ongoing concerns about cross-lake mobility as a challenge to disease control. The camp at Rumonge had been designed to situate the campaign in a historically well-connected place, proximate to the town's significant market that brought together highland and lowland populations as well as traders from Uvira, Ujiji, and Ubwari. Postwar Belgian documents show that the camp also had a good vantage of landings on Rumonge's wide, sheltered bay and was surrounded by Rundi homes and farms.[60] In 1910, frequent traffic from Ubwari, where tsetse were common but Belgian clearing work was reportedly minimal, continued to arrive at Rumonge. Many people settled in to live nearby.[61] Later that year, a group of very ill people arrived at Rumonge from Baraka without travel passes or any evidence of screening from the Belgian side, they told Dr. Wittrock, the supervising official at the Rumonge camp, that they had been neither examined nor treated before their departure. This, combined with Wittrock's discovery of people landing elsewhere on the Imbo lakeshore outside of surveillance—certainly no real surprise to the doctor—induced German authorities to try to enlist Belgian assistance with surveillance of travelers and influencing the direction of travelers from the western shore. Wittrock appealed to Belgian political authorities at Uvira, requesting that they direct people to come to shore only at Kigwena, Rumonge, and Urambi.[62] Directing people to particular locations for inspection was one of the few measures aimed at individuals around the lake, where colonial authorities admitted that they had neither the manpower nor the technology for surveillance of all travel on the lake. But reports leading up to the outbreak of World War I underscore that German officers were constantly finding evidence of travel through and use of "fly" zones or areas they wished to keep closed—here and there a fishing camp, a site of palm oil production, or a home near a landing on the lake. Their main recourse was destruction of anything valuable in the hope that this would deter further activity.[63]

Doctors at each camp generally did not attempt to restrict access to or use of the lake or to closely monitor all activity on the water. Staff regularly noted that such work was untenable, given the size of the population and the great expanse of the lakeshore.[64] But while controlling individuals was difficult, camp doctors could aim for technologies of mobility—rowed larger canoes and punted smaller boats—and therefore limit certain activities. Attention to boats and threats to destroy them were an aspect of contemporary British policy at Lake Victoria by 1908, but new to Lake Tanganyika and German anti–sleeping sickness work.[65] Confiscating or destroying boats represented a shift in the campaign's approach to Imbo populations and had potentially serious implications for those who fished for food and trade. Canoes, traps, and nets represented investments of collective and household labor. Proposals early on in the campaign had tried to keep the sleeping sickness campaign's field of intervention in fishing work narrow. Campaign leader Friedrich Kleine had suggested in 1908 that only daytime fishing be prohibited, given the frequent use of night-fishing with torches; this would not interfere with food supply, he argued.[66] The next year, the head of the Nyanza camp argued conversely that "policing measures," such as prohibiting fishing in areas where people would come into contact with fly vectors, could only work in "small, limited districts."[67] Fishing by canoe took groups of men onto the lake from their homes, regardless of whether the nearest stretch of lakeshore had been denuded of tsetse-harboring vegetation. The campaign's focus on canoes, boats, and discrete kinds of fishing work operated within particularly gendered ideas about labor and a model of fishing that considered local consumption only. Fishing was a crucial productive activity that could involve the labor of men, women, and children with different technologies during day or night; restricting certain kinds of fishing work would drive increases in others and likely also have wider-ranging economic impacts. Beginning in 1911, Imbo camp officials therefore focused on regulating only "*Stakverkehr*"—punted boats—and canoes on the basis of when and where they were used.[68] In their coordination meeting that same year, German and Belgian authorities agreed in theory to confiscate any canoes or boats discovered in uncleared areas on the lakeshore.[69] Daytime fishing was the primary worry because of tsetse feeding habits; officials mooted confiscating or destroying boats of people who continued to frequent cleared or "forbidden" areas, in cooperation with the resident.[70] While policies aimed at punted fishing boats were never fully in force, camp staff destroyed the boats and houses of fishermen discovered between Kigoma and Nyanza in 1913 and, around camps near larger populations, attempted to prohibit daytime fishing.[71] Attention to fishing boats was one area of the Tanganyika campaign where German officials were not preoccupied with perceived differences within Imbo populations: all who fished were implicated by 1911, regardless of provenance.

Parity in intercolonial anti–sleeping sickness measures and their enforcement, particularly along the Rusizi River, posed a constant source of tension and frustration for campaign staff stationed there.[72] Officials in Berlin and East Africa acknowledged the problem of managing movement across the colonial borders with the Belgian Congo in the Rusizi valley, as well as regarding traffic originating from the Belgian side; part of the work of colonial public health, here, was pursuing diplomatic solutions and bilateral commitments to monitor travelers' health.[73] In a 1911 meeting, German and Belgian medical officers met and agreed that, ideally, those who were sick would be identified before traveling and receive a high-dose, one-gram atoxyl injection and a paper pass, upon which atoxyl treatments would be noted.[74] In addition to the multicolored bands issued to laborers and travelers at major commercial centers and the treatment passes issued at the colonial border were health passes issued from the sleeping sickness camps. These latter passes were particularly relevant for Imbo populations, as they affected free local movement rather than only travel in long-distance caravans or large dhows. These health passes forced people to circulate through the sleeping sickness camps, the only places locally where passes could be obtained, a process that some resisted. Passes issued by the sleeping sickness camps fit into other health surveillance and labor regimes used by the German colonial government, chiefly regarding caravan porters, but they largely targeted individual travel and smaller-scale mobilities in Imbo. While campaign officials on the Urundi shore kept in touch periodically with their Belgian counterparts and had a general awareness of Belgian work to destroy fly habitats and monitor infected people, coordination between the two campaigns remained sporadic at best leading up to the outbreak of World War I.

CLEARING WORK AND COLONIAL AUTHORITY: IMBO COMMUNITIES CHALLENGE THE CAMPAIGN

In the campaign's field of work between Nyanza and Cape Magala—that is, near Kigwena, Rumangu, and Rumonge—surveillance and case-finding were an initial priority and clearing work emerged as a parallel concern as the campaign wore on. Success was measured in areas cleared of flies and the number of people completing regimens of double injections with atoxyl. Medical officers at the camps defined the work to be done at a given time seasonally, prioritizing certain tasks and demanding work on particular projects, with a fair measure of latitude granted by the colonial administration and campaign leadership. Both clearing work and searching for the sick developed a pattern that followed the ebb of seasonal rains. This seasonal variation was largely out of necessity—it was difficult to maneuver in the countryside during the rains. But it also allowed camp doctors to shift how they deployed their own energies and the energies of those whose labor they demanded, according to staffing shortages, planting and harvest times, and expansion into new areas. Colonial categorizations of mobility and illness, which pathologized particular populations and activities, generated tension with colonial economic needs when the work of sleeping sickness prevention began in earnest. Throughout, doctors and sanitation officers did not have a free hand in implementing disease prevention measures: latitude in what camp doctors on the Urundian coast could do remained limited, chiefly through the engagement or resistance of local Rundi and Congolese communities, which defined what was practicable and possible.

Processes of finding people and surveying them for signs of sleeping sickness were contentious and frequently involved punitive violence and destruction of home and property. In dealing with Bwari populations, efforts at enticement and persuasion soon gave way to threats and officials employed punitive measures readily. Officials characterized their responses as *leichteren Gewaltsmassregeln* (light disciplinary force), which often took the form of burning down the homes and sometimes villages of people who evaded surveillance. In mid-1908, soldiers burned down multiple locations in the southern Imbo, the resident reported, hoping to make the fear of "these measures" prevail over the fear of examination. The Bwari population in the area became a particular target of colonial violence. "Emigrated and constituting no blessing for the country," the resident groused to his superiors, the Bwari were a source of particular frustration as they did not appear at all when summoned. But near Nyanza, he noted, camp doctors also burned down Rundi villages that exhibited particular "passive resistance" or whose inhabitants fled into the nearby mountains.[75] German colonial doctors, accompanied by a contingent of *askari*, here employed the brutal scorched-earth tactics that had brought widespread mortality and immiseration to the repression of the Maji Maji rebellion between 1905 and 1907.[76] Colonial punitive strategies and the deployment of violence recapitulated recent warfare and raiding tactics around the lake. Certainly on the Ubwari peninsula and likely on the Imbo coast, wars were made and won through strategies targeting villages in just such ways, and, in the era of Rumaliza's expanding sphere of influence from Uvira, such violence produced captives, too.

As threats and realizations of retaliatory violence played out, doctors' early appraisals of particular Bwari recalcitrance gave way over time to more positive appraisals of compliance. Doctors apparently began to recognize that the challenge of attracting and retaining patients was a more general one. Throughout, however, German officers maintained distinctions between Bwari and Rundi potential patients. German officials continued to assert that Bwari and other migrants had brought sleeping sickness to Imbo, and that movement to and from the Congo continued to import new cases. In 1910, news circulated in the southern Imbo camps that the epidemic still raged on the Ubwari peninsula, Belgian measures having failed to check its spread.[77] But doctors stationed at Urambi and Rumonge at the time also noted that, while Bwari patients constituted most of the new cases, they also sought treatment early and voluntarily, for which the doctors praised Bwari populations as being of a "higher intelligence" than their Rundi counterparts.[78] A sense of ease in dealing with some Bwari likely had much to do with ease of communication in Kiswahili and association with coastal culture—reports make clear that doctors and Bwari patients spoke the coastal lingua franca (and preferred German colonial language) of Kiswahili to one another.[79] Thus, treatments with atoxyl were purportedly more easily negotiated and, even in cases of dispute, resolved with greater clarity from the doctors' perspective. But simply seeking treatment on German terms also raised colonial estimations of certain populations, and historic mobility and resulting patterns of settlement made Bwari and other western littoral populations paradoxically easier to engage. Doctors cast the larger Rundi population in Imbo as more immiserated by the disease, but also more difficult to reach. With connections to villages in the bordering mountains of southwestern

Urundi, where German colonial power was barely relevant, Imbo Rundi populations had an ever-present escape—simply leaving the area where doctors were searching for cases. Doctors effectively lost these departed populations until and only *if* they returned voluntarily. Chasing people to the hinterlands of the Imbo region was wholly impractical: neither camp staff capacity nor wider German colonial authority extended into the nearby highlands. Bwari communities settled near the lakeshore had more constrained options to evade surveillance. Fleeing across the lake required time, planning, resources, and manpower, and moving into the adjacent hills appeared to be an infrequently used option.

As the campaign wore on into 1911, the practicalities of bringing people into consistent treatment with various drugs continued to stymie camp staff. Clearing work firmly took center stage.[80] This dual brief of the camps led to tensions with Imbo populations, as demands for regular inputs of intensive labor fit uneasily with demands for participation within treatment regimes. The idea behind large-scale *Abholzung* (deforestation) projects was to clear vegetation that could harbor tsetse flies and offer them breeding places, especially at locations that people also frequented: riverbanks, landings on the lake, marshes, footpaths, forests, and so on. Ultimately, too, cassava fields, banana groves, and oil palm forests would come under scrutiny.[81] Kleine as campaign director preferred to emphasize clearing tsetse habitats over resettlement or relocation, because it had the longer-term benefit of preventing infection among future inhabitants.[82] Adding an additional layer of complexity was the matter of fitting the campaign's need for labor into the Urundi Residency's hut tax and forced-labor requirements. Linking cooperation with or participation in the campaign's diverse projects to new tax obligations had been mooted from Oskar Feldmann's original 1908 budget for the Tanganyika campaign. There, Feldmann proposed that a wide range of activities—from caring for the sick to carrying wood to clearing forests—could serve as substitute tax work.[83] The work of bush clearing, through a designated number of days per month, had been linked with forgiveness of the hut tax around Nyanza and Kigwena since at least mid-1909.[84] But just how campaign labor demands would fit into wider colonial forced-labor and taxation regimes in the region remained contentious.

Dynamics surrounding the Kigwena camp and its work illustrate the difficult situation that the camps' dual brief created. German officials had sited a camp there because they saw it as an entrepôt of Congolese mobility and because its palm forests attracted people involved in oil production and trade. Dense oil palm groves nearby, which provided palm kernels for the area's lucrative oil trade, also harbored tsetse flies and therefore presented a perennial, intractable challenge to cutting transmission of sleeping sickness. The forests could not be cut down or closed off, particularly if people lived nearby; surveillance and case detection, along with deforestation of all other possible areas that people frequented, became the only possible tactical focus. At the same time, ambulatory treatment of positive cases with atoxyl was the only chance, however slim, to break the transmission cycle by killing parasites in the peripheral blood, although its effects were not durable and its side effects remained significant. The disease remained fatal for the vast majority of cases; detection and diagnosis did little of lasting consequence for those in advanced stages of the disease, and atoxyl could not cure it. The camp asked much and offered little to Imbo populations: clearing work was arduous, injections were painful, and the impositions of surveillance were not borne without attempts to evade and avoid.

The new relationships that sleeping sickness prevention efforts created between people living in the southern Imbo and the German administration around taxation, clearing work, and compensation touched off a significant controversy between the resident, Rumonge and Kigwena camp staff, and "Congolese" populations. While the Rundi population around Kigwena proved elusive in efforts to detect cases and bring people to camp for injections, doctors found that the "foreigners with sleeping sickness living in this district [the Wabwari] were, in the course of time, all pretty much in treatment."[85] Their participation in clearing work was, however, another matter entirely. A turn of events in 1911–12 ran counter to colonial expectations and experience with the Bwari population; trouble brewed. A series of urgent memoranda narrates how migrants deployed their economic power and perceived distinctive status to oppose the demands of the sleeping sickness campaign. In late 1911, Bwari populations rejected German demands for laborers for bush clearing around the Tanganyika camps in the southern Imbo. Per Resident von Langenn-Steinkeller's order of December 13, 1911, the Bwari population had refused to work for the camps at Kigwena and Rumonge based on claims of having already paid taxes to the residency. His frustration with the *Kongoleute* in response was deep: "It has been brought to the attention of the Rumonge sleeping sickness camp that the Wabwari living on the Tanganyika coast refuse to carry out any work for the camps or to combat sleeping sickness. Apart from the fact that I do not know of a special arrangement . . . people have also lied, claiming that they were freed from work because they paid taxes. Firstly, they pay taxes from January 1912 and, secondly, they have to follow the orders of the sleeping sickness camps unconditionally because they have been paid for their services."[86] Von Langenn-Steinkeller made known that he would travel within a fortnight to visit Urambi, Rumonge, and Nyanza in order to speak with local watwale as well as all the Bwari, Vira, and Sanse people there ("in short, all Congo people") and settle the controversy. Marshalling labor for clearing work around Kigwena and Rumonge was difficult enough for several reasons. Sleeping sickness affected many there, so the pool of healthy workers was proportionally diminished. The environmental and ecological conditions in the area created ideal fly habitats, which made the field of work extensive. Further, regardless of their reasons for refusing, people could easily evade being mustered by leaving homes and villages temporarily. Evasion of camp work by leaving for the Congolese shore, however, was not common enough to register with German officials up to this point; the Bwari population had seemed inclined to stay put in the southern Imbo.

The resident's intervention did not diffuse tension between Vorwerk, the supervising doctor, and the Bwari population, nor did it resolve the problem of mustering enough workers to execute planned bush clearance. A flurry of reports from Vorwerk in May 1912 reveal a sharp escalation around continued demands for clearing work near the Mulembwe River, between Rumonge and Kigwena.[87] With the end of the rains, as was becoming typical, Vorwerk set out to resume aggressive clearing work, but remarked, "I have constantly run into direct resistance from the emigrated Congolese people."[88] In subsequent encounters that spanned several days and were recorded in detail by Vorwerk, his pressure on Bwari populations to follow the orders of the resident and the campaign leadership met with flat refusal. In refusing, however, these men also negotiated the monetization of their work and asserted the higher status brought by their wealth. When Vorwerk informed them that they had not *all* paid the required tax—effectively, a petty charge that one or the other had been sick or absent, and thus that the obligation was not fulfilled—they countered that they would simply pay the remaining tax later. One man said, simply, "I have no need to work," an assertion of independence to which Vorwerk responded violently, with a sentence of twenty-five lashes of flogging. The askari responsible for the beating reported to Vorwerk that the same man fought against him, saying "*er sei kein Mshenzi*"—"he was not a slave"—an invocation of his economic and social status as *mngwana*, a free-born man, as opposed to washenzi, the slave-born.[89] The group of men were summoned to come again the following day. On the subsequent afternoon, Vorwerk's auxiliaries at Rumonge then intercepted a group of thirty-five men—all of the people summoned to work the previous day, save six—who were ostensibly on the way to Usumbura, "in order to bring their rupees to the resident's till and in this way, with no further difficulty to themselves, free themselves from the work."[90] But they lacked required travel passes, which had to be acquired at a camp. Vorwerk, reporting to the resident, wished to expel the troublesome Bwari, fearing their influence on the purportedly more cooperative Rundi living nearby; he called in reinforcements, requesting additional askari to assist him, an unusual reaction within the camp's typical activities.

The group of men abjured negotiation with the camp authorities, instead seeking to engage directly with the official they considered their political superior and holding colonial authorities to the calculus of wage labor and taxation they had been offered—that work was equivalent to money. The political and economic stakes of distinctive categorizations and identities within the colonial regime were thus heightened amid the pressure of controlling sleeping sickness. That the men sought to travel to Usumbura overland from Rumonge may have been an elegant tale to distract the askari and the doctor—the men, after all, had been caught without a necessary pass in an area where they were not supposed to be. But the resident's appearance six months prior to speak to "Congolese" populations directly, while superficially bolstering the campaign's authority, also suggested that local campaign officials might be strategically superseded. Appealing directly to the regional authority to whom their taxes were paid bypassed the changeable personnel of the local camp. Further, there was the matter of the nature of the work—hard physical labor—the prospect of which had generated vehement opposition. The man who asserted in Kiswahili that he was no slave articulated resistance against physical restraint and violent coercion, doing so based on an assertion of waungwana status. Washenzi, the slave-born, could be thus treated and thus compelled to physical labor, but he could not. This assertion of being waungwana, free-born, and not washenzi, slave-born, echoes the initial German characterization of the Bwari and "foreign" populations in Imbo, which German observers had used initially to differentiate them from their Rundi neighbors. Here, however, waungwana and its inverse washenzi were deployed to signal economic and social status, not provenance. Waungwana were now taxpayers to the German regime, an economic and political arrangement, and they expected obligation in return.[91]

The conflict over laboring for the camp continued, freighted as it was with consequences for Bwari and other Congolese men's social and economic status, as well as their political position within the colonial regime. The doctor encountered the Congolese men again three days later. They had again attempted to travel overland to Usumbura and had been turned back again by a camp auxiliary for traveling without the required pass. Vorwerk's back-and-forth with one man indicates that Vorwerk's nominal power to compel and direct African labor was slipping rapidly away, as his increasing pressure met with equal resistance. "*Kazi ya malale* they would never do," the Bwari people told him, "but rather would return again to the Congo."[92] Threatening a return to the Congolese side of the lake served here to emphasize the fundamental disdain for clearing work as much as it suggested a viable alternative to life under the German colonial public health regime. A station doctor would remark that "in Rumonge there is a constant traffic with the Congo coast that is close by, just opposite," and that traffic had a distinguishable vector: "many people immigrate completely from there to Rumonge."[93] Return was a proposition full of risk amid sleeping sickness mortality, with its toll on social connections, political stability, and agricultural production following on years of disruption from warfare and raiding.

For waungwana men wealthy enough to pay taxes, their identity opened up space to deploy mobility and independence as another aspect of their status. Further, "kazi ya malale" was a distinctive and telling shorthand. *Malale* was the Congolese-Kiswahili name for disease in circulation locally that the Germans glossed as *Schlafkrankheit*/sleeping sickness. For the men, it defined the *kazi*—in Kiswahili, the work—of strenuous labor hacking down brush and reeds along the lakeshore and roads for days at a time. Here, as elsewhere, Bwari interlocutors spoke Kiswahili with Vorwerk; he retained it in his text in a manner that reinforced the speaker's emphasis and passion, as when his primary Bwari interlocutor told him "er tun keine Arbeit '*hata kidogo*'"—he would not work *at all*.[94] Retaining their interlocutors' Kiswahili words at multiple critical moments in their interaction is rare in extant reports from the Imbo camps. It underscores that articulations of status and obligation were embedded in Bwari arguments with campaign officials; Kiswahili, the colonial language and language of economic power in the region, was a language of negotiation.[95]

After his argument with Vorwerk, the man was seized by an African camp policeman and detained. Vorwerk sent him and the twenty-five other men, accompanied by an askari, directly to the resident on a colonial steamer. Thirteen of the men, a mix of Sanse and Bwari migrants, were also present at the first refusal to work under Vorwerk's direction months before. Their names (as transcribed by a German speaker): Kajungiro, Takuti, Ismahili, Ramazani, Mbuasa, Muyema, Mthunatjake, Tanambele, Marombe, Makina, Feruzi bin Soso, Simatara, and Amari.[96] The majority of the group Vorwerk identified as Bwari migrants, though Sanse, Bembe, and Vira men's provenance were also noted. Swahili, Quranic, and Arabic names indicate that some had durable connections to coastal culture, if not personal Islamic religious practice. Waungwana indeed, these people and their countrymen, Vorwerk asserted somewhat breathlessly, had lived "since their emigration almost without contact with the administration in complete independence," constituting a small, self-governing "republic."[97]

That republic seemed to stand little chance against the colonial government's ordinances against unregulated travel, not to mention its interest in maintaining, visibly, its monopoly on coercion and violence. The twenty-six men were each punished at Usumbura with twenty-one days in chains and twenty-five lashes. "Doubtless," wrote the resident, "we already have these people to thank for the spread of the disease, which they have brought to us from over there."[98] And yet, the resident's suggestion for resolving the overall conflict was to recognize the practical independence and the differing economic means of the Bwari population. Rather than be subject to compulsory clearing work at the camp doctor's discretion, they would be responsible for a particular territory near their own dwellings, clearing either on their own or by paying others to do the work. And they would, moving forward, be subject to an expanded regime of taxes on trade and profit, as well as taxed per head or hut. While Bwari and other western littoral populations could and must take up some of the work of the campaign, von Langenn-Steinkeller argued, their economic value to the colony necessitated careful management. In threatening a return to the Congo, the Bwari men also held forward the prospect of lost tax and trade revenue for the German colony, a prospect of little concern to camp doctors but pertinent for the Urundi resident. After all, "these people supply a not insignificant income through their trade and smuggling with the Congo State [*sic*]," he argued, "which should not be allowed to diminish because of too abrupt measures."[99] Such economic concerns would consistently influence how the campaign pursued its work in the southern Imbo. Up to the very end of German colonial rule in the region, the Urundi Residency afforded Bwari a different responsibility in the work of sleeping sickness control than the Rundi populace at large.

In Urundi, Imbo constituted the primary area of the sleeping sickness epidemic's outbreak and was also the main area of German colonial concern, due to the perceived impact of lacustrine mobilities on the health of populations on the Ubwari peninsula and in the southern Imbo. Mobility had significant meaning for ideas about the epidemiology of sleeping sickness, while also holding practical importance for the southern Imbo populations contending with the German campaign. Beliefs about the eastward path of sleeping sickness and fixations on the epidemic's provenance focused colonial attention on particular Congolese migrants. The behavior of the populations of diverse origins that circulated around and through the camps in the southern Imbo was often understood by doctors as quintessentially unpredictable—a puzzle that they had trouble solving. Neither in the Imbo region nor in the Rusizi valley could camp doctors consistently compel the cooperation of populations in a camp's immediate environs as well as further afield. The campaign's purview remained limited to a strip along the lakeshore, encroaching on villages in foothills and mountains nearby only shortly before the campaign was disrupted by the outbreak of the First World War. People in lakeshore villages, particularly those within a network of Bwari immigrants, could move or travel to avoid being drawn into the campaign, if identified as sick; people with connections to settlements higher into the mountains could—and did—move beyond the campaign's reach. Plans for the campaign in future years, left unattempted because of the war, focused on controlling Rundi settlement and cultivation patterns, effectively using farming and maintenance of oil palm forests as a means to limit tsetse fly habitats.[100]

And yet, disease control programs at Lake Tanganyika and campaign staff at the scattered camps—the majority of which were more than a day's walk from the Usumbura or Ujiji stations—constituted a regular and relatively permanent colonial presence otherwise lacking in the Tanganyika valley. Sleeping sickness interventions, particularly around bush clearing work, triggered direct engagement between Bwari migrants and the resident and provided a focal point for negotiations over labor, taxes, and political obligation. Differing modes of engagement with colonial authorities and the sleeping sickness campaign stemmed from leveraging the power, economy, and labor of Bwari migrants, as opposed to their Rundi neighbors. Bwari and other Congolese populations retained economic power, important for the big picture of colonial success in the Urundi Residency, and their relatively free mobility and political autonomy was central to that power. They reasserted that autonomy and demanded accommodation, invoking mobility and wealth as part of being "Bwari" or waungwana. Claims to engage directly with the resident were an assertion of parity and mutual obligation, however diminished by the colonial administrator's use of violence. Rejection of colonial demands on the basis of understandings of obligation or grounded in assertions of distinctive status, as articulated by Bwari men, drive home how important colonial public health work was for defining relations between colonial officials and their subjects.

Conclusion

THE GERMAN sleeping sickness campaign in eastern Africa did not have a deliberate conclusion. The disruptions of the First World War halted the German campaign and deferred any deeper reckoning with the sustainability of its strategies and tactics. Some camps continued to use a combination of monetary incentives, surveillance by African auxiliaries, and coercion via local political authorities to keep people with sleeping sickness circulating through them. But others changed their work completely: Kigarama was already a polyclinic by the outbreak of war, while Kigwena's populations were slated for resettlement into unnamed fly-free locations, away from their homes in an area that was to be "closed."[1] The ongoing use of atoxyl had a horizon already visible to some camp doctors by 1912–13, and the immense commitment that bush clearing work at Lake Tanganyika truly entailed had also become evident as German officials moved deeper into the Rusizi River watershed. New savannah foci of sleeping sickness in central and southern German East Africa discovered in 1911 promised to dramatically expand the campaign's scope.[2] Stories of the campaign and its workers percolated occasionally into the postwar world. Some of the Ziba gland-feelers became "Bukoba boys" at work on the Imbo campaign; others employed at the Kigarama camp were remembered to have gone to work at the Bukoba hospital after British takeover.[3] A few men trained during the sleeping sickness campaign reconnected with German doctors during British-German collaborative research on trypanosomiasis in the region in the 1920s.[4] For Belgian officials, whose assumption of control around Lake Tanganyika after World War I involved dealing with the persistent, acute problem of sleeping sickness on the lakeshore, evaluations of the German campaign were ongoing. They approvingly surveyed cleared portions of the Imbo shore, mapped the camps and their environs, and assessed how to manage mobility that now moved within a single administrative unit (after the transfer of Rwanda and Urundi to Belgium as Mandate territories) but remained vigorous and difficult to control.[5] The Ssese islanders whose experiences of illness and death had so profoundly shaped German colonial strategies were displaced by a newly energetic British administration that sought to depopulate fly-dense areas along the shores of Lake Victoria. The islands were officially closed to their inhabitants for decades, though clandestine traffic and fishing continued. The populations affected by the German campaign moved through the upheaval and the lean years of World War I to confront new colonial impositions and concerns in the interwar years—some still contending with the dangers of tsetse bites and trypanosomes, some not. The contingencies that shaped the campaign's design and execution were enduring, and little argument for long-term success in suppressing sleeping sickness can be made for the campaign on the whole. Further, given the harm inherent to atoxyl use, we can assume that affected populations who circulated through the sleeping sickness camps not only continued to deal with sleeping sickness but with blindness and vision impairment as well.

The history of epidemic illness and political change that I have offered, situated and embedded in complex social contexts, is concerned with continuity as much as with rupture. It is a history that invites consideration of generations and the consistencies and alterations of experiences lived by and discussed within families and communities across time. I have focused on the specific places where sleeping sickness camps were located to explore how ideas about health, prosperity, and political power collided amid an epidemic, and, further, to understand how the ideas and actions of diverse people interacted to shape measures taken to mitigate its impact.

The interlacustrine societies in my case studies had particular ways of approaching different environments, including domestic and social spaces and those marginal to daily life, and places that provided different opportunities for prosperity or safeguards from illness or misfortune. Social and political relationships were embedded in both micro-geographies of the home, farm, or village and macro-geographies of the polity or of the lakeshore. Powers to heal and mitigate misfortune were historically rooted in place; such places, how they were oriented in particular environments, and how they were used mattered for their therapeutic power and who could access it. Economies of land and labor impacted how generations and genders related to one another and to higher political authorities. These durable systems were also focused on control and use of particularly beneficial and useful places—good ground for cultivation, good pasture for cattle. In the interlacustrine kingdoms, very few kinds of places were neutral or without political or social consequence. Embedded politics, historical touchstones, and situated experiences still provided durable points of reference for understanding one's world and making decisions about one's life, even as the social, political, and economic dynamics that were part of people's intellectual worlds were changing rapidly in the early twentieth century.

Intersecting with this complexity were the impositions of new European colonial administrations, which sought to change how people worked, farmed, built families, and healed illness, among other aspects of life. Medical and public health concerns constituted a key impetus for engagement between Africans and Europeans, few more urgently and ambitiously than epidemics of sleeping sickness at the turn of the century. Contemporary European approaches to sleeping sickness were also preoccupied with place, but in anxious modalities of surveillance and knowledge production, concerned with where the sick were, where fly vectors were, in which zones transmission could potentially occur, and which areas could serve as bulwarks against further spread of the disease.[6] The activities of European scientists and colonial administrators prioritized vector habitats and conditions of exposure as they oriented to African populations and environments. These activities inevitably became entangled in the pervasive, encompassing complexity of interlacustrine worlds. Medical officers might have recognized the need to engage, negotiate with, or coerce their African interlocutors to bring about a desired outcome, but they frequently did not recognize how extant political structures, social relations, or forms of community shaped how they did so. But, as this book demonstrates, the contexts within which camps were located truly determined how their fortunes rose and fell. The intellectual worlds of people living around Lake Victoria, and particularly of Ssese islanders, influenced the efforts of German scientists as they sought to develop sleeping sickness prevention strategies and homed in on the establishment of isolation camps as a course of action. Subsequently, as the German campaign unfolded around both Lake Victoria and Lake Tanganyika and sleeping sickness camps were established across a wide territory, medical officers had to adapt to conditions on the ground in order to make the progress they desired, even if they did not recognize explicitly that they were making accommodations or adaptations at the time. The people whom sleeping sickness research and control targeted had an impact on both the conceptualization of sleeping sickness campaigns and on the horizons of their feasibility.

Several arguments about the nature of disease control efforts have animated this book's examination of sleeping sickness prevention in the Great Lakes region, in seeking to understand how and why people have engaged, or not, with public health programs in the past. First, that people avail themselves of treatment at some times and not others within complex and locally specific logics. Second, that the placement and setting of an intervention (like a dispensary or camp) affects the way people use it. Finally, that historical experiences and intergenerational points of reference shape how people engage with particular treatments or intervention strategies. The early, formative period of colonial-era attempts to control sleeping sickness illuminates how deeply integral that local processes of healing, mitigating illness, preventing disease, and political engagement were to colonial public health interventions.

The history I have explored here holds lessons for fundamental elements of public health and medical practice, particularly how we understand the nature of "community participation" and the factors that drive or deter it today. In this respect, the history of sleeping sickness interventions has bearing on the ideas and practices of global health which still seek, as Packard puts it, to make "interventions into the lives of other peoples."[7] This book cannot provide a handbook of instructions for what public health practitioners should consider today. It can, however, identify the kinds of factors that might ultimately shape a program's success or failure, as well as the kinds of questions practitioners might begin to ask in their work: What was here before this intervention (and, implicitly, "we") came? Why situate a program here, or there, and not elsewhere? What are the political and social implications of doing so, and the broader consequence for a community's health? Some interventions take these matters up with energy, foresight, and creativity, but many others do not. There is still work to be done.

The people across the African continent who could fall sick with sleeping sickness now and be examined for trypanosomes likely have parents or grandparents who experienced similar processes of screening and testing.[8] Consider how Binyavanga Wainaina, in *Beyond River Yei*, reflected on sleeping sickness surveillance he observed in the early 2000s: "I saw the team in full operation first last week, unpacking their lab—several large metal trunks—and placing the two or three hundred villagers into a conveyer [*sic*] belt of activity: checking blood, checking lymph nodes, checking lymphatic fluid for tryps."[9] What Robert Koch sought to do on the Ssese Islands, and what subsequent efforts in German East Africa before World War I, Belgian Congo in the interwar years, and French Equatorial Africa at mid-century (to name three well-studied examples), was to align expertise and people on just such a "conveyor belt of activity." These continuities across a century of history are not incidental. For, although much about the world has changed for the rural African populations that live in sleeping sickness foci or "tsetse belts" on the continent, the technologies and tactics of investigation and diagnosis have remained remarkably consistent; likewise, the challenges of treatment and difficulty of cure.

One of my touchstones in this book is Lyons' masterful study of trypanosomiasis and its prevention in the Belgian Congo. Lyons covers a vast territory, intellectually and geographically, and succinctly describes one of the main challenges of sleeping sickness historically, both in our understandings of it and in populations' experiences with it. "Many historical foci," Lyons wrote, "have continued to flare up in spite of concentrated efforts since the 1930s on the part of surveillance and prevention personnel."[10] While her implication of consistent effort should be approached with caution, given the under-prioritization of sleeping sickness in drug development and many states' limited capacities to do prevention and control work, the attention to persistence over time and in space is crucial. While epidemics might flare up and die down, and populations considered at risk may change over time, the disease's connection to particular ecologies and environments creates circumstances of persistent, enduring experience and exposure.

As a focal, vector-borne disease, sleeping sickness is fixed in certain spaces because of the tsetse fly vector's habitat. Thus, scientists look for the disease in particular places and target it there. This idea of focality resonates with large-scale national or international public health interventions as well as late twentieth-century disease eradication agendas. Such modern programs deploy a language of coverage and comprehensiveness—of *reaching* all affected people—or conversely of *wiping out* or erasure of all relevant pathogens, in articulating their priorities and goals. What we often lose in such global views and biomedical or environmental thinking about epidemics is a sense of history and the durability of experience—precisely what Lyons, Lachenal, and others have sought to recover about sleeping sickness. Ideas of focality necessitate a consideration of how place and emplacement matter for public health, as I have argued, but must also account for the lives of affected people and both rupture and continuity in their lives over time. For it is not simply the fly in place, but also the human (and animal) hosts of the parasite, that must overlap for an epidemic to occur; complex processes of political and economic change and of mobility drive those epidemics.

In the twentieth century experience of sleeping sickness, people have coped with periodic and often devastating episodes of weakness, tremors, frailty, sleepiness, and death, often corresponding with hunger or famine, social dislocation, unanticipated mobilities, warfare, or insecurity. After 1902, when the parasite-fly nexus of trypanosomiasis gained coherence, African people's experiences of illness became intertwined with experiences of attempts at treatment and prevention from successive different medical authorities: missionaries, colonial health officers, itinerant international researchers, national ministry of health officials, WHO-sponsored teams, still more missionaries, nongovernmental organization workers, and more. These efforts occurred as affected populations did their own work of treatment, prevention, and mitigation, engaged with and evolving in dialogue with the solutions on offer as they changed over time. By identifying links between responses to sleeping sickness and extant strategies of managing illness, this book suggests how we might restore a sense of continuity and a focus on people's experiences of interventions in our studies of public health. How, for instance, are the consistencies in sleeping sickness treatment and prevention viewed, understood, discussed, and remembered over an extended period of time? How do these experiences and memories affect the way that people orient to a community, to a state, or to particular ecosystems or environments? Once again, connectivity across eras is important, particularly as medical and pharmaceutical approaches to sleeping sickness have changed far more slowly than political regimes. The long sweep of history pulls into dialogue the successive generations of people living where sleeping sickness epidemics can flare up, and therefore presents a further opportunity for practitioners and scholars to engage with—to learn from—their experiences.

In the interwar years, biomedical strategies to combat epidemic sleeping sickness advanced into new territory in some areas (with the advent of suramin/Germanin as a treatment after 1917) but remained remarkably consistent in others, particularly in the dominance of environmental approaches to sleeping sickness control and in the continued reliance on mass drug administration campaigns.[11] Throughout, the problem of incentivizing or coercing use of colonial sites and strategies remained. Particularly given the persistent toxicity of anti-trypanosomal drugs and the glaring lack of development of more effective and tolerable alternatives for most of the twentieth century, these colonial troubles were effectively the troubles of the twentieth century on the whole. No watershed moment occurred when treating or preventing sleeping sickness suddenly became easier or more straightforward, as, for example, with the advent of mectizan in the 1970s to treat another vector-borne, focal disease, river blindness (onchocerciasis) with a single pill.[12] Treating people, but also reaching people for testing and treatment in the first place, was a consistent and sometimes intractable challenge, shaped by the complexity of the trypanosome parasite and its evasion of pharmaceutical interventions, but more damningly by matters outside of the laboratory: insufficient resources for public health colonially and then nationally, inattention to seriously addressing the social and economic factors that kept some populations vulnerable to infection for generations, and a failure to prioritize a disease that attacked exclusively rural and generally impoverished African populations. These matters are still the crux of the persistence of sleeping sickness into the present day, despite ambitions to eradicate it.

Sleeping sickness is no longer a disease that is impossible to cure because of a lack of appropriate therapies.[13] Rather, sleeping sickness is a disease that is seen as very difficult to cure because effective therapies are few, on the one hand, and often remain difficult to administer because of the chemical combinations and lengthy duration of treatment they require. On the other hand, new therapies have been slow in development, because they are prohibitively expensive to produce compared to their potential market.[14] Some of this difficulty, particularly in the latter case of combination treatments, is due to the complex nature of the trypanosome parasite and its ability to evade the chemicals deployed to kill it. But it is also rooted in the structures of global markets of health care and pharmaceuticals, in pervasive racism, and in the ongoing precarity and marginalization of people living in areas where sleeping sickness can, and does, strike. These are perennial issues that biomedical approaches alone cannot resolve. The history of sleeping sickness invites us to turn a critical eye on approaches to diseases that are cast in our current parlance as urgent, emerging, or resurgent, and query why others remain "neglected."[15] This history pieced together through stories of epidemics, narratives of research, maps and charts, explanations of changing interventions, photographs, and statistics—offers a lens through which we can see more clearly the dynamics of political, social, economic, and environmental change in the twentieth century and gauge the nature, scope, and root causes of health disparities globally today.

Notes

INTRODUCTION

1. Georgina H. Endfield et al., "'The Gloomy Forebodings of This Dread Disease': Climate, Famine and Sleeping Sickness in East Africa," *Geographical Journal* 175, no. 3 (September 2009): 181–95; Fèvre et al., "Reanalyzing the 1900–1920 Sleeping Sickness Epidemic in Uganda," *Emerging Infectious Diseases* 10, no. 4 (April 2004): 567–73.

2. Evidence strongly suggests that not all deaths attributed to the epidemic were caused by trypanosomiasis, per Endfield et al., "Gloomy Forebodings"; Jennifer Lee Johnson, "Fishwork in Uganda: A Multispecies Ethnohistory about Fish, People, and Ideas about Fish and People" (PhD diss., University of Michigan, 2014), 194–95. On the novelty of epidemic trypanosomiasis, see John Ford, *The Role of the Trypanosomiases in African Ecology: A Study of the Tsetse-Fly Problem* (Oxford: Clarendon, 1971), 239–43; James Giblin, "Trypanosomiasis Control in African History: An Evaded Issue?," *Journal of African History* 31, no. 1 (1990): 59–80; and Fèvre et al., "Reanalyzing the 1900–1920 Sleeping Sickness Epidemic." The debate about the newness of human trypanosomiasis, as well as the discussion of which species of trypanosome caused the 1900–20 Lake Victoria epidemic, remains active; see Geoff Hide, "History of Sleeping Sickness in East Africa," *Clinical Microbiology Reviews* 12, no. 1 (1999): 112–25. See Luise White, "Tsetse Visions: Narratives of Blood and Bugs in Colonial Northern Rhodesia, 1931–9," *Journal of African History* 36, no. 2 (1995): 222–26, on historical ramifications of debates over tsetse and trypanosome species.

3. Maryinez Lyons, *The Colonial Disease: A Social History of Sleeping Sickness in Northern Zaire, 1900–1940* (Cambridge: Cambridge University Press, 1992); Kirk A. Hoppe, *Lords of the Fly: Sleeping Sickness Control in British East Africa, 1900–1960* (Westport, CT: Praeger, 2003).

4. Hoppe, *Lords of the Fly*; Helen Tilley, "Ecologies of Complexity: Tropical Environments, African Trypanosomiasis, and the Science of Disease Control in British Colonial Africa, 1900–1940," *Osiris*, 2nd ser., 19 (2004): 21–38.

5. Notable exceptions regarding the pre–World War I era are brief discussions in Lyons, *Colonial Disease*, chap. 9; Hoppe, *Lords of the Fly*, chap. 3; and Myriam Mertens, "Chemical Compounds in the Congo: Pharmaceuticals and the 'Crossed History' of Public Health in Belgian Africa (ca. 1905–1939)" (PhD diss., University of Ghent, 2014), chap. 9.

6. Frederick Cooper, "Conflict and Connection: Rethinking African Colonial History," *American Historical Review* 99, no. 5 (December 1994): 1533.

7. Donna Haraway, "Situated Knowledges: The Science Question in Feminism and the Privilege of Partial Perspective," *Feminist Studies* 14, no. 3 (Fall 1988): 575–99; Petra Tschakert et al., "Situated Knowledge of Pathogenic Landscapes in Ghana: Understanding the Emergence of Buruli Ulcer through Qualitative Analysis," *Social Science & Medicine* 150 (2016): 161–62.

8. Victor W. Pentreath and Peter G. E. Kennedy, "Pathogenesis of Human African Trypanosomiasis," in *The Trypanosomiases*, ed. Ian Maudlin, Peter H. Holmes, and Michael A. Miles (Cambridge, MA: CABI Publishing, 2004), 283, 289–90.

9. Thorsten Koerner, Peter de Raadt, and Ian Maudlin, "The 1901 Uganda Sleeping Sickness Epidemic Revisited: A Case of Mistaken Identity?," *Parasitology Today* 11, no. 8 (August 1995): 304.

10. Susan C. Welburn et. al., "Sleeping Sickness: A Tale of Two Diseases," *Trends in Parasitology* 17, no. 1 (January 2001): 20; Philippe Büscher and Veerle Lejon, "Diagnosis of Human African Trypanosomiasis," in *The Trypanosomiases*, ed. Ian Maudlin, Peter H. Holmes, and Michael A. Miles (Cambridge, MA: CABI Publishing, 2004), 204.

11. Welburn et al., "Sleeping Sickness," 20–21, 23.

12. Maryinez Lyons, "African Trypanosomiasis," in *Cambridge World History of Human Disease*, ed. Kenneth Kiple et al. (Cambridge: Cambridge University Press, 1993), 556.

13. Jan Kuhanen, *Poverty, Health, and Reproduction in Early Colonial Uganda* (Joensuu, Finland: Joensuu University Publications, 2005), 110–34; Shane Doyle, *Before HIV: Sexuality, Fertility and Mortality in East Africa, 1900–1980* (Oxford: Oxford University Press, 2013), 63–74; Roger Botte, "Rwanda and Burundi, 1889–1930: Chronicle of a Slow Assassination, Part I," *International Journal of African Historical Studies* 18, no. 1 (1985): 53–91.

14. Lyons, "African Trypanosomiasis," 555–56.

15. Lyons, *Colonial Disease*; Deborah Neill, *Networks in Tropical Medicine: Internationalism, Colonialism, and the Rise of a Medical Specialty, 1890–1930* (Stanford, CA: Stanford University Press, 2012); Michael Worboys, "The Comparative History of Sleeping Sickness in East and Central Africa," *History of Science* 32, no. 1 (1994): 89–102

16. Steven Riethmiller, "From Atoxyl to Salvarsan: Searching for the Magic Bullet," *Chemotherapy* 51, no. 5 (August 2005): 234–42.

17. Susan C. Welburn, Eric Fèvre, and Paul G. Coleman, "Sleeping Sickness Rediscovered," review article, *Parasitology Today* 15, no. 8 (August 1999): 303–5.

18. See Julie M. Weiskopf, "Living in 'Cold Storage': An Interior History of Tanzania's Sleeping Sickness Concentrations, 1933–1946," *International Journal of African Historical Studies* 49, no. 1 (2016): 1–22 for an exploration of later British efforts in Buha that parallel early German work.

19. Mari K. Webel, "Mapping the Infected Landscape: Sleeping Sickness Prevention and the African Production of Colonial Knowledge in the Early

Twentieth Century," *Environmental History* 20, no. 4 (October 2015): 722–35; Adam Jones and Isabel Voigt, "'Just a First Sketchy Makeshift': German Travellers and Their Cartographic Encounters in Africa, 1850–1914," *History in Africa* 39 (2012): 9–39.

20. Considering the dearth of reliable or comprehensive information about disease at Lake Victoria or Lake Tanganyika, these maps present a limited and incomplete view of circumstances at the time. On colonial mapping of eastern Africa as an imaginative endeavor, discursive practice, and political move, see Michael Pesek, "The Boma and the Peripatetic Ruler: Mapping Colonial Rule in German East Africa, 1889–1903," *Western Folklore* 66, no. 3/4 (2007): 233–57. See also Webel, "Mapping the Infected Landscape."

21. David L. Schoenbrun, *A Green Place, a Good Place: Agrarian Change, Gender, and Social Identity in the Great Lakes Region to the 15th Century* (Portsmouth, NH: Heinemann, 1998), chap. 3.

22. On spaces carved out for global health and research and their multiple meanings, see Johanna Tayloe Crane, *Scrambling for Africa: AIDS, Expertise, and the Rise of American Global Health Science* (Ithaca, NY: Cornell University Press, 2013), introduction and chap. 4; see also Nancy Rose Hunt, *A Colonial Lexicon: Of Birth Ritual, Medicalization, and Mobility in the Congo* (Durham, NC: Duke University Press, 1999), 84–88.

23. Schoenbrun, *Green Place,* chaps. 4 and 5.

24. Neil Kodesh, *Beyond the Royal Gaze: Clanship and Public Healing in Buganda* (Charlottesville: University of Virginia Press, 2010); on the territorial ranges of *kubándwa* deities, their origin sites, and principal shrines, see Schoenbrun, *A Green Place,* 107–12, 236–45; Iris Berger, *Religion and Resistance: East African Kingdoms in the Precolonial Period* (Tervuren, Belgium: Musée royal de l'Afrique centrale, 1981); Shane Doyle, "The Cwezi-Kubandwa Debate: Gender, Hegemony and Pre-Colonial Religion in Bunyoro, Western Uganda," *Africa* 77, no. 4 (2007): 559–81; Henri Médard, *Le royaume du Buganda au XIXe siècle: Mutations politiques et religieuses d'un ancien état d'Afrique de l'Est* (Paris: Karthala, 2007), chap. 4, see maps 23 and 24.

25. Michael G. Kenny, "The Powers of Lake Victoria," *Anthropos* 72, no. 5/6 (1977): 721–22; Johnson, "Fishwork," chaps. 4 and 5.

26. Kim Picozzi et al., "Sleeping Sickness in Uganda: A Thin Line between Two Fatal Diseases," *British Medical Journal* 331, no. 7527 (November 26, 2005): 1238–41; Pere P. Simarro et al., "Estimating and Mapping the Population at Risk of Sleeping Sickness," *PLoS Neglected Tropical Diseases* 6, no. 10 (October 2012): e1859.

27. Fèvre et al., "Reanalyzing the 1900–1920 Sleeping Sickness Epidemic."

28. Endfield et al., "Gloomy Forebodings."

29. Lyons, *Colonial Disease.*

30. Myron J. Echenberg, *Africa in the Time of Cholera: A History of Pandemics from 1817 to the Present* (Cambridge: Cambridge University Press, 2011); Henri Médard, "La peste et les missionnaires: Maladies et syncrétisme médical au royaume du Buganda à la fin du XIXe siècle," *Outre-Mers Revue*

d'Histoire no. 346–347 (First Semester, 2005): 79–102; William H. Schneider, "Smallpox in Africa during Colonial Rule" *Medical History* 53, no. 2 (April 2009): 193–227; Marc H. Dawson, "Socioeconomic Change and Disease: Smallpox in Colonial Kenya, 1880–1920," in *The Social Basis of Health and Healing in Africa*, ed. Steven Feierman and John M. Janzen (Berkeley: University of California Press, 1992); Eugenia W. Herbert, "Smallpox Inoculation in Africa," *Journal of African History* 16, no. 4 (1975): 547–50.

31. Hoppe, *Lords of the Fly*, 134, 199.

32. Schoenbrun, *Green Place*; Jean-Pierre Chrétien, *The Great Lakes of Africa: Two Thousand Years of History*, trans. Scott Strauss (New York: Zone Books, 2003); Steven Feierman, "A Century of Ironies in East Africa (ca. 1780–1890)," in *African History from Earliest Times to Independence*, ed. Philip Curtin, Steven Feierman, Leonard Thompson, and Jan Vansina (London: Longman, 1995); Kodesh, *Royal Gaze*; Rhiannon Stephens, *A History of African Motherhood: The Case of Uganda, 700–1900* (Cambridge: Cambridge University Press, 2013); Brad Weiss, *Sacred Trees, Bitter Harvests: Globalizing Coffee in Northwest Tanzania* (Portsmouth, NH: Heinemann, 2003); Brad Weiss, *The Making and Unmaking of the Haya Lived World: Consumption, Commoditization, and Everyday Practice* (Durham, NC: Duke University Press, 1996); Henri Médard, *Le royaume du Buganda*; Richard J. Reid, *Political Power in Pre-colonial Buganda: Economy, Society, and Warfare in the Nineteenth Century* (Athens: Ohio University Press, 2002); Richard J. Reid, "The Ganda on Lake Victoria: a Nineteenth-Century East African Imperialism," *Journal of African History* 39, no. 3 (1998): 349–63; Shane Doyle, *Crisis and Decline in Bunyoro: Population and Environment in Western Uganda 1860–1955* (Athens: Ohio University Press, 2006); Randall M. Packard, *Chiefship and Cosmology: An Historical Study of Political Competition* (Bloomington: Indiana University Press, 1981); Jan Vansina, *Antecedents to Modern Rwanda: The Nyiginya Kingdom* (Madison: University of Wisconsin Press, 2004); David Newbury, *Kings and Clans: Ijwi Island and the Lake Kivu Rift, 1780–1840* (Madison: University of Wisconsin Press, 1991); Holly Elisabeth Hanson, *Landed Obligation: The Practice of Power in Buganda* (Portsmouth, NH: Heinemann, 2003).

33. Schoenbrun, *Green Place*, 195–207; Berger, *Religion and Resistance*; Kodesh, *Royal Gaze*.

34. Steven Feierman, *The Shambaa Kingdom: A History* (Madison: University of Wisconsin Press, 1974); Packard, *Chiefship and Cosmology*; Doyle, *Crisis and Decline*; Stephens, *African Motherhood*; James L. Giblin, *The Politics of Environmental Control in Northeastern Tanzania, 1840–1940* (Philadelphia: University of Pennsylvania Press, 1992); Weiss, *Making and Unmaking*; Weiss, *Sacred Trees*; Hanson, *Landed Obligation*; Feierman, *Peasant Intellectuals: Anthropology and History in Tanzania* (Madison: University of Wisconsin Press, 1990).

35. Steven Feierman, "Struggles for Control: the Social Roots of Health and Healing in Modern Africa, *African Studies Review* 28, no. 2–3 (June–September 1985): 73–147; Julie Livingston, "Productive Misunderstandings

and the Dynamism of Plural Medicine in Mid-century Bechuanaland," *Journal of Southern African Studies* 33, no. 4 (December 2007): 801–10; Livingston, *Debility and the Moral Imagination in Botswana* (Bloomington: Indiana University Press, 2005); Hunt, *Colonial Lexicon*; Stacey A. Langwick, *Bodies, Politics, and African Healing: The Matter of Maladies in Tanzania* (Bloomington: Indiana University Press, 2011); Patrick Thomas Malloy, "Holding [Tanganyika] by the Sindano": Networks of Medicine in Colonial Tanganyika" (PhD diss., University of California–Los Angeles, 2003).

36. See, for example, F. Courtin et al., "Sleeping Sickness in West Africa (1906–2006): Changes in Spatial Repartition and Lessons from the Past," *Tropical Medicine and International Health* 13, no. 3 (2008): 334–44.

37. Lyons, *Colonial Disease*; Giblin, *Politics of Environmental Control*; Maryinez Lyons, "The Power to Heal: African Auxiliaries in Colonial Belgian Congo and Uganda," in *Contesting Colonial Hegemony: State and Society in Africa and India*, ed. Dagmar Engels and Shula Marks (London: British Academic Press, 1994).

38. Lyons, *Colonial Disease*; Hoppe, *Lords of the Fly*; Mwelwa C. Musambachime, "The Social and Economic Effects of Sleeping Sickness in Mweru-Luapula 1906–1922," *African Economic History* 10 (1981): 151–73; Rita Headrick and Daniel R. Headrick, *Colonialism, Health and Illness in French Equatorial Africa, 1885–1935* (Atlanta, GA: African Studies Association Press, 1994).

39. Musambachime, "Social and Economic Effects"; Heather Bell, *Frontiers of Medicine in the Anglo-Egyptian Sudan, 1899–1940* (Oxford: Clarendon Press, 1999); Wolfgang Eckart, *Medizin und Kolonialimperialismus, Deutschland 1884–1945* (Paderborn, Germany: Schöningh, 1997).

40. Abena Dove Osseo-Asare, *Bitter Roots: The Search for Healing Plants in Africa* (Chicago: University of Chicago Press, 2014); Melissa Graboyes, *The Experiment Must Continue: Medical Research and Ethics in East Africa, 1940–2014* (Athens: Ohio University Press, 2015); Neill, *Networks in Tropical Medicine.*

41. On littorals as a frame of analysis in recent historical scholarship, see "Tracks and Trails: Indian Ocean Worlds as Method," a special edition of *History Compass* with contributions by Antoinette Burton, Madhavi Kale, Isabel Hofmeyr, Clare Anderson, Christopher J. Lee, and Nile Green, *History Compass* 11, no. 7 (2013): 497–535; "AHR Forum: Oceans of History," with contributions by Kären Wigen, Peregrine Horden, Nicholas Purcell, Matt K. Matsuda, and Alison Games, *American Historical Review* 111, no. 3 (June 2006): 717–57.

42. On the mutual impacts of colony and metropole on cultural production, see Frederick Cooper and Ann Laura Stoler, eds., *Tensions of Empire: Colonial Cultures in a Bourgeois World* (Berkeley: University of California Press, 1997); Frederick Cooper, *Colonialism in Question: Theory, Knowledge, History* (Berkeley: University of California Press, 2005).

43. Daniel Lord Smail and Andrew Shryock, "History and the 'Pre,'" *American Historical Review* 118, no. 3 (June 2013): 709–37. I deliberately avoid the use of "precolonial" to describe nineteenth-century eastern Africa throughout this book, following Smail and Shryock.

44. Nicola A. Batchelor et al., "Spatial Predictions of Rhodesian Human African Trypanosomiasis (Sleeping Sickness) Prevalence in Kaberamaido and Dokolo, Two Newly Affected Districts of Uganda," *PLoS Neglected Tropical Diseases* 3, no. 12 (2009): e563; Lyons, *Colonial Disease*, chaps. 7 and 10.

45. Lea Berrang-Ford et al., "Sleeping Sickness in Southeastern Uganda: A Systems Approach," *Ecohealth* 2, no. 3 (2005): 183–94 provides a sense of a typical spatialized approach to sleeping sickness as well as trenchant critiques of its narrowness.

46. Crane, *Scrambling for Africa*; Susan Reynolds-Whyte, ed., *Second Chances: Surviving AIDS in Uganda* (Durham, NC: Duke University Press, 2014); P. Wenzel Geissler and Catherine Molyneux, eds., *Evidence, Ethos, and Experiment: The Anthropology and History of Medical Research in Africa* (New York: Berghahn, 2011); Jennifer Tappan, *The Riddle of Malnutrition: The Long Arc of Biomedical and Public Health Interventions in Uganda* (Athens: Ohio University Press, 2017).

47. Several excellent collections embody this trend: Tamara Giles-Vernick and James L. A. Webb, Jr., eds., *Global Health in Africa: Historical Perspectives on Disease Control* (Athens: Ohio University Press, 2013); Ruth J. Prince and Rebecca Marsland, eds., *Making and Unmaking Public Health in Africa: Ethnographic and Historical Perspectives* (Athens: Ohio University Press, 2014); João Biehl and Adriana Petryna, eds., *When People Come First: Critical Studies in Global Health* (Princeton, NJ: Princeton University Press, 2013); Adriana Petryna, Andrew Lakoff, and Arthur Kleinman, eds., *Global Pharmaceuticals: Ethics, Markets, Practices* (Durham, NC: Duke University Press, 2006); P. Wenzel Geissler, ed., *Para-States and Medical Science: Making African Global Health* (Durham, NC: Duke University Press, 2015).

48. Stephen L. Cochi and Walter R. Dowdle, *Disease Eradication in the 21st Century: Implications for Global Health* (Cambridge, MA: MIT Press, 2013); Amy Moran-Thomas, "A Salvage Ethnography of the Guinea Worm: Witchcraft, Oracles and Magic in a Disease Eradication Program," in *When People Come First: Critical Studies in Global Health*, ed. João Biehl and Adriana Petryna (Princeton, NJ: Princeton University Press, 2013); Elisha P. Renne, *The Politics of Polio in Northern Nigeria* (Bloomington: Indiana University Press, 2010); Nancy Leys Stepan, *Eradication: Ridding the World of Diseases Forever?* (Ithaca, NY: Cornell University Press, 2011); Randall M. Packard, *A History of Global Health: Interventions into the Lives of Other Peoples* (Baltimore: Johns Hopkins University Press, 2016); Kirsten Moore-Sheeley, "'Nothing but Nets': The History of Insecticide-Treated Nets in Africa, 1980s–Present" (PhD diss., Johns Hopkins University, 2017).

49. Guillaume Lachenal, "A Genealogy of Treatment as Prevention (TasP): Prevention, Therapy, and the Tensions of Public Health in African History," in *Global Health in Africa: Historical Perspectives on Disease Control*, ed. Tamara Giles-Vernick and James L. A. Webb, Jr. (Athens: Ohio University Press, 2013).

50. Vincanne Adams, ed., *Metrics: What Counts in Global Health* (Durham, NC: Duke University Press, 2016); Peter Redfield, *Life in Crisis: The Ethical Journey of Doctors without Borders* (Berkeley: University of California Press, 2013); Hansjörg Dilger, Abdoulaye Kane, and Stacey A. Langwick, eds., *Medicine, Mobility, and Power in Global Africa: Transnational Health and Healing* (Bloomington: Indiana University Press, 2012).

51. Melissa Parker and Tim Allen, "De-Politicizing Parasites: Reflections on Attempts to Control the Control of Neglected Tropical Diseases," *Medical Anthropology* 33, no. 3 (2014): 223–39.

52. Here, I follow Schoenbrun, *Green Place*.

53. Peter Robertshaw and David Taylor, "Climate Change and the Rise of Political Complexity in Western Uganda," *Journal of African History* 41, no. 1 (2000): 1–28; David L. Schoenbrun, "Cattle Herds and Banana Gardens: The Historical Geography of the Western Great Lakes Region, ca. AD 800–1500," *African Archaeological Review* 11 (December 1993): 39–72.

54. Schoenbrun, *Green Place*, chap. 2.

55. For Burundi-specific examples, see Chrétien's "Calendrier agricole du Burundi/Plateau Central," in Jean-Pierre Chrétien, *Burundi: L'histoire retrouvée: 25 ans de métier d'historien en Afrique* (Paris: Karthala, 1993), 83; and fig. 3, "Most common cropping systems in the agrarian history of Burundi," in Hubert Cochet, "Agrarian Dynamics, Population Growth and Resource Management: The Case of Burundi," *GeoJournal* 60, no. 2 (2004): 114.

56. Chrétien, *Burundi*, 80–82; Schoenbrun, *Green Place*.

57. Chrétien, *Burundi*, 83.

58. Stephens cautions against overemphasizing the gendered aspects of agricultural labor or presuming their stability over time in *African Motherhood*, 69.

59. On cattle in the interlacustrine kingdoms, see Schoenbrun, *Green Place*, Schoenbrun, "Cattle Herds," and David L. Schoenbrun, "We Are What We Eat: Ancient Agriculture between the Great Lakes," *Journal of African History* 34, no. 1 (1993): 1–31; Doyle, *Crisis and Decline*; Jacques Depelchin, "From Pre-Capitalism to Imperialism: A History of Social and Economic Formations in Eastern Zaïre (Uvira Zone, c. 1800–1965)" (PhD diss., Stanford University, 1974), chap. 5; Cochet, "Agrarian Dynamics," 113–14. See also David M. Anderson, "Cow Power: Livestock and the Pastoralist in Africa," review article, *African Affairs* 92, no. 366 (1993): 121–33; Dorothy L. Hodgson, ed., *Rethinking Pastoralism in Africa: Gender, Culture, and the Myth of the Patriarchal Pastoralist* (Athens: Ohio University Press, 2000).

60. Stephens, *African Motherhood*, 71–73.

61. Schoenbrun, *Green Place*, 168–71.

62. Peter R. Schmidt, *Historical Archaeology in Africa: Representation, Social Memory, and Oral Traditions* (Lanham, MD: AltaMira, 2006).

63. David W. Cohen, "Peoples and States of the Great Lakes Region," in *Africa in the Nineteenth Century until the 1880s*, ed. J. F. Ade Ajayi, UNESCO General History of Africa 6 (Paris: UNESCO, 1989); Chrétien, *Great Lakes*;

David W. Cohen, *Womunafu's Bunafu: A Study of Authority in a Nineteenth-Century African Community* (Princeton, NJ: Princeton University Press, 1977); David Newbury, *The Land beyond the Mists: Essays on Identity and Authority in Precolonial Congo and Rwanda* (Athens: Ohio University Press, 2009); Christopher Wrigley, *Kingship and State: The Buganda Dynasty* (Cambridge: Cambridge University Press, 1996); Médard, *Le royaume du Buganda*; Doyle, *Crisis and Decline*; Chrétien, *Burundi*; David Newbury, *Kings and Clans*.

64. Feierman, *Shambaa Kingdom*; Kodesh, *Royal Gaze*.

65. Schoenbrun, *Green Place*, 185–95.

66. Kodesh, *Royal Gaze*, 10; Berger, *Religion and Resistance*; David L. Schoenbrun, "Conjuring the Modern in Africa: Durability and Rupture in Histories of Public Healing between the Great Lakes of East Africa," *American Historical Review* 111, no. 5 (December 2006): 1403–39; David Newbury, "The Clans of Rwanda: a Historical Hypothesis," chap. 8 in *The Land beyond the Mists*.

67. Kodesh, *Royal Gaze*, 7–11.

68. Richard J. Reid, *Pre-colonial Buganda*, 244–45.

69. Hanson, *Landed Obligation*; Kodesh, *Royal Gaze*.

70. Kodesh, *Royal Gaze*, 85–86; Berger, *Religion and Resistance*, 17, 52.

71. Doyle, "Cwezi-Kubandwa Debate," 565; Johnson, "Fishwork," 228–29. *Misambwa* could be mobile and might also be linked to ancestors. See Johnson's discussion of Bega in "Fishwork," 216–17.

72. Schoenbrun, *Green Place*, 112, 204–6, 268–69; Doyle, "Cwezi-Kubandwa Debate," 560–61.

73. Cohen, "Peoples and States," 274.

74. Cohen, "Peoples and States"; Richard J. Reid, *Pre-colonial Buganda*.

75. Catharine Newbury, *The Cohesion of Oppression: Clientship and Ethnicity in Rwanda, 1860–1960* (New York: Columbia University Press, 1988); Alison L. Des Forges, *Defeat Is the Only Bad News: Rwanda under Musinga, 1896–1931*, ed. David Newbury (Madison: University of Wisconsin Press, 2011); Michael Twaddle, *Kakungulu and the Creation of Uganda, 1868–1928* (Athens: Ohio University Press, 1993); Wrigley, *Kingship and State*.

76. Cohen, "Peoples and States," 273; Michael G. Kenny, "Mutesa's Crime: Hubris and the Control of African Kings," *Comparative Studies in Society and History* 30, no. 4 (October 1988): 595–612.

77. Charles M. Good, "Salt, Trade, and Disease: Aspects of Development in Africa's Northern Great Lakes Region," *International Journal of African Historical Studies* 5, no. 4 (1972): 543–86; Michele D. Wagner, "Trade and Commercial Attitudes in Burundi before the Nineteenth Century," *International Journal of African Historical Studies* 26, no. 1 (1993): 149–66; Gerald W. Hartwig, "The Victoria Nyanza as a Trade Route in the Nineteenth Century," *Journal of African History* 11, no. 4 (1970): 535–52. Uvira in particular was a key regional market for iron wares; Richard Burton, *The Lake Regions of Central Africa: A Picture of Exploration* (New York: Harper and Brothers, 1860), 325.

78. Henri Médard and Shane Doyle, eds., *Slavery in the Great Lakes Region of East Africa* (Athens: Ohio University Press, 2007).

79. Kodesh, *Royal Gaze*, 122–24; Botte, "Rwanda and Burundi, 1889–1930: Chronicle of a Slow Assassination, Part I."

80. Stephen J. Rockel, *Carriers of Culture: Labor on the Road in Nineteenth-Century East Africa* (Portsmouth, NH: Heinemann, 2006); Thomas F. McDow, *Buying Time: Debt and Mobility in the Western Indian Ocean* (Athens: Ohio University Press, 2018).

81. Erik Gilbert, *Dhows and the Colonial Economy of Zanzibar, 1860–1970* (Athens: Ohio University Press, 2004), chap. 2.

82. Rockel, *Carriers of Culture*; Jonathon Glassman, *Feasts and Riot: Revelry, Rebellion, and Popular Consciousness on the Swahili Coast, 1856–1888* (Portsmouth, NH: Heinemann, 1995), chaps. 1–2; Michelle R. Moyd, *Violent Intermediaries: African Soldiers, Conquest, and Everyday Colonialism in German East Africa* (Athens: Ohio University Press, 2014), 213.

83. Frederick Cooper, *Plantation Slavery on the East Coast of Africa* (Portsmouth, NH: Heinemann, 1997); McDow, *Buying Time.*

84. Feierman, "Century of Ironies"; Abdul Sheriff, *Slaves, Spices, and Ivory in Zanzibar: Integration of an East African Commercial Empire into the World Economy, 1770–1873* (Athens: Ohio University Press, 1987); and Edward A. Alpers, *Ivory and Slaves: Changing Pattern of International Trade in East Central Africa to the Later Nineteenth Century* (Berkeley: University of California Press, 1975).

85. Sheryl McCurdy, "Transforming Associations: Fertility, Therapy, and the Manyema Diaspora in Urban Kigoma, Tanzania, c. 1850–1993" (PhD diss., Columbia University, 2000); Rockel, *Carriers of Culture*; Stephen J. Rockel, "Enterprising Partners: Caravan Women in Nineteenth Century Tanzania," *Canadian Journal of African Studies/Revue canadienne des études africaines* 34, no. 3 (2000): 748–78; Philip Gooding, "Lake Tanganyika: Commercial Frontier in the Era of Long-Distance Commerce, East and Central Africa, c. 1830–1890" (PhD diss., SOAS University of London, 2017).

86. Médard and Doyle, *Slavery in the Great Lakes Region*; see also Marcia Wright, *Strategies of Slaves and Women: Life-Stories from East/Central Africa* (London: James Currey, 1993).

87. C. F. Holmes, "Zanzibari Influence at the Southern End of Lake Victoria: The Lake Route," *African Historical Studies* 4, no. 3 (1971): 477–503.

88. Richard J. Reid, *Pre-colonial Buganda*, 151–71; Chrétien, *Great Lakes*, 195–97.

89. David Newbury, *The Land beyond the Mists*, chap. 12.

90. Johannes Fabian, *Out of Our Minds: Reason and Madness in the Exploration of Central Africa* (Berkeley: University of California Press, 2000); Jones and Voigt, "Sketchy Makeshift"; Donald H. Simpson, *Dark Companions: The African Contribution to the European Exploration of East Africa* (New York: Barnes and Noble, 1976).

91. D. A. Low, *Fabrication of Empire: The British and the Uganda Kingdoms, 1890–1902* (Cambridge, MA: Cambridge University Press, 2009); Chrétien, *Great Lakes*, 217–19; Ralph A. Austen, *Northwest Tanzania under German*

and British Rule: Colonial Policy and Tribal Politics, 1889–1939 (New Haven, CT: Yale University Press, 1968), chap. 3.

92. Chrétien, *Great Lakes*, 217–18.

93. Chrétien, *Great Lakes*, 244, 247–50.

PART I: THE SSESE ISLANDS, C. 1890–1907

THE SSESE ISLANDS, C. 1890: AN OVERVIEW

1. The Ssese archipelago is made up of eighty-four islands, covering 166 square miles; the largest island, Bugala, covers 103 square miles, a majority of the islands' total landmass. Johnson recounts narratives from islanders that derived the *Ssese* name from a Luganda word meaning "islands that have moved." Jennifer Lee Johnson, "Fishwork in Uganda: A Multispecies Ethnohistory about Fish, People, and Ideas about Fish and People" (PhD diss., University of Michigan, 2014), 228.

2. "Letters from Uganda," Letter from R. H. Walker to his father, Buganda, 24 Oct. 1889, *Church Missionary Intelligencer and Record* 15 (June 1890): 361.

3. M. Afr.–Rome, Sese Correspondence (Nyanza Septentrional), Dossier 87, 087457–59, Fr. Reynès, 25 July 1904; Andrew Reid, "The Lake, Bananas and Ritual Power in Buganda," in *A History of Water: Series* 3, vol. 3, *Water and Food*, ed. Terje Tvedt, Terje Oestigaard, and R. Coopey (London: I. B. Tauris, 2010).

4. See, for instance, CMS, ACC 558 F5, Journal of Aileen Weatherhead, 11 Jan. 1902 (17), and CMS, ACC 588 F6, Journal of Aileen Weatherhead, 14 Mar. 1903. Weatherhead's description accords with European photographs of smaller Ssese homes in the first decade of the twentieth century.

5. John Roscoe's "typical ganda home" in *The Baganda: An Account of Their Native Customs and Beliefs* (London: Macmillan, 1911), is a large structure. Contrasted with a photograph from the Lake Victoria islands (like Buvuma) from 1902 in the Christy collection of photographs on Africa (RCS, Y304A/51), Roscoe's "typical" home is large by comparison.

6. Production of these perennial crops could fit into the margins of forested areas at the islands' centers, where fast-growing yams complemented slower-maturing banana trees, per CMS, ACC 558 F5, 26 Jan. 1902, 2. See David L. Schoenbrun, "We Are What We Eat: Ancient Agriculture between the Great Lakes," *Journal of African History* 34, no. 1 (1993): 20–21, and Schoenbrun, *A Green Place, a Good Place: Agrarian Change, Gender, and Social Identity in the Great Lakes Region to the 15th Century* (Portsmouth, NH: Heinemann, 1998), 79–83, on banana cultivation development. On legume crops, see Schoen-brun, *Green Place*, 73; Rhiannon Stephens, *A History of African Motherhood: The Case of Uganda, 700–1900* (Cambridge: Cambridge University Press, 2013), 68–70; RCS, RCMS 124/4, Diary of Cuthbert Christy, 13 Oct. 1902; Richard J. Reid, *Political Power in Pre-colonial Buganda: Economy, Society, and Warfare in the Nineteenth Century* (Athens: Ohio University Press, 2002), 37.

7. Ndyabahika Matete and Bakama B. BakamaNume, "Climate of Uganda," in *Contemporary Geography of Uganda*, ed. Bakama B. BakamaNume (Dar es Salaam, Tanzania: Mkuki na Nyota, 2010), 6–7, 9–10.

8. "Extracts from the Annual Letters of the Missionaries," Rev. E. C. Gordon, Bukasa, Sesse [*sic*] Islands, 28 Dec. 1896, *Church Missionary Intelligencer and Record* 21 (June 1896): 531.

9. M. Afr.–Rome–EE, Diaire de Rubaga, 24 Mar. 1892, *CTSMA*, no. 56 (October 1892). Per my Note on Digitized and Manuscript Archival Materials above, sources accessed only in the *Éditions électronique* database of the Society of the Missionaries of Africa (White Fathers) Archives housed in the Generalate in Rome have the designation M. Afr.–Rome–EE.

10. Richard J. Reid, *Pre-colonial Buganda*, 64; Johnson, "Fishwork."

11. CMS, ACC 611 F3, Martin J. Hall, "Through My Spectacles in Uganda" (1898), 67.

12. These larger, plank-constructed canoes were distinct from dugout canoes ranging from ten to twenty feet long, created by hollowing out an entire tree trunk. Richard J. Reid, *Pre-colonial Buganda*, 229; Michael G. Kenny, "The Powers of Lake Victoria," *Anthropos* 72, no. 5/6 (1977): 725–26.

13. Richard J. Reid, *Pre-colonial Buganda*, 230; CMS ACC 611-F3, Martin J. Hall, "Through My Spectacles in Uganda," 1898, 63–65. See also CMS ACC 276, Harold B. Lewin Papers, F2—Photograph Album; Letter from Rev. R. H. Walker, 24 Oct. 1889, *Church Missionary Intelligencer and Record* 15 (June 1890): 361.

14. Kenny, "Powers of Lake Victoria," 723–28.

15. Roscoe, *Baganda*; Apolo Kagwa, *The Customs of the Baganda*, trans. Ernest B. Kalibala, ed. May M. Edel (New York: Columbia University Press, 1934); Richard J. Reid, *Pre-colonial Buganda*; and Richard J. Reid, "The Ganda on Lake Victoria: A Nineteenth-Century East African Imperialism," *Journal of African History* 39, no. 3 (1998): 349–63.

16. Richard J. Reid, *Pre-colonial Buganda*, 5; Richard J. Reid, *A History of Modern Uganda* (Cambridge: Cambridge University Press, 2017), 128–35, 141–46.

17. Richard J. Reid, *Pre-colonial Buganda*; Neil Kodesh, *Beyond the Royal Gaze: Clanship and Public Healing in Buganda* (Charlottesville: University of Virginia Press, 2010), 46.

18. Richard J. Reid, *Pre-colonial Buganda*, 228–31, 238–39.

19. Richard J. Reid, 240.

20. "Extracts from the Annual Letters of the Missionaries," Rev. H. T. C. Weatherhead, Bugala, Sese [*sic*] Islands, 30 Oct. 1902, *Church Missionary Intelligencer and Record* 27 (October 1902). 170.

21. "Letters from Uganda," Letter from Bishop Tucker, Island of Bukassa, Victoria Nyanza, 25 Jan. 1896, *Church Missionary Intelligencer and Record* 21 (June 1896): 422; Richard J. Reid, *Pre-colonial Buganda*, 66.

22. Richard J. Reid, *Pre-colonial Buganda*, 66.

23. Holly Elisabeth Hanson, *Landed Obligation: The Practice of Power in Buganda* (Portsmouth, NH: Heinemann, 2003), 234; Stephens, *African Motherhood*, 107–8; Johnson, "Fishwork," 143–45.

24. Roscoe, *Baganda*, 294; Kagwa, *Customs*, 114–15.

25. Kouzi was described as chief of Sseses in 1890, per M. Afr.–Rome–EE, Diaire de Buganda, *CTSMA*, no. 50 (April 1891).

26. Kodesh, *Royal Gaze*, 46–47; Richard J. Reid, *Pre-colonial Buganda*, 238; Johnson, "Fishwork," 168.

27. Richard J. Reid, *Pre-colonial Buganda*, 244–45.

28. Richard J. Reid, 234.

29. Richard J. Reid, *Modern Uganda*, 154–55. See also Michael Twaddle, "The Emergence of Politico-Religious Groupings in Late Nineteenth-Century Buganda," *Journal of African History* 29, no. 1 (1988): 81–92; D. A. Low, *Fabrication of Empire: The British and the Uganda Kingdoms, 1890–1902* (Cambridge, MA: Cambridge University Press, 2009), chap. 3; Henri Médard, *Le royaume du Buganda au XIXe siècle: Mutations politiques et religieuses d'un ancien état d'Afrique de l'Est* (Paris: Karthala, 2007).

30. See Clive A. Spinage, *African Ecology: Benchmarks and Historical Perspectives* (New York: Springer, 2012), chap. 22; James L. Giblin pushes against narratives of generalized devastation in *The Politics of Environmental Control in Northeastern Tanzania, 1840–1940* (Philadelphia: University of Pennsylvania Press, 1992).

31. Richard J. Reid, *Modern Uganda*, 155. See also Richard J. Reid, *Pre-colonial Buganda*, 34–39.

32. Here, I follow Reid's synthesis in *Modern Uganda*, 153–60.

33. M. Afr.–Rome–EE, Diaire de Rubaga, 24 Mar. 1892, *CTSMA*, no. 56 (October 1892). Admittedly, given the polemical nature of White Fathers' diaries regarding religious wars, we must take the "total" destruction of communities with a grain of salt.

34. Richard J. Reid, *Pre-colonial Buganda*, 245–48.

CHAPTER 1: FINDING SLEEPING SICKNESS ON THE SSESE ISLANDS

1. Michael G. Kenny, "The Powers of Lake Victoria," *Anthropos* 72, no. 5/6 (1977): 721–22; see Nyambura Mpesha, *Mugasha: Epic of the Bahaya* (Nairobi: East African Educational Publishers, 2000) for a narrative of the life and powers of Mugasha (Mukasa) within Haya cosmology.

2. Jennifer Lee Johnson, "Fishwork in Uganda: A Multispecies Ethnohistory about Fish, People, and Ideas about Fish and People" (PhD diss., University of Michigan, 2014), chaps. 1 and 4.

3. Jennifer Lee Johnson, "Fish, Family, and the Gendered Politics of Descent along Uganda's Southern Littoral," *History in Africa* 45 (June 2018): 459.

4. Neil Kodesh, *Beyond the Royal Gaze: Clanship and Public Healing in Buganda* (Charlottesville: University of Virginia Press, 2010), 46–47. The lubaale's home is not a settled issue, however. David L. Schoenbrun discusses the vexed history of locating Mukasa on either Bubembe or Bukasa Islands in early twentieth-century historical writing on Buganda in "Ethnic Formation with Other-than-Human Beings: Island Shrine Practice in Uganda's Long Eighteenth Century," *History in Africa* 4 (June 2018): 405–9.

5. John Roscoe, *The Baganda: An Account of Their Native Customs and Beliefs* (London: Macmillan, 1911), 290; Apolo Kagwa, *The Customs of the Baganda*, trans. Ernest B. Kalibala, ed. May M. Edel (New York: Columbia University Press, 1934), 114, 222–23.

6. Christopher Wrigley, "The Story of Rukidi," *Africa: Journal of the International African Institute* 43, no. 3 (July 1973): 232.

7. Roscoe, *Baganda*, 300–301; and Diane Leinwand Zeller, "The Establishment of Western Medicine in Buganda" (PhD diss., Columbia University, 1974), 32.

8. Schoenbrun, "Ethnic Formation," 400.

9. Schoenbrun, 422; A. Mukanga, "The Traditional Belief in Balubaale," *Occasional Research Papers in African Religions and Philosophies* [Makerere University, Department of Religious Studies and Philosophy] 17, no. 167 (1974): 5.

10. Johnson, "Fishwork," 8–9, 163–64.

11. Roscoe, *Baganda*, 300; M. Afr.–Rome–EE, "Rapport du P. Brard sur les Tribus Insulaires du Nyanza Méridional," *CTSMA*, no. 73 (January 1897).

12. Hermann Rehse, *Kiziba: Land und Leute* (Stuttgart: Strecker and Schröder, 1910), 128–29; M. Afr.–Rome–EE, Fr. Lévesque, Letter from Kadouma, 1 Nov. 1880, *CTSMA*, no. 10 (April 1881).

13. M. Afr.–Rome–EE, Journal de voyage du Père Lévesque, *CTSMA*, no. 19 (July 1883), entries for 26 Nov. and 25 Dec. 1882; Roscoe, *Baganda*, 300; M. Afr.–Rome–EE, "Rapport du P. Brard sur les Tribus Insulaires du Nyanza Méridional," *CTSMA*, no. 73 (January 1897).

14. Henri Médard, *Le royaume du Buganda au XIXe siècle: Mutations politiques et religieuses d'un ancien état d'Afrique de l'Est* (Paris: Karthala, 2007), maps 23 and 24.

15. Richard J. Reid, *Political Power in Pre-colonial Buganda: Economy, Society, and Warfare in the Nineteenth Century* (Oxford: James Currey, 2002), 65; Kodesh, *Royal Gaze*, 46–48; Kenny, "Powers of Lake Victoria," 723–24.

16. Richard J. Reid, *Pre-colonial Buganda*, 65; Holly Elisabeth Hanson, *Landed Obligation: The Practice of Power in Buganda* (Portsmouth, NH: Heinemann, 2003), 73. Regarding relations between the kabaka and Mukasa's mediums, see Michael G. Kenny, "Mutesa's Crime: Hubris and the Control of African Kings," *Comparative Studies in Society and History* 30, no. 4 (October 1988): 595–612. Hall recounts specific items sent periodically to the kabaka: large amounts of dried *nkeje* (fish), coffee berries, goats, cowries, and several wooden stools "peculiar to the islands." CMS, ACC 611 F3, Martin J. Hall, "Through My Spectacles in Uganda" (1898), 67.

17. Roscoe, *Baganda*, 299; Michael W. Tuck, "Kabaka Mutesa and Venereal Disease: An Essay on Medical History and Sources in Precolonial Buganda," *History in Africa* 30 (2003): 309–25.

18. M. Afr.–Rome–EE, Missions des Lacs Nyanza, Fr. Lévesque, Letter from Kadouma, 1 Nov. 1880, *CTSMA*, no. 10 (April 1881).

19. Kenny, "Mutesa's Crime," 598; Richard J. Reid, *A History of Modern Uganda* (Cambridge: Cambridge University Press, 2017), 166.

20. Hanson, *Landed Obligation*, 72.

21. These titled chiefs were also historically associated with particular territories. A map of the Sseses marks the northern half of present-day Bugala Island as "Sewaia" and the southern half as "Semagala," and Bukasa as "Kaganda Island or Bukasa." See NAB MR 1/1395, "Map no. 1. Sleeping sickness and Glossina palpalis in the Uganda Protectorate, Sleeping Sickness Extended Investigation," 1906. Johnson, "Fishwork," 159–60.

22. Roscoe, *Baganda*, 294; Kagwa, *Customs*, 114–15.

23. M. Afr.–Rome–EE, Rapport—Buyango, *CTSMA*, no. 104 (March 1904). Per Hans Cory and M. M. Hartnoll, *Customary Law of the Haya Tribe: Tanganyika Territory* (London: Percy Lund, Humphries, 1945), the clan of the firelighters of the king had Wamara and Mugasha (Mukasa) as their associated spirit and originated in Buganda; Appendix, 19.

24. Johnson, "Gendered Politics of Descent."

25. Zeller, "Establishment," 24, 54, 379.

26. Zeller, "Establishment," 42–44.

27. Steven Feierman, "Struggles for Control: The Social Roots of Health and Healing in Modern Africa, *African Studies Review* 28, no. 2–3 (June–September 1985): 79–83; Kodesh, *Royal Gaze*, 15; David L. Schoenbrun, "Conjuring the Modern in Africa: Durability and Rupture in Histories of Public Healing between the Great Lakes of East Africa," *American Historical Review* 111, no. 5 (December 2006): 1403–39.

28. Roscoe, *Baganda*, mentions but does not detail particular plants and substances used in healing. Zeller, "Establishment," Appendix A: "Kiganda Pharmacopoeia," 397–409.

29. Luise White, "'They Could Make Their Victims Dull': Genders and Genres, Fantasies and Cures in Colonial Southern Uganda," *American Historical Review* 100, no. 5 (December 1995): 1390.

30. Roscoe, *Baganda*, 100. Roscoe mentions cupping used for headache, coughs, abscesses, and sharp pain in the sides. Kagwa mentions cupping used by women for headaches in *Customs*, 126, and Aileen Weatherhead recounts it among elite Ssese women at roughly the same time Kagwa and Roscoe were writing; CMS, ACC 558 F6, Journal of E. A. Weatherhead, 5 Feb. 1903. Extracts of a particular beetle known as *kauka kawumpuli* were used to raise blisters on people affected with kaumpuli (understood as bubonic plague by European observers), per Henri Médard, "La peste et les missionnaires: Maladies et syncrétisme médical au royaume du Buganda à la fin du XIXe siècle," *Outre-Mers Revue d'Histoire* no. 346–347 (First Semester, 2005): 96–97.

31. Roscoe, *Baganda*, 103; Johnson, "Gendered Politics of Descent," 464.

32. Roscoe, *Baganda*, 100–103; Kagwa, *Customs*, 125–26. On generalized "traditional" etiologies of illness and therapeutic practice observed in the 1970s, see L. Sseguya, "The Concept of Death and Future among the Ganda," *Occasional Research Papers in African Religions and Philosophies* [Makerere University, Department of Religious Studies and Philosophy] 32, no. 30 (September 1974): 7–8. Sseguya aligns roughly with descriptions from Roscoe and Kagwa.

33. See, for example, Johnson, "Gendered Politics of Descent," 462; and Johnson, "Fishwork."

34. Feierman, "Struggles for Control," 79–83.

35. Kenny, "Mutesa's Crime," 599, 602.

36. David William Cohen, "Review: The Cwezi Cult," *Journal of African History* 9, no. 4 (October 1968): 655.

37. Kagwa, *Customs*, 113–14; Roscoe, *Baganda*, 313–14; James Francis Cunningham, *Uganda and Its Peoples: Notes on the Protectorate of Uganda, Especially the Anthropology and Ethnology of Its Indigenous Races* (London: Hutchinson, 1905), 89, 253, 255.

38. Kagwa, *Customs*, 114.

39. Kagwa, 121.

40. Kagwa, 122–23; Kodesh, *Royal Gaze*, 44, 146.

41. Kodesh, *Royal Gaze*, 85.

42. David Schoenbrun, *The Historical Reconstruction of Great Lakes Bantu Cultural Vocabulary: Etymologies and Distributions* (Cologne: Rüdiger Köppe, 1997), 249.

43. I thank David Schoenbrun for this insight.

44. The extent to which Islamic medicine was available to Ssese islanders is not clear from extant sources, likewise the extent of Islamic religious practice on the islands. On Islamic medicine historically in Buganda, see Zeller, "Establishment," 58–60; Médard notes Islamic medicine in the court of Mutesa in "La peste," 83.

45. The Bugoma mission (Our Lady of Good Help) was founded in early 1890; its westerly location on the Sseses placed it on the route from the White Fathers' base at Rubaga and their mission south of Lake Victoria at Bukumbi. It was occupied for six months, then abandoned and refounded in September 1891, abandoned again in 1892, and ultimately transferred to Bumangi. The station at Bumangi was officially "founded," and Bugoma given up, in November 1893. M. Afr.–Rome–EE, Diaire de Rubaga, 9 Jan. 1890, *CTSMA*, no. 50, 1st trimester 1891 (April 1891); M. Afr.–Rome–EE, Notice IV.—Stations en fin 1906, *RA*, no. 1 (1905–6).

46. M. Afr.–Rome–EE, Diaire de Rubaga, July 1891, *CTSMA*, no. 54, 1st trimester 1892 (April 1892).

47. CMS ACC 611 F3, Hall, "Through My Spectacles," 66–67. Aileen Weatherhead discusses expanding CMS presence at Bugala in late 1901 and early 1902 in her journals; see CMS, ACC 558 F5, Journal of Aileen Weatherhead, Nov. 1901–Jan. M. Afr.–Rome, Diary of Sese-Bumangi Mission, 1902, 125.

48. The Ssese mission was understood as a potential refuge from war on the mainland. M. Afr.–Rome–EE, "Ouganda," Livinhac, Letter, 23 Feb. 1890, *CTSMA*, no. 47, 2nd Trimester 1890 (July 1890).

49. Megan Vaughan, *Curing Their Ills: Colonial Power and African Illness* (Cambridge: Polity Press, 1991), 58–62.

50. Vaughan, *Curing Their Ills*; John Iliffe, *East African Doctors: A History of the Modern Profession* (Cambridge: Cambridge University Press, 1998).

51. CMS work centered on Bukasa Island until 1900, when European missionaries were moved to Bugala. "Extracts of the Annual Letters of the Missionaries," Rev. E. C. Gordon, Bukasa, Sese [*sic*] Islands, 23 Nov. 1900, *Church Missionary Intelligencer and Record* 25 (November 1900): 202–4.

52. Yolanda Pringle, "Crossing the Divide: Medical Missionaries and Government Service in Uganda, 1897–1940," in *Beyond the State: The Colonial Medical Service in British Africa*, ed. Anna Greenwood (Manchester, UK: Manchester University Press, 2015), 19–24, 34n2, 35n10; Vaughan, *Curing Their Ills*.

53. Elizabeth Prevost, "Married to the Mission Field: Gender, Christianity, and Professionalization in Britain and Colonial Africa, 1865–1914," *Journal of British Studies* 47, no. 4 (October 2008): 816; CMS, ACC 558 F6, Journal of A. Weatherhead, 9 Feb. 1903, 14 Mar. 1903.

54. M. Afr.–Rome, "La maladie du sommeil aux îles Sésé. (Lettre d'un missionaire de Boumangi)," *Mission d'Alger*, no. 177 (May 1905–January 1906): 289.

55. M. Afr.–Rome–EE, Bumangi—Notre Dame de Bon Secours, "Commission pour l'étude de la maladie du sommeil," Nov. 1906, *CTSMA*, no. 138 (May–June 1907).

56. On saltpeter, see M. Afr.–Rome–EE, Diaire de Notre-Dame de Kamoga, 4 May 1887, *CTSMA*, no. 36 (October 1887). On calaya, see, for instance, M. Afr.–Rome–EE, Diaire de Ushirombo—Notre-Dame Auxiliatrice, Apr. 1900, *CTSMA*, no. 88 (October 1900). Calaya was a drug derived from extract of the fruit of *Anneslea febrifuga*, a plant native to tropical Africa, according to contemporary pharmacists' manuals, and lauded in the early twentieth century as an exciting, novel antipyretic; "New Remedies," *Practical Druggist and Pharmaceutical Review of Reviews* (April 1903), 63. On citric acid, see M. Afr.–Rome–EE, Diaire de Île de Komé—Notre-Dame du Perpétuel Secours, 28 Oct. 1901, *CTSMA*, no. 94 (April 1902).

57. At Rubaga in 1885, for example, priests recorded distributing remedies for smallpox and plague; M. Afr.–Rome–EE, Diaire de Rubaga, 28 Oct. 1885, *CTSMA*, no. 36 (October 1887).

58. Cantharidin, a substance extracted from blister beetles, was known as a blistering agent and poison, with diverse uses in folk medicine. Priests appear to have used it for plague and for an unidentified illness that included fever, weakness, and cough in the early 1890s. See Quy Pham and Sharon E. Jacob, "Cantharidin: Relevance of an Ancient Remedy," *Journal of the Dermatology Nurses' Association* 8, no. 2 (March 2016): 141; Lisa Moed, Tor A. Shwayder, and Mary Wu Chang, "Cantharidin Revisited: A Blistering Defense of an Ancient Medicine," *Archives of Dermatolology* 137, no. 10 (2001): 1357–60. Regarding cantharides as treatment for kaumpuli, see M. Afr.–Rome–EE, Diaire, Ste. Marie de Rubaga, 5 May 1891, *CTSMA*, no. 54 (April 1892); as a topical treatment, see M. Afr.–Rome–EE, Diaire de Buganda, 9–10 May 1890, *CTSMA*, no. 46 (April 1890).

59. Aylward Shorter, *Cross and Flag in Africa: The "White Fathers" during the Colonial Scramble (1892–1914)* (Maryknoll, NY: Orbis Books, 2006), 115.

60. Missions provided a crucial potential refuge for people who found themselves in marginal or tenuous social positions at times of insecurity, such as women, orphans, and slaves. Marcia Wright, *Strategies of Slaves and Women: Life-Stories from East/Central Africa* (London: James Currey, 1993); Kathleen R. Smythe, "African Women and White Sisters at the Karema Mission Station, 1894–1920," *Journal of Women's History* 19, no. 2 (June 2007): 59–84; Vaughan, *Curing Their Ills*, 61–64.

61. Kaumpuli's mother was the goddess Naku; Naku was, according to Kagwa (*Customs*, 116, 121), a second or third wife of lubaale Mukasa. See also Roscoe, *Baganda*, 309. Kodesh notes that, by the nineteenth century, Ndahura (Kawali) was associated with irruptions of the skin equated with smallpox or yaws in *Royal Gaze*, 122–24.

62. Kagwa, *Customs*, 121–22.

63. George L. Pilkington, *Luganda-English and English-Luganda Vocabulary* (London: Society for Promoting Christian Knowledge, 1892), 36; Cuthbert Christy, "Bubonic Plague ('Kaumpuli') in Central East Africa," *British Medical Journal* 2, no. 2237 (November 14, 1903): 1265; RCS, RCMS 124/4, Cuthbert Christy MSS, Diary of Cuthbert Christy, vol. 2, 13 Nov. 1902.

64. See M. Afr.–Rome–EE, Diaire de Rubaga, 14 Apr. 1881, 24 Apr. 1881, *CTSMA*, no. 18 (April 1883).

65. Pilkington, *Luganda-English Vocabulary*, 46, 113; W. A. Crabtree, *Elements of Luganda Grammar* (London: Society for Promoting Christian Knowledge, 1902), 169, 182, 209. See also M. Afr.–Rome, Sese-Bumangi Diary, 18 Oct. 1904, 21 Nov. 1904, 27 Nov. 1904. Dysentery, by comparison, is mentioned in a letter, sent from Villa Mariya (Buddu) by Fr. P. Bresson to Mgr. Hirth, 26 May 1895, M. Afr.–Rome–EE, *CTSMA*, no. 68, 3rd Trimester 1895 (October 1895).

66. Pilkington, *Luganda-English Vocabulary*, 36. George L. Pilkington, a CMS missionary, further noted, "blisters on the swollen parts, and an emetic, is the treatment which has had most success" for kaumpuli.

67. Andrew Cunningham, "Transforming Plague: The Laboratory and the Identity of Infectious Disease," in *The Laboratory Revolution in Medicine*, ed. Andrew Cunningham and Perry Williams (Cambridge: Cambridge University Press, 1992); Monica H. Green, "Taking 'Pandemic' Seriously: Making the Black Death Global," in *Pandemic Disease in the Medieval World: Rethinking the Black Death*, vol. 1, ed. Monica H. Green (Kalamazoo, MI: Arc Medieval Press, 2015), 35–45.

68. Christy, "Bubonic Plague," 1265.

69. Philip D. Curtin, "The White Man's Grave: Image and Reality, 1780–1850," *Journal of British Studies* 1, no. 1 (November 1961): 94–110; Curtin, *Death by Migration: Europe's Encounter with the Tropical World in the Nineteenth Century* (Cambridge: Cambridge University Press, 1989); and K. David Patterson, "Disease and Medicine in African History: A Bibliographical Essay," *History in Africa* 1 (1974): 141–48 are fundamental works on Africa. See also Alfred W. Crosby, *The Columbian Exchange: Biological and Cultural*

Consequences of 1492 (Westport, CT: Greenwood, 1972); William H. McNeil, *Plagues and Peoples* (New York: Anchor Books, 1989); James L. A. Webb, Jr., "Historical Epidemiology and Infectious Disease Processes in Africa," *Journal of African History* 54, no. 1 (March 2013): 3–10; Webb, "The Historical Epidemiology of Global Disease Challenges," *Lancet* 385, no. 9965 (January 24, 2015): 322–23; Monica H. Green, ed., *Pandemic Disease in the Medieval World: Rethinking the Black Death*, vol. 1 (Kalamazoo, MI: Arc Medieval, 2015); Tamara Giles-Vernick and Stephanie Rupp, "People, Great Apes, Disease, and Global Health in the Northern Forests of Equatorial Africa," in *Global Health in Africa: Historical Perspectives on Disease Control*, ed. Tamara Giles-Vernick and James L. A. Webb, Jr. (Athens: Ohio University Press, 2013); and K. David Patterson, *Pandemic Influenza, 1700–1900: A Study in Historical Epidemiology* (Totowa, NJ: Rowman and Littlefield, 1986).

70. See, for example, Shane Doyle, *Before HIV: Sexuality, Fertility and Mortality in East Africa, 1900–1980* (Oxford: Oxford University Press, 2013), chap. 2; Jan Kuhanen, *Poverty, Health, and Reproduction in Early Colonial Uganda* (Joensuu, Finland: Joensuu University Publications, 2005), chaps. 2–6; Médard, *Le royaume du Buganda*.

71. Myron J. Echenberg, *Africa in the Time of Cholera: A History of Pandemics from 1817 to the Present* (Cambridge: Cambridge University Press, 2011), chap. 3.

72. Michael Worboys, *Spreading Germs: Disease Theories and Medical Practice in Britain, 1865–1900* (Cambridge: Cambridge University Press, 2000); Bruno Latour, *The Pasteurization of France* (Cambridge, MA: Harvard University Press, 1988); Latour, "Give Me a Laboratory and I Will Raise the World," in *Science Observed: Perspectives on the Social Study of Science*, ed. Karin Knorr-Cetina and Michael Mulkay (London: Sage, 1983); Andrew Cunningham, "Transforming Plague."

73. This subordination, I emphasize, was not successful or complete. Julie Livingston, "Productive Misunderstandings and the Dynamism of Plural Medicine in Mid-century Bechuanaland," *Journal of Southern African Studies* 33, no. 4 (December 2007): 801–10; Julie Livingston, *Debility and the Moral Imagination in Botswana* (Bloomington: Indiana University Press, 2005), chap. 4; Nancy Rose Hunt, *A Colonial Lexicon: Of Birth Ritual, Medicalization, and Mobility in the Congo* (Durham, NC: Duke University Press, 1999).

74. Here, I differ from Médard's argument in "La peste," 88.

75. Kodesh, *Royal Gaze*; Steven Feierman, *The Shambaa Kingdom: A History* (Madison: University of Wisconsin Press, 1974).

76. Pilkington, *Luganda-English Vocabulary*, 61. Kodesh works with Walumbe as the "personification of death" in *Royal Gaze*, 51–59.

77. See M. Afr.–Rome–EE, Diaire de Rubaga, 24 Apr. 1881, *CTSMA*, no. 18 (April 1883).

78. M. Afr.–Rome–EE, Diaire de Rubaga, 24 Apr. 1881, *CTSMA*, no. 18 (April 1883).

79. Kagwa, *Customs*, 121–22; Roscoe, *Baganda*, 102.

80. Kagwa, *Customs*, 121–22. Kagwa asserts Kaumpuli's medium would say, after an epidemic, "Come! Now I have forgiven you and I will take you back to your homes." Roscoe describes extensive processes of marking homes of the sick, removing the unaffected, and purification of homes by priests of Kaumpuli, *Baganda*, 309–11.

81. Kagwa, *Customs*, 121–22.

82. Kagwa, 121–22; Roscoe, *Baganda*, 102.

83. Kagwa, *Customs*, 120. The concept and definition of "plague" itself was slippery in the late nineteenth century. Kagwa's reference to glands and Roscoe's parallel mention of rat involvement indicate that, in their time, kaumpuli had begun to cohere as an analog to bubonic plague. See John Roscoe, "Further Notes on the Manners and Customs of the Baganda," *Journal of the Anthropological Institute of Great Britain and Ireland* 32 (January–June 1902): 27; and A. B. K. Kasozi, "The Impact of Islam on Ganda Culture, 1844–1894," *Journal of Religion in Africa* 12, no. 2 (1981): 128. John C. Ssekamwa, a Ganda ethnographer, notes *kawumpuli* as a lubaale in "Witchcraft in Buganda Today," *Transition* 30 (April–May 1967): 30. English and French interpretations of Kaumpuli as the "god of plague" seem to have shifted over time as meanings of *peste/fléau* or plague also changed. Meanings of *fléau* or plague by Francophone and Anglophone missionaries appear to have grown more rigid in the late nineteenth century, shifting from a generalized scourge, epidemic, or disaster to mean a specific biological and medical entity, bubonic plague.

84. NAB CO 685/1, "Annual Report of the Principal Medical Officer, Uganda Protectorate, 1907," 1 Jan. 1908, 33.

85. Médard, "La peste," 88.

86. M. Afr.–Rome–EE, Diaire de Rubaga, 14 Apr. 1881, *CTSMA*, no. 18 (April 1883).

87. J. Howard Cook, "Notes on Cases of 'Sleeping Sickness' Occurring in the Uganda Protectorate," *Journal of Tropical Medicine* 4 (July 15, 1901): 236–39.

88. M. Afr.–Rome, Mission de Notre Dame de Bon Secours, Bumangi, Pére Ramond; Nyanza Septentrional, Sese Correspondence, Dossier 87, 087469–70.

89. M. Afr.–Rome, Nyanza Septentrional, Sese Correspondence, Dossier 87, 087470, Mission de Notre Dame de Bon Secours, Bumangi, Pére Ramond, Questionnaire, 1901–1902. Some of the supposed six thousand patients could have been repeat visits by the same individuals.

90. "Extracts from the Annual Letters of the Missionaries," Rev. H. T. C. Weatherhead, Bugala, Nov. 1903, *Church Missionary Intelligencer and Record* 28 (November 1903): 177.

91. Pilkington, *Luganda-English Vocabulary*, 6; H. Lyndhurst Duke, "On the Zoological Status of the Polymorphic Mammalian Trypanosomes of Africa and Their Relation to Man," *Parasitology* 13, no. 4 (November 1921): 358; Cook, "Sleeping Sickness," 236.

92. Pilkington, *Luganda-English Vocabulary*, 6.

93. See, for example, WL, RST/G27/8, George Wilson, Confidential Report, 3 Nov. 1904.

94. Kuhanen, *Poverty, Health, and Reproduction*, 251; Zeller, "Establishment," 163.

95. *Mongota*, in Luganda, also appeared in later-noted Lunyoro names for the illness, such as *endwala ya mongota* (mongota illness/disease). A. R. Dunbar, *A History of Bunyoro-Kitara* (Nairobi: Oxford University Press, 1965), 1.

96. I thank Neil Kodesh for this insight.

97. Duke, "On the Zoological Status," 358. Duke's writing on the historical epidemiology of human African trypanosomiasis in Uganda is somewhat convoluted, reflecting his own attempts to research the epidemic's origins and surmise its chronology. He appears to rely heavily on Kagwa and Dr. A. H. Cook, as well as some conversations with Ganda-speaking residents; these latter conversations may indeed have been mediated by Kagwa himself.

98. In modern Luganda, *bongota* (rel. to mongota) and *simagira* are translated as "to sleep." *Mongota* (Luganda) and *isimagira* (Oluhaya) presently imply sleepiness, drowsiness, and sleep and are currently glosses for sleeping sickness (human African trypanosomiasis). Further research might reveal whether shifts in the meaning of *bongota* and *simagira* could be traced to their association with particular diseases or symptoms. Ronald A. Snoxall, *Luganda–English Dictionary* (London: Oxford University Press, 1967), glosses *kù-bongoota* as both to "be drowsy" and to "have sleeping sickness" (25).

99. M. Afr.–Rome, Sese-Bumangi Diary, 18 Mar. 1902.

100. WL, RST/G27/4, E. D. W. Greig to David Bruce, 14 Apr. 1904. This point is echoed by Robert Koch in BArch R1001/5895, draft, "Bericht über die Tätigkeit," 15 Oct. 1906.

101. Christy, "Bubonic Plague," 1265.

102. "Extracts from the Annual Letters of the Missionaries," Rev. H. T. C. Weatherhead, Bugala, 18 Nov. 1904, *Church Missionary Intelligencer and Record* 29 (November 1904): 524.

103. Richard J. Reid, *Modern Uganda*, 160.

104. Lyons argues in *Colonial Disease* that people in Uele associated sleeping sickness with the advent of Belgian colonization and Free State labor and resource demands.

105. Johnson ("Fishwork," 239) argues that historically on the Sseses family (and even household) could encompass large numbers of people. I anticipate that the reconfiguration of life amid mongota would have involved these wider networks of kinship and relation.

106. M. Afr.–Rome–EE, Sacrés-Cœurs de Jésus et de Marie–Entebbé, Letter from Fr. Puel to Mgr. Livinhac, 3rd trimester 1902, *CTSMA*, no. 97 (April 1903).

107. CMS, ACC 558 F6, Journal of A. Weatherhead, 13 Nov. 1902; RCS, RCMS 124/4, Diary of Cuthbert Christy, 12 Nov. 1902.

108. CMS, ACC 558 F4, Memoirs of Henry Weatherhead, 57–8; CMS, ACC 558 F6, Journal of A. Weatherhead, 10 Dec. 1902.

109. CMS, ACC 558 F4, Memoirs of Henry Weatherhead, 57.

110. M. Afr.–Rome, Sese and Bumangi Correspondence, Dossier 87, 087455–56, Letter from Fr. Ramond to Mgr. (Hirth?), 24 May 1903.

111. M. Afr.–Rome, Sese-Bumangi Diary, June/July/Aug. 1903.

112. Uganda National Archives/Entebbe Secretariat Archives/Buganda Correspondence Inward A-8, Ebitegeza Bwebanaziizanga Mongoto Okubana (Sleeping Sickness Native Regulations and Precaution), Mengo, May, 1902, translated by and quoted in Zeller, "Establishment," 163n2. Zeller also cites "Endwadde eyali Ettula Mongota," *Munno* 7 (January 1917), regarding belief that sleeping sickness was caused by fishes, because the sick lived near the lake shore or on islands (163n2).

113. NAB FO 2/828, 328, David Bruce to Commissioner Hayes-Sadler, 28 Apr. 1903.

114. Kirk A. Hoppe, *Lords of the Fly: Sleeping Sickness Control in British East Africa, 1900–1960* (Westport, CT: Praeger, 2003), chap. 3.

115. On bigenge, see NAB CO 685/1, Appendix to the Annual Medical Report for the Uganda Protectorate, 1908, Van Someren, "On the Treatment of Leprosy by Soamin," 115. On leprosy in colonial Uganda in the interwar period, see Kathleen Vongsathorn, "'First and Foremost the Evangelist'? Mission and Government Priorities for the Treatment of Leprosy in Uganda, 1927–1948," *Journal of Eastern African Studies* 6, no. 3 (August 2012): 544–60.

116. M. Afr.–Rome, Sese-Bumangi Diary, 20–21 Nov. 1904.

117. Kagwa, *Customs*, 115.

118. RKI Fotoarchiv 6105, Fotoalbum Koch in Sese, 6105036, "Krankenlager bei Bugala."

119. See RCS, Y304A, Christy Collection of Photographs on Africa, Image 51.

120. The root verb *-byam* is glossed as "to sleep, to lie down"; I thank Neil Kodesh for his insight on parsing *lubyamira/lubyamwa*. Schoenbrun, *Historical Reconstruction*, 205; Roger Botte, "Rwanda and Burundi, 1889–1930: Chronicle of a Slow Assassination, Part I," *International Journal of African Historical Studies* 18, no. 1 (1985): 54n7. See M. Afr.–Rome–EE, Letter from Fr. A. Brard to Mgr. Hirth, 15 Nov. 1892, *CTSMA*, no. 58, 1st Trimester 1893 (April 1893).

121. Lyons, *Colonial Disease*, chap. 4.

122. Feierman, "Struggles for Control," and *Peasant Intellectuals: Anthropology and History in Tanzania* (Madison: University of Wisconsin Press, 1990), chaps. 3 and 4; Kodesh, *Royal Gaze*.

123. M. Afr.–Rome, Sese Correspondence (Nyanza Septentrional), Dossier 87, 087457–59, Fr. Reynès, 25 July 1904.

124. M. Afr.–Rome–EE, Fr. Reynès, Bumangi Annual Report, *RA*, no. 1 (1905–6).

125. M. Afr.–Rome–EE, Fr. Reynès, Bumangi Annual Report, *RA*, no. 1 (1905–6).

126. It is possible that work to meet accelerating demands for an Italian concession focused on rubber, hides, and coffee might have increased opportunities for exposure to the parasite. See, for instance, M. Afr.–Rome–EE, Report, Bumangi—N.-D. de Bon-Secours, *CTSMA*, no. 103 (February 1904). The potential impact of colonial-era pressure to exploit rubber in changing a population's exposure to tsetse flies is strongly suggested by the history of sleeping sickness in the Congo Free State, despite the differences in Leopoldian and British regimes.

127. M. Afr.–Rome–EE, Fr. Reynès, Bumangi Annual Report, *RA*, no. 1 (1905–6).

128. M. Afr.–Rome–EE, Fr. Reynès, Bumangi Annual Report, *RA*, no. 1 (1905–6).

129. M. Afr.–Rome–EE, Fr. Reynès, Bumangi Annual Report, *RA*, no. 1 (1905–6).

130. Setting people apart in the forest raises interesting questions about demarcation of village and home from forests, some of which were purportedly used for the burial of the dead by non-elites and were reputedly spaces where ancestral and other spirits resided. See M. Afr.–Rome–EE, Sese-Bumangi Annual Report, 1906, *RA*, no. 2 (1906–7).

131. M. Afr.–Rome, "La maladie du sommeil aux îles Sésé. (Lettre d'un missionaire de Boumangi)," *Mission d'Alger*, no. 177 (May 1905–January 1906): 289.

132. M. Afr.–Rome, "La maladie du sommeil aux îles Sésé. (Lettre d'un missionaire de Boumangi)," *Mission d'Alger*, no. 177 (May 1905–January 1906): 290.

133. M. Afr.–Rome, "La maladie du sommeil aux îles Sésé. (Lettre d'un missionaire de Boumangi)," *Mission d'Alger*, no. 177 (May 1905–January 1906): 290. See Hoppe, *Lords of the Fly*, 74n69, regarding use of British camps as a last resort.

134. Hoppe, *Lords of the Fly*, 75–76, suggests that care at the mission was a possible solution for coping with sick relatives when they displayed the "madness" that occasionally accompanied the late stages of the disease.

135. WL, RST/G27/4, E. D. W. Greig to David Bruce, 4/14/1904.

136. BArch R1001/5895, Koch, Draft of "Bericht über die Tätigkeit," 15 Oct. 1906, 2. On the Sese-Bumangi hospital, see M. Afr.–Rome, "La maladie du sommeil aux îles Sésé. (Lettre d'un missionaire de Boumangi)," *Mission d'Alger*, no. 177 (May 1905–January 1906), 289.

137. Quoted in BArch R1001/5895, Koch, Draft of "Bericht über die Tätigkeit," 15 Oct. 1906, 3.

138. M. Afr.–Rome–EE, Fr. Reynès, Bumangi Annual Report, 1905, *RA*, no. 1 (1905–6).

139. "Extracts from the Annual Letters of the Missionaries," Hugh Savile, Bugala, 18 Feb. 1905, *Church Missionary Intelligencer and Record* 30 (February 1905): 525. Richard J. Reid also notes that, with the coming of Ganda suzerainty on the islands, the Ganda likely "appointed chiefs from among their own governing class to manage Sesse affairs"; *Pre-colonial Buganda*, 238.

140. M. Afr.–Rome, Sese-Bumangi Diary, 23 July 1906.

CHAPTER 2: HEALING *MONGOTA*, TREATING TRYPANOSOMIASIS

1. Jan Kuhanen, *Poverty, Health, and Reproduction in Early Colonial Uganda* (Joensuu, Finland: Joensuu University Publications, 2005), 254–55.

2. Deborah Neill, *Networks in Tropical Medicine: Internationalism, Colonialism, and the Rise of a Medical Specialty, 1890–1930* (Stanford, CA: Stanford University Press, 2012).

3. Mari K. Webel, "Borderlands of Research: Medicine, Empire, and Sleeping Sickness in East Africa, 1902–1914" (PhD diss., Columbia University, 2012), 37–42, 55. On nagana, see B. I. Williams, "African Trypanosomiasis," in *The Wellcome Trust Illustrated History of Tropical Medicine*, ed. F. E. G. Cox (London: Wellcome Trust, 1996), 184–87. Drs. David Bruce and David Nabarro and Capt. Edward Greig concluded that infection with a trypanosome parasite caused sleeping sickness, and that a local species of biting fly, *Glossina palpalis*, carried the trypanosome. Bruce and Nabarro arrived in Uganda March 16, 1903, overlapping for several weeks with Dr. Aldo Castellani, one of the Sleeping Sickness Commission's original three members; Greig joined them on May 25, 1903. Though acknowledging Castellani's prior work, Bruce's team claimed discovery of the trypanosome as the causative agent of sleeping sickness, touching off a controversy that would persist until Castellani's death in 1971. See John Boyd, "Sleeping Sickness: The Castellani-Bruce Controversy," *Notes and Records of the Royal Society of London* 28, no. 1 (June 1973): 93–110.

4. Thomas D. Brock, *Robert Koch: A Life in Medicine and Bacteriology* (Washington, DC: ASM Press, 1999), chap. 18; Christoph Gradmann, *Laboratory Disease: Robert Koch's Medical Bacteriology* (Baltimore: Johns Hopkins University Press, 2009), chap. 3.

5. In the last sixteen years of his life, Koch spent nearly eight years on research expeditions away from Germany, the majority in southern and eastern Africa. After remarrying in 1893, Koch was alive for another sixteen years, eight months. Of that time, he spent seven years, eleven months abroad. Measured after he began his research in Africa in 1896, he was in Germany for only five years, eight months in the subsequent twelve years until his health likely began to fail in 1908. I have tabulated these figures using data from Bernhard Möllers, *Robert Koch: Persönlichkeit und Lebenswerk* (Hannover: Schmorl and Von Seefeld, 1951); Brock, *Robert Koch*; and items from the *Lancet* and *British Medical Journal*.

6. Brock, *Robert Koch*, 262–63; Robert Koch, "Vorläufige Mitteilungen über die Ergebnisse einer Forschungsreise nach Ostafrika," in *Gesammelte Werke von Robert Koch*, ed. Georg Gaffky and Eduard Pfuhl (Leipzig: Thieme, 1912), vol. 2, part 1, 477, 486; Koch to Gaffky, 3 May 1905, and Koch to Gaffky, 25 June 1905, quoted in Möllers, *Robert Koch*, 298.

7. Neill, *Networks in Tropical Medicine*; Gradmann, *Laboratory Disease*.

8. Webel, "Borderlands," 79–80, 157–63; Neill, *Networks in Tropical Medicine*.

9. My narrative of Koch's expedition follows the reports he drafted and published in a series between October 1906 and September 1907, found in

Koch's *Gesammelte Werke*; see Friedrich Kleine, *Ein deutscher Tropenarzt* (Hannover: Schmorl and Von Seefeld Nachf., 1949); Möllers, *Robert Koch*; Brock, *Robert Koch*; Gradmann, *Laboratory Disease*. Gradmann argues that travel to Africa late in life was important for Koch both personally and professionally, 171–78.

10. BArch R1001/5895, Koch, Draft of "Bericht über die Schlafkrankheits-Expedition während des Aufenthalts in Muansa," 31 July 1906, 13; BArch R901/20872, Posadowsky (Ministry of Interior) to Minister of the Foreign Office, 3 Mar. 1906.

11. Gradmann, *Laboratory Disease*, 208–10.

12. NAB/CO 885/9/11, A. H. Milne (Liverpool School of Tropical Medicine) to Colonial Office, 2 May1906, 42–44; NAB/CO 885/9/11, A. H. Milne to Colonial Office, 20 July 1906; NAB/CO 885/9/11, Colonial Office to Liverpool School of Tropical Medicine, 3 Aug. 1906, 50; NAB/CO 885/9/11, R. N. Moffat to Colonial Office, 9 Aug. 1906, 51; WL, RST/G27/40, A.D.P Hodges to David Bruce, 31 Oct. 1907; Kirk A. Hoppe, *Lords of the Fly: Sleeping Sickness Control in British East Africa, 1900–1960* (Westport, CT: Praeger, 2003), 55–79.

13. On the expedition's initial itinerary, see BArch R1001/5895, R. Koch, Draft, "Bericht über die Tätigkeit," 15 Oct. 1906, 1. On the laboratory, see NAB/CO 885/9/11, Hesketh-Bell to Colonial Office, 23 Nov. 1906, 104.

14. LSHTM, Hodges MSS, GB 0809 Hodges/01/10, A.D.P. Hodges Diary, 11 Aug. 1906, 16 Aug. 1906.

15. Dietmar Steverding, "The Development of Drugs for Treatment of Sleeping Sickness: A Historical Review," *Parasites and Vectors* 3, no. 15 (2010): 1–9; Enrique Raviña, *The Evolution of Drug Discovery: From Traditional Medicines to Modern Drugs* (Weinheim, Ger.: Wiley-VCH Verlag, 2011), 40–41; Monika Mölders, "Formeln gegen Flagellaten—Der pharmakologische Kampf gegen die afrikanische Trypanosomiasis, 1900 bis 1950" (PhD diss., Ruprecht-Karls-Universität Heidelberg, 2003), 75–84.

16. Steverding, "Development," 3.

17. Neill, *Networks in Tropical Medicine*; Deborah Neill, "Paul Ehrlich's Colonial Connections: Scientific Networks and Sleeping Sickness Drug Therapy Research, 1900–1914," *Social History of Medicine* 22, no. 1 (April 2009): 61–77.

18. On methylene blue, see WL, RST/G27/2, Greig to Bruce, 10 Sept. 1903; on using prisoners for sleeping sickness drug experiments, see WL, RST/G27/4, Greig to Bruce, 10 June 1904, Greig to Bruce, 7 July 1904; WL, RST/G27/9, Gray to Bruce, 19 Jan. 1905; NAB CO 885/9/11, Royal Society to Colonial Office, 3 Apr. 1905, 1–2; NAB CO 885/9/11, Hayes-Sadler to Colonial Office, 13 June 1905, 13; Neill, "Paul Ehrlich's Colonial Connections," 61–77.

19. NAB/CO 885/9/7, A. H. Milne (Liverpool School of Tropical Medicine) to Colonial Office, 9 June 1906, 72–73. See, for example, John L. Todd, "The Treatment of Human Trypanosomiasis by Atoxyl," *British Medical Journal* 1, no. 2366 (1906): 1037; Gustave Martin, Alexis Leboeuf, and Émile

Roubaud, *Rapport de la Mission d'Études de la Maladie du Sommeil au Congo Français*, 1906–08 (Paris: Masson, 1909).

20. NAB CO 885/18/7, George Will (British East Africa Principal Medical Officer) to Colonial Office, 22 Dec. 1907, 10.

21. Christoph Gradmann, "'It Seemed About Time to Try One of Those Modern Medicines': Animal and Human Experimentation in the Chemotherapy of Sleeping Sickness 1905–1908," in *Twentieth Century Ethics of Human Subjects Research: Historical Perspectives on Values, Practices, and Regulations*, ed. Volker Roelcke and Giovanni Maio (Stuttgart: Fritz Steiner, 2004), 83–99.

22. Georgina H. Endfield et al., "'The Gloomy Forebodings of This Dread Disease': Climate, Famine and Sleeping Sickness in East Africa," *Geographical Journal* 175, no. 3 (September 2009): 181–95.

23. RCS, RCMS 124/4, Cuthbert Christy MSS, Diary of Cuthbert Christy, vol. 2, 12 Nov. 1902; M. Afr.–Rome–EE, Annual Report for Bumangi, 1907–1908, *RA*, no. 3 (1907–8). The White Fathers mission at Bumangi was founded in 1893; M. Afr.–Rome–EE, "IV-Stations en fin 1906," *RA*, no. 1 (1905). The Church Missionary Society (Anglican) mission at Bubembe was founded in 1896, and the mission at Bugala dates from at least 1900; Anne Clark, University of Birmingham, personal correspondence, August 26, 2010.

24. M. Afr.–Rome, *Mission d'Alger*, no. 177 (May–June 1906), Letter from "missionary of Bumangi," 289.

25. M. Afr.–Rome, Sese Correspondence (Nyanza Septentrional), Dossier 87, 087461, Delévaux to "Pierre-Marie," 29 Oct. 1906.

26. RCS, RCMS 124/4, Cuthbert Christy MSS, Diary of Cuthbert Christy, vol. 2, 12 Nov. 1902; M. Afr.–Rome–EE, Annual Report for Bumangi, 1907–1908, *RA*, no. 3 (1907–8).

27. BArch R1001/5895, R. Koch, Draft of "Bericht über die Tätigkeit," 15 Oct. 1906.

28. Gradmann, *Laboratory Disease*; Myriam Mertens, "Chemical Compounds in the Congo: Pharmaceuticals and the 'Crossed History' of Public Health in Belgian Africa (ca. 1905–1939)" (PhD diss., University of Ghent, 2014); Guillaume Lachenal, "A Genealogy of Treatment as Prevention (TasP): Prevention, Therapy, and the Tensions of Public Health in African History," in *Global Health in Africa: Historical Perspectives on Disease Control*, ed. Tamara Giles-Vernick and James L. A. Webb, Jr. (Athens: Ohio University Press, 2013), and Lachenal, *The Lomidine Files: The Untold Story of a Medical Disaster in Colonial Africa* (Baltimore: Johns Hopkins University Press, 2017).

29. See Aylward Shorter, *Cross and Flag in Africa: The "White Fathers" during the Colonial Scramble (1892–1914)* (Maryknoll, NY: Orbis Books, 2006), 115; regarding the hospital at Kisubi (nearest Entebbe), see NAB CO 885/9/11, Hesketh-Bell to Colonial Office, 23 Nov. 1906, 103–4. On relationships between researchers and missions on the Sseses, see, for instance, RCS, RCMS 124/4, Cuthbert Christy MSS, Diary of Cuthbert Christy, 10 Oct. 1902, 11 Nov. 1902. Koch had visited Kisubi and the Sseses in 1905, per BArch R1001/5895, Koch, Draft of "Bericht über die Tätigkeit," 15 Oct. 1906.

30. M. Afr.–Rome, Sese Correspondence (Nyanza Septentrional), Dossier 87, 087461, Delévaux to "Pierre-Marie," 29 Oct. 1906.

31. While missionaries generally welcomed the expedition, friction between the White Fathers and German doctors developed periodically in September and October 1906 regarding the camp potentially superseding the Bumangi hospital. See M. Afr.–Rome, Sese-Bumangi Diary entries for October–November 1906.

32. Koch's reports from the 1906–7 expedition were published serially in the *Deutsche medizinische Wochenschrift* and collected in Gaffky and Pfuhl, *Gesammelte Werke von Robert Koch*.

33. Cunningham refers to the high ground as "Mount Bugalla" and "Bugalla hill." Fort Stanley, the British station, sat below on the Lutoboka peninsula. James Francis Cunningham, *Uganda and Its Peoples: Notes on the Protectorate of Uganda, Especially the Anthropology and Ethnology of Its Indigenous Races* (London: Hutchinson, 1905), 76; CMS, ACC 558 F6, Journal of A. Weatherhead, 13 Nov. 1902, indicates that the missionaries at Bugalla village could see canoes arriving from the station and communicated readily with those stationed at the "fort." See also RCS, Y3045C, Fisher Collection, "Sketch Map of Western Uganda" (1904).

34. "Extracts of the Annual Letters of the Missionaries," Rev. E. C. Gordon, Bukasa, Sese [sic] Islands, 23 Nov. 1900, *Church Missionary Intelligencer and Record* 25 (November 1900): 203.

35. CMS, ACC 558 F6, Journal of A. Weatherhead, 9 Feb. 1903.

36. BArch R1001/5895, "Lager Bugalla" map, with report dated 27 Nov. 1906.

37. M. Afr.–Rome, Sese-Bumangi Diary, 11 Aug. 1906. "Semugala" (or Semagala) was noted as the title of one of the islands' principal chiefs by John Roscoe, *The Baganda: An Account of Their Native Customs and Beliefs* (London: Macmillan, 1911), 294. Semagala complained to the Bumangi priests about Koch's incursion onto his land to and traveled to Kampala to bring complaints to authorities there. The "Baswahili" men employed by the Germans were also accused shortly after their arrival of "debauching" three local women. Bumangi diarists recorded these controversies accompanying the arrival of Koch's expedition in detail, until receiving a letter from their father superior with encouragement to assist and cooperate on 22 August, following which accounts of the expedition's work turned generally more favorable (if measured as to atoxyl's efficacy). Koch also offered the priests German wine and cigars, engaged with them socially, and explained his research in the context of successes at the Pasteur Institute.

38. BArch R1001/5895, R. Koch, "Bericht über die Tätigkeit der Schlafkrankheits-Expedition bis zum 25. November 1906," 27 Nov. 1906.

39. BArch R1001/5895, Koch, Draft of "Bericht über die Tätigkeit," 15 Oct. 1906, 9.

40. RKI Fotoarchiv 6105, Fotoalbum Koch in Sese, 6105005.

41. BArch R1001/5895, Koch, Draft of "Bericht über die Tätigkeit," 15 Oct. 1906, 9.

42. RKI Fotoarchiv 6105, Fotoalbum Koch in Sese, 6105037, "Durch Abholzen freigemachter Landungsplatz / Provisorisches Kranken-Lager bei Bugala."

43. BArch R1001/5896, Koch, Draft, "Bericht über die Tätigkeit der Schlafkrankheits-Expedition bis zum 25. April 1907," 25 Apr. 1907.

44. Ziba fishermen occupied camps around the Sseses, particularly on the northern peninsula of Bugala Island (called Buninga). BArch R1001/5896, Koch, "Bericht über die Tätigkeit," 25 April 1907; RKI AS/w6/005, Draft, Letter from Koch, Ssese Islands, 12 May 1907.

45. M. Afr.–Rome, Sese-Bumangi Diary, 1 Dec. 1906. Mission access to boats or rowers may have facilitated travel in groups.

46. BArch R1001/5895, Koch, Draft of "Bericht über die Tätigkeit," 15 Oct. 1906, 3; see RKI Fotoarchiv 6105, Fotoalbum Koch in Sese, 6105003, 6105004, 6105005, 6105006 and RKI Friedrich Karl Kleine Nachlass, Box 2.

47. BArch R1001/5895, Koch, Draft of "Bericht über die Tätigkeit," 15 Oct. 1906, 9.

48. BArch 1001/5895, Koch, Draft of "Bericht über die Tätigkeit," 15 Oct. 1906, 7.

49. BArch R86/ 2613, Appendix to "Betrifft Expedition zur Erforschung der Schlafkrankheit. Auf den Erlass vom 1. Februar 1906," Max Beck to Bumm (Director, Imperial Health Institute), 29 Feb. 1906. Atoxyl was transported in powdered form and then mixed with distilled water for subcutaneous injection.

50. M. Afr.–Rome, Sese-Bumangi Diary, 19 Sept. 1906.

51. BArch R1001/5895, Koch, Draft of "Bericht über die Tätigkeit," 15 Oct. 1906, 7–8.

52. Webel, "Borderlands," chapter 3.

53. BArch R1001/5895, Koch, Draft of "Bericht über die Tätigkeit," 15 Oct. 1906, 7–8; BArch R1001/5896, Koch, Draft, "Bericht über die Tätigkeit," 26 Apr. 1907, 6. Gray offers an extensive description of atoxyl injections in NAB CO 885/19/14, "Report on the Sleeping Sickness Camps, Uganda, from December, 1906, to November, 1907," 6 Dec. 1907; collected in Gray, "Reports on the Sleeping Sickness Camps, Uganda, and on the Medical Treatment of Sleeping Sickness Patients at the Segregation Camps, from December, 1906 to January, 1908," in *Reports of the Sleeping Sickness Commission of the Royal Society* 9, no. 24 (1908), see 73.

54. Atoxyl was given along an increasing gradient from the first 0.06 gram dose; Koch indicates the first few doses increased by 0.02 grams each time, but does not detail whether this same increase was maintained between 0.10 and 0.50 grams. BArch R1001/5895, Koch, Draft of "Bericht über die Tätigkeit," 15 Oct. 1906, 7.

55. BArch R86/2613, Koch, Draft, "Bericht über die Tätigkeit," 27 Nov. 1906, 6–8.

56. BArch R86/2613, Koch, Draft, "Bericht über die Tätigkeit," 27 Nov. 1906, 10.

57. Enthusiasm for injections presents its own set of questions, touched upon by Luise White in *Speaking with Vampires: Rumor and History in Colonial Africa* (Berkeley: University of California Press, 2000), 99–100. More specifically, on injections and healing in colonial Tanganyika, see Patrick

Thomas Malloy, "Holding [Tanganyika] by the Sindano": Networks of Medicine in Colonial Tanganyika" (PhD diss., University of California–Los Angeles, 2003). On injections with atoxyl for sleeping sickness, see Nancy Rose Hunt, *A Colonial Lexicon: Of Birth Ritual, Medicalization, and Mobility in the Congo* (Durham, NC: Duke University Press, 1999), 92–96; on injections for yaws, 96–99.

58. BArch R1001/5895, Koch, Draft of "Bericht über die Tätigkeit," 15 Oct. 1906, 9. NAB CO 885/9/11, Hesketh-Bell to Colonial Office, 23 Nov. 1906, 105.

59. Neill, *Networks in Tropical Medicine*, cites earlier evidence of people clamoring for a cure to A.D.P. Hodges' station at Jinja in 1902, 108–9.

60. M. Afr.–Rome, Nyanza Septentrional, Dossier 87, Sese and Bumangi Correspondence, 087460-62, Letter from Delévaux, 29 Oct. 1906.

61. BArch R1001/5895, Koch, Draft of "Bericht über die Tätigkeit," 15 Oct. 1906; M. Afr.–Rome, Sese-Bumangi Diary, 31 Oct. 1906.

62. M. Afr.–Rome, Nyanza Septentrional, Dossier 87, Sese and Bumangi Correspondence, 087464, Letter from Fr. Bec, 12 Dec. 1906.

63. BArch R1001/5896, Koch, "Bericht über die Tätigkeit," 25 Apr. 1907.

64. BArch R86/2613, Koch, Draft, "Bericht über die Tätigkeit," 27 Nov. 1906, 1.

65. M. Afr.–Rome, Nyanza Septentrional, Dossier 87, Sese and Bumangi Correspondence, 087460-62, Letter from Delévaux, 29 Oct. 1906; M. Afr.–Rome–EE, "Diaires arriérés—Mitala-Mariya," September 1906, *CTSMA*, no. 138 (May–June 1907).

66. BArch R1001/5896, Koch, "Bericht über die Tätigkeit," 25 April 1907, 11.

67. M. Afr.–Rome–EE, Diaires du 4e Trimestre 1906 (suite), Bumangi, 1st Diaire, 6 Jan. 1907 and Feb. 1907, *CTSMA*, no. 140 (August 1907); Bumangi, Diary for May 1907, *CTSMA*, no. 143 (November 1907).

68. M. Afr.–Rome–EE, Sese-Bumangi Diary, November 1906, *CTSMA*, no. 138 (May–June 1907).

69. M. Afr.–Rome, Sese-Bumangi Diary, 10 Nov. 1906.

70. M. Afr.–Rome, Sese-Bumangi Diary, 15 Nov. 1906.

71. M. Afr.–Rome, Sese-Bumangi Diary, 15 Nov. and 18 Dec. 1906.

72. Kenny refers to the difficulties in understanding the function of Mukasa's shrine in particular in "Mutesa's Crime: Hubris and the Control of African Kings," *Comparative Studies in Society and History* 30, no. 4 (October 1988): 603.

73. Luise White, "'They Could Make Their Victims Dull': Genders and Genres, Fantasies and Cures in Colonial Southern Uganda," *American Historical Review* 100, no. 5 (December 1995): 1396.

74. M. Afr.–Rome–EE, Diaire de Bumangi (N.-D. de Bon Secours), Diaire November 1906, *CTSMA*, no. 138 (May–June 1907).

75. While the extent of co-infection with malaria and other illness is beyond the scope of my research, malaria was endemic in the region and would have had wide-ranging impacts on health. Koch observed high rates of infection among adults at the Bugalla camp in an April 25, 1907, report. Generally,

adults likely acquired functional immunity after repeated exposure to the parasite, but would have suffered active infections in times of immune stress; the impact on children would have been more acute and severe. See James L. A. Webb, Jr., *Humanity's Burden: A Global History of Malaria* (Cambridge: Cambridge University Press, 2009).

76. M. Afr.–Rome, Sese-Bumangi Diary, 16 Sept. 1906.

77. BArch R1001/5895, Koch, Draft of "Bericht über die Tätigkeit," 15 Oct. 1906, 5.

78. WL, RST/G27/4, E. D. W. Greig to David Bruce, 14 Apr. 1904, quoting White Fathers at Kisubi; Gray and Grieg, "Continuation Report on Sleeping Sickness Uganda," *Reports of the Sleeping Sickness Commission of the Royal Society* 4, no. 11 (1905): 12; BArch R1001/5895, Koch, Draft of "Bericht über die Tätigkeit," 15 Oct. 1906, 5.

79. A. C. H Gray and F. M. G. Tulloch, "Continuation Report on Sleeping Sickness in Uganda," *Reports of Sleeping Sickness Commission of the Royal Society* 8, no. 16 (February 1907): 7.

80. M. Afr.–Rome, Sese-Bumangi Diary, 5 Aug. 1905, 24 Sept. 1906. Tuck argues that kabotongo should be understood as "endemic syphilis"; his exploration of kabotongo relates to evidence for or against venereal syphilis in precolonial Buganda, and the matter of kabaka Mutesa's illnesses late in his reign; Michael W. Tuck, "Kabaka Mutesa and Venereal Disease: An Essay on Medical History and Sources in Precolonial Buganda," *History in Africa* 30 (2003): 314–18. Similarities between endemic syphilis and yaws are such that differentiation in the modern era remains challenging.

81. M. Afr.–Rome, Sese-Bumangi Diary, 3 Feb. 1905, regarding Sewoya, and entries 7 Jan., 6 Mar., 22 May 1905.

82. M. Afr.–Rome, Sese-Bumangi Diary, 16 Sept. 1906; Randall M. Packard, *Chiefship and Cosmology: An Historical Study of Political Competition* (Bloomington: Indiana University Press, 1981); Steven Feierman, *The Shambaa Kingdom: A History* (Madison: University of Wisconsin Press, 1974).

83. RCS, RCMS 124/4, Cuthbert Christy MSS, Diary of Cuthbert Christy, Nov. 11–13, 1902.

84. M. Afr.–Rome–EE, Sese-Bumangi Annual Report, 1905, *RA*, no. 1 (1905–6); M. Afr.–Rome, Letter from "a missionary of Bumangi," *Mission d'Alger*, no. 177 (May–June 1906), vol. 1, 289.

85. M. Afr.–Rome, Sese-Bumangi Diary, 12 Feb. 1906. Kabotongo was, at the time, read as venereal syphilis but is now understood to have more likely been non venereal syphilis or yaws.

86. Missionaries pointedly struggled with providing for people with kabotongo, despite their willingness to minister to the sick, because they associated kabotongo with undesirable behavior and sexual immorality.

87. M. Afr.–Rome, Sese-Bumangi Diary, 14 Mar. 1902.

88. Gray and Greig, "Continuation Report," 15–17.

89. Of those living on the Church shamba on Bugala, a cohort of seven women had the given last name "wa Muzungu" (of the European or white

person), suggesting dependence on the mission. From Bukasa, all those surveyed were on the shamba of Danieli Kaganda, a chief who actively supported the CMS. Gray and Greig, "Continuation Report," 15–17. Historically, the chief bearing the name *Kaganda* was one of the principal chiefs of the islands involved in Mukasa's ceremonies, per Roscoe, *Baganda*, 294.

90. A. C. H. Gray and F. M. G. Tulloch, "Continuation Report on Sleeping Sickness in Uganda," *Reports of Sleeping Sickness Commission of the Royal Society* 8, no. 16 (February 1907): 9.

91. David Schoenbrun, "Pythons Worked: Constellating Communities of Practice with Conceptual Metaphor in Northern Lake Victoria, ca. A.D. 800 to 1200," in *Knowledge in Motion: Constellations of Learning across Time and Place*, ed. Andrew P. Roddick and Ann B. Stahl (Tucson: University of Arizona Press, 2016), 218.

92. David L. Schoenbrun, "Ethnic Formation with Other-Than-Human Beings: Island Shrine Practice in Uganda's Long Eighteenth Century." *History in Africa* 45 (June 2018): 400–401.

93. M. Afr.–Rome, Sese-Bumangi Diary, June/July/Aug. 1903.

94. NAB CO 885/9/11, no. 100, Hesketh Bell to Colonial Office, 23 Nov. 1906, 108. Following Luise White's influential *Speaking with Vampires*, the recent work of Tappan and Graboyes points to the additional importance of understanding rumor and accusations of blood-sucking or cannibalism as, per Graboyes, "accurately describ[ing] what African participants were seeing" (132). Jennifer Tappan, *The Riddle of Malnutrition: The Long Arc of Biomedical and Public Health Interventions in Uganda* (Athens: Ohio University Press, 2017), 23–29; Melissa Graboyes, *The Experiment Must Continue: Medical Research and Ethics in East Africa, 1940–2014* (Athens: Ohio University Press, 2015), 132–33.

95. See *Reports of the Sleeping Sickness Commission of the Royal Society*, nos. 1–8.

96. Zeller argues that chiefs were involved in the siting of subsequent sleeping sickness camps, but it is unclear whether this was true of Buwanuka; Diane Leinwand Zeller, "The Establishment of Western Medicine in Buganda" (PhD diss., Columbia University, 1974), 171.

97. M. Afr.–Rome–EE, "Diaires du 2e trimestre 1907 (suite)," Rubaga (Sainte-Marie), 1 June 1907, *CTSMA* 143 (Nov. 1907); A. C. H. Gray, "Reports on the Sleeping Sickness Camps, Uganda, and on the Medical Treatment of Sleeping Sickness Patients at the Segregation Camps, from December, 1906 to January, 1908," in *Reports of the Sleeping Sickness Commission of the Royal Society* 9, no. 24 (1908): 65.

98. A. C. H. Gray, "Report on the Sleeping Sickness Camps, Uganda, from December, 1906 to November, 1907," *Reports of the Sleeping Sickness Commission of the Royal Society* 9, no. 24 (1908): 66.

99. Neill, *Networks in Tropical Medicine*, 126.

100. Gray, "Report on the Sleeping Sickness Camps, Uganda, from December, 1906 to November, 1907," 70.

101. Apolo Kagwa, *The Customs of the Baganda*, trans. Ernest B. Kalibala, ed. May M. Edel (New York: Columbia University Press, 1934), 122; A. B. K. Kasozi, "The Impact of Islam on Ganda Culture, 1844–1894," *Journal of Religion in Africa* 12, no. 2 (1981): 128; Jennifer Lee Johnson, "Fish, Family, and the Gendered Politics of Descent along Uganda's Southern Littorals," *History in Africa* 45 (June 2018): 458.

102. Holly Elisabeth Hanson, *Landed Obligation: The Practice of Power in Buganda* (Portsmouth, NH: Heinemann, 2003), 75.

103. Kiwanuka has also been understood to have powers of fertility, and particularly granting twin children, when appealed to along with his twin brother Musoke (*lubaale* of rainbows). See Peter Hoesing, "Kusamira Ritual Music and the Social Reproduction of Wellness in Uganda" (PhD diss., Florida State University, 2011), 71–77, 120–25.

104. Schoenbrun, "Ethnic Formation," 422–23.

105. Fred Espenak, "Phases of the Moon: 1901 to 1910, Universal Time," last updated December 21, 2014, http://astropixels.com/ephemeris/phasescat/phases1901.html.

106. BArch R1001/5895, Koch, Draft of "Bericht über die Tätigkeit," 15 Oct. 1906, 9.

107. M. Afr.–Rome–EE, Bumangi (N.-D. de Bon Secours), Diaire, Nov. 1906 and Dec. 1906, *CTSMA*, no. 138 (May June 1907); BArch R1001/5895, Koch, Draft of "Bericht über die Tätigkeit," 15 Oct. 1906.

108. M. Afr.–Rome, Sese-Bumangi Diary, 15 Nov. 1906 and 4 Mar. 1907.

109. M. Afr.–Rome, Sese-Bumangi Diary, 18 Nov. 1906.

110. M. Afr.–Rome–EE, Bumangi (N.-D. de Bon Secours), Diaire, Dec. 1906, *CTSMA*, no. 138 (May–June 1907); neither Koch's reports nor the mission diaries discuss a regular market in the area, but subsequent sleeping sickness camps offered markets within or nearby.

111. Endfield et al., "'Gloomy Forebodings,'" asserts the importance of famine for the severity of the Uganda epidemic.

112. M. Afr.–Rome–EE, Bumangi (N.-D. de Bon Secours), Diaire, Dec. 1906, *CTSMA*, no. 138 (May–June 1907); M. Afr.–Rome, Sese-Bumangi Diary, 1 Dec. 1906.

113. This is the same period during which the doctor posted at Bumangi treated only particular cases whose course was relevant for scientific publications, according to priests, and offered quinine to the rest. Koch mentions stocks of atoxyl being nearly exhausted two to three months after most patients began treatment, i.e., December–January, in BArch R1001/5896, "Bericht über der Tätigkeit," 25 Apr. 1907, 1.

114. M. Afr.–Rome, Sese-Bumangi Diary, 7–14 Jan. 1907.

115. BArch R1001/5896, Koch, "Bericht über der Tätigkeit," 25 Apr. 1907; M. Afr.–Rome, Sese-Bumangi Diary, 19 Sept. 1906.

116. BArch R1001/5896, Koch, "Bericht über der Tätigkeit," 25 Apr. 1907.

117. Neill, *Networks in Tropical Medicine*, 114.

118. BArch R1001/5896, Koch, "Bericht über der Tätigkeit," 25 Apr. 1907, 5.

119. M. Afr.–Rome–EE, Bumangi (N.-D. de Bon Secours), February 1907, *CTSMA*, no. 140 (August1907).

120. BArch R1001/5896, Koch, "Bericht über der Tätigkeit," 25 Apr. 1907, 6.

121. M. Afr.–Rome, Sese-Bumangi Diary, 14 Mar. 1907.

122. BArch R1001/5896 Koch, "Bericht über der Tätigkeit," 25 Apr. 1907, 2.

123. BArch R1001/5896, Koch, "Bericht über der Tätigkeit," 25 Apr. 1907, 8; Mülders, "Formeln gegen Flagellaten," 59–61 (regarding Trypanrot) and 59–60 (regarding Trypanblau); Anthony S. Travis, "Anilines: Historical Background," in *The Chemistry of Anilines*, part 1, ed. Zvi Rappoport (Chichester, UK: Wiley, 2007), 55–56; Steverding, "Development."

124. Steverding, "Development," 3, notes that Trypanblau caused the skin of experimental animals to turn blue and was judged "unacceptable for use in patients," but research in Uganda by German and British teams used these dyestuffs in African patients and subjects.

125. BArch R1001/5896, Koch, "Bericht über der Tätigkeit," 25 Apr. 1907, 4–6.

126. Hoppe documented mass burial in trenches at Bussu camp, *Lords of the Fly*, 74.

127. M. Afr.–Rome–EE, Bumangi (N.-D. de Bon Secours), May 1907, *CTSMA*, no. 143 (November 1907).

128. M. Afr.–Rome–EE, Bumangi (N.-D. de Bon Secours), May 1907, *CTSMA*, no. 143 (November 1907).

129. Gradmann and Eckart's respective works make a powerful ethical critique of Koch's methods. See Gradmann, *Laboratory Disease*; Wolfgang U. Eckart, "The Colony as Laboratory: German Sleeping Sickness Campaigns in German East Africa and in Togo, 1900–1914," *History and Philosophy of the Life Sciences* 24, no 1. (February 2002): 69–89; Christoph Gradmann, "Modern Medicines."

130. WL, Dutton/Todd MSS, MS 2248/4792, Letter from A. Broden to J. L. Todd, 19 Dec. 1908, looseleaf in Congo Expedition Case book, vol. 1.

131. M. Afr.–Rome–EE, Sese-Bumangi Annual Report, 1905, *RA*, no. 1 (1905).

132. BArch R1001/5896, Koch, Draft of "Bericht über die Tätigkeit der Schlafkrankheits-Expedition bis zum 5. September 1907," 5 Sept. 1907, 7; Webel, "Borderlands," 156–66.

133. BArch R1001/5896, Koch, Draft of "Bericht über die Tätigkeit," 5 Sept. 1907, 9.

134. For broader context on experimentation, see Neill, *Networks in Tropical Medicine*, 176–80; Neill, "Paul Ehrlich's Colonial Connections," 70–71; Lenny A. Ureña Valerio, "The Stakes of Empire: Colonial Fantasies, Civilizing Agendas, and Biopolitics in the Prussian-Polish Provinces (1840–1914)" (PhD diss., University of Michigan, 2010), 203, 192n302, 196–98; Manuela Bauche, "Medizin und staatliche Herrschaft im deutschen Kaiserreich: Die Bekämpfung von Malaria in Kamerun, Ostafrika und Ost-Friesland, 1890–1919" (PhD diss., Universität Leipzig, 2015).

135. British camps were Koch's primary reference point; see BArch R1001/5876, "Aufzeichnung über die Sitzung des Reichsgesundheits-rats," 18 Nov. 1907, 10. Sleeping sickness camps were also called variously *Isolierstationen* (isolation stations), *Isolationslager* (isolation camps), *Sammlungslager* (collection camps), *Krankenlager* (sick camps), or *Schlafkrankheitslager* throughout the course of the *Schlafkrankheitsbekämpfung* (sleeping sickness campaign). On British camps, see Hoppe, *Lords of the Fly*, 73–76.

136. Alison Bashford and Carolyn Strange, eds., *Isolation: Places and Practices of Exclusion* (London: Routledge, 2003); Maryinez Lyons, *The Colonial Disease: A Social History of Sleeping Sickness in Northern Zaire, 1900–1940* (Cambridge: Cambridge University Press, 1992), 38–40.

137. BArch R86/2613, Koch, Draft, "Bericht über die Tätigkeit der Schlafkrankheits-Expedition bis zum 25. November 1906," 27 Nov. 1906, 11.

138. BArch R1001/5896, Koch, "Bericht über der Tätigkeit," 25 Apr. 1907, 3.

139. Hoppe, *Lords of the Fly*, 74.

140. Though he remained involved in metropolitan discussions about sleeping sickness, Koch would not return to Africa after his departure from East Africa in 1907; he died of heart failure in 1910. Brock, *Robert Koch*, 266, 285.

141. Eckart, "Colony as Laboratory," 69–89.

PART II: THE KINGDOM OF KIZIBA, C. 1890–1914

THE KINGDOM OF KIZIBA, C. 1890: AN OVERVIEW

1. I follow Michele Wagner's treatment of Buha in "Environment, Community, and History: 'Nature in the Mind' in Nineteenth- and Early Twentieth-Century Buha, Tanzania," in *Custodians of the Land: Ecology and Culture in the History of Tanzania*, ed. Gregory Maddox, James Giblin, and Isaria N. Kimambo (Athens: Ohio University Press, 1996), 176; Frederick J. Kaijage, "AIDS Control and the Burden of History in Northwestern Tanzania," *Population and Environment* 14, no. 3 (January 1993), 279.

2. Brad Weiss, *The Making and Unmaking of the Haya Lived World: Consumption, Commoditization, and Everyday Practice* (Durham, NC: Duke University Press, 1996), 16; Peter R. Schmidt, *Historical Archaeology in Africa: Representation, Social Memory, and Oral Traditions* (Lanham, MD: AltaMira, 2006), 12–13, suggests a relatively consistent 1,250 people per square mile dating at least from 1930s onward, with historic population density the greatest in areas closest to Lake Victoria.

3. BArch R1001/5936, Eggel, Report from Bukoba, 18 Apr. 1901.

4. Interview with Odilia Kokujwara, Bugombe, Tanzania, 21 Aug. 2008; Bethel Mission Archives, M194, Andrea Kajarero, "Aus meinem Leben"; Brad Weiss, *Sacred Trees, Bitter Harvests: Globalizing Coffee in Northwest Tanzania* (Portsmouth, NH: Heinemann, 2003), and Weiss, *Making and Unmaking*; Schmidt, *Historical Archaeology*; Library of Congress Geography and Map Division, Republic of Tanzania survey maps, "Kashozi" and "Mwiziro" (1965). BArch R1001/5936, Gallus, Report from Bukoba, 8 June 1901.

5. David L. Schoenbrun, *A Green Place, a Good Place: Agrarian Change, Gender, and Social Identity in the Great Lakes Region to the 15th Century* (Portsmouth, NH: Heinemann, 1998), 169–76; Schmidt, *Historical Archaeology*, chap. 3.

6. Hermann Rehse, *Kiziba: Land und Leute* (Stuttgart: Strecker and Schröder, 1910), 53; Weiss, *Sacred Trees*, 115–25. Weiss points to continuities between early twentieth-century descriptions of cultivation and late-twentieth century practices.

7. Rehse, *Kiziba*, 43–45; Rehse also notes common forms of basket traps in Kiziba and Buganda.

8. BArch R1001/5936, Robert Ahlbory, Report from Bukoba, 12 Mar. 1902; Charles M. Good, "Salt, Trade, and Disease: Aspects of Development in Africa's Northern Great Lakes Region," *International Journal of African Historical Studies* 5, no. 4 (1972): 564; Bethel Mission Archives, M194, Andrea Kajarero, "Aus meinem Leben"; Weiss, *Sacred Trees*, 166, quoting White Fathers Diary for Bwanja.

9. Bartolomeo Tibawa (b. 1910) discussed marketing fish caught around the Kagera River, as well as dry fish, to the Ganda in his youth; interview with Bartolomeo Tibawa, Bugombe Kashekere, Tanzania, 25 Aug. 2008. Ralph A. Austen notes that the name "Haya" was "originally applied by foreigners to the fishermen of the lake coast," in *Northwest Tanzania under German and British Rule: Colonial Policy and Tribal Politics, 1889–1939* (New Haven, CT: Yale University Press, 1968), 10n6, citing Hans Cory, *Historia ya Wilaya Bukoba* (Mwanza, Tanzania: Lake Printing Works, 1956); Gerald W. Hartwig, "The Victoria Nyanza as a Trade Route in the Nineteenth Century," *Journal of African History* 11, no. 4 (1970): 546.

10. M. Afr.–Rome–EE, Diaire d'Usui, Letter from P. Brard, *CTSMA*, no. 83 (July 1899); Tanzania National Archives, GM/118, "Militärische Wegekarte von Deutsch-Ostafrika in 9 Blättern, Blatt 1.2, Usumbura (Urundi, Ruanda, Kigali, Kissenyi, Bukoba, Ussuwi, Kifumbiro, Muanza)," undated, indicates rough marching times and state of roads.

11. Edmond Cesard, "Le Muhaya (l'Afrique orientale) (fin)," *Anthropos* 32, no. 1–2 (January–April 1937): 22, 27–29; Hans Cory and M. M. Hartnoll, *Customary Law of the Haya Tribe: Tanganyika Territory* (London: Percy Lund, Humphries, 1945), 277. German officials seem to have conflated the role of *mwami* and *mukungu*, and both with that of *katikiro*, a head elder or minister of the royal court, and applied *katikiro* widely to refer to a local African authority figure.

12. Schoenbrun, *Green Place*, 181–94.

13. See Francis X. Lwamgira, *Amakuru ga Kiziba: The History of Kiziba and Its Kings*, trans. E. R. Kamuhangire (1949; Kampala, Uganda: Makerere University College, Department of History, 1969), chaps. 22–25; Abel G. M. Ishumi, *Kiziba, the Cultural Heritage of an Old African Kingdom* (Syracuse, NY: Foreign and Comparative Studies Program, Syracuse University, 1980), 47.

14. Priscilla Copeland Reining, "The Haya: The Agrarian System of a Sedentary People" (PhD diss., University of Chicago, 1967), 59–61; Schoenbrun, *Green Place*, 222–25.

15. Kiziba accords with a model of political, social, and economic organization focused on banana cultivation and cattle husbandry, per David L. Schoenbrun, "Cattle Herds and Banana Gardens: The Historical Geography of the Western Great Lakes Region, ca. AD 800–1500," *African Archaeological Review* 11 (December 1993): 39–72.

16. Schmidt, *Historical Archaeology*, 101; Rehse, *Kiziba*, 1. See also Randall M. Packard, *Chiefship and Cosmology: An Historical Study of Political Competition* (Bloomington: Indiana University Press, 1981), 67–71.

17. Schmidt, *Historical Archaeology*, 29–30; Cory and Hartnoll, *Customary Law*, 125, 261–62; Hans Cory, *Historia ya Wilaya Bukoba* (Mwanza, Tanzania: Lake Printing Works, 1956); Rehse, *Kiziba*, 54.

18. Schmidt, *Historical Archaeology*, 28–29; Reining, "The Haya," 65–66. Schmidt argues that processes of consolidation were complete by the eighteenth century.

19. Weiss, *Sacred Trees*, 111; *Nyarubanja* referred to a system of landholding and tenant farming of a group of banana plantations, historically under control of clan heads and chiefs but subject to the favor of the king, per Cory and Hartnoll, *Customary Law*, 123–26. See Priscilla Copeland Reining, "Haya Land Tenure: Landholding and Tenancy," *Anthropological Quarterly* 35, no. 2 (April 1962): 65–66, on conferring a nyarubanja.

20. Rehse, *Kiziba*, 80, 107; Weiss, *Sacred Trees*, 51–54. On kibanja, see Schoenbrun, *Green Place*, 176–78.

21. Rehse, *Kiziba*, 1–3; M. Afr.–Rome–EE, Vicariat apostolique du Nyanza Méridional, "Notice sur la pays," Fr. Van Thiel to Mgr., *CTSMA*, no. 79 (July 1898).

22. Lwamgira, *Amakuru ga Kiziba*; Ishumi, *Kiziba*, chaps. 2 and 3.

23. BArch R1001/1029, Carl Hermann, Report, 30 Aug. 1892.

24. Austen, *Northwest Tanzania*, chaps. 3–4.

25. Austen, *Northwest Tanzania*, 37–41.

26. On Ismaili Indian traders at Bukoba see John Iliffe, *A Modern History of Tanganyika* (New York: Cambridge University Press, 1979), 139; Juhani Koponen, *Development for Exploitation: German Colonial Policies in Mainland Tanzania, 1884–1914* (Münster: Lit Verlag, 1994), 152. Koponen (254, 608) notes that a German coffee farm was established in 1908, with only seven European farms in Bukoba Residency in 1912. The White Fathers founded missions at Kashozi/Marienberg (1892), Mugana (1902), Kagondo (1903), and a mission and school at Rubya (1904), per *Catholic Directory of Tanzania 2006* (Dar es Salaam: Tanzania Episcopal Conference, 2006), held in the Missionaries of Africa (White Fathers) Provincial Archives in Dar es Salaam, Tanzania. German missionaries did not arrive until 1910, Koponen, *Development*, 165–66.

27. Austen, *Northwest Tanzania*, 85, 90–91; Koponen, *Development*, 215–24.

28. Austen, *Northwest Tanzania*, 38, 42, 46.

29. Koponen, *Development*, 116.

30. Jean-Pierre Chrétien, *The Great Lakes of Africa: Two Thousand Years of History*, trans. Scott Strauss (New York: Zone Books, 2003), 251–60.

31. Austen, *Northwest Tanzania*, 88, 254.

32. See BArch R1001/5898, Gudowius, "Bericht über die Ausbreitung der Schlafkrankheit im Bezirk Bukoba," 31 May 1908; BArch R1001/1029, Haber, "Die innerpolitischen Verhältnisse im Bezirke Bukoba," 30 June 1904.

33. Kenneth Curtis, "Capitalism Fettered: State, Merchant, and Peasant in Northwestern Tanzania" (PhD diss., University of Wisconsin–Madison, 1989); Reining, "The Haya."

34. Francis X. Lwamgira, *Amakuru ga Kiziba: The History of Kiziba and Its Kings*, trans. E. R. Kamuhangire (1949; Kampala, Uganda: Makerere University College, Department of History, 1969), 74–77; Rehse, *Kiziba*, 197–215.

CHAPTER 3: THE PRINCE AND THE PLAGUE

1. *Rubunga* and *lubunga* are interchangeable transliterations in English and French sources of an Oluhaya word denoting a particular illness.

2. Hermann Rehse, *Kiziba: Land und Leute* (Stuttgart: Strecker and Schröder, 1910), 45; Priscilla Copeland Reining, "The Haya: The Agrarian System of a Sedentary People" (PhD diss., University of Chicago, 1967), 30.

3. Peter R. Schmidt, *Historical Archaeology in Africa: Representation, Social Memory, and Oral Traditions* (Lanham, MD: AltaMira, 2006), 14.

4. M. Afr.–Rome–EE, "À Bikira-Maria," from "Un Coin du Buganda (Extraits d'un rapport du P. Gorju)," *CTSMA* no. 110 (November 1904).

5. BArch R1001/5936, Gallus, Report, 8 June 1901.

6. Rehse, *Kiziba*, 2.

7. Carl Hermann, "Die Wasiba und ihr Land," *Mitteilungen von Forschungsreisenden und Gelehrten aus den deutschen Schutzgebieten* 7 (1894): 47.

8. Edmond Cesard, "Le Muhaya (l'Afrique orientale) (fin)," *Anthropos* 32, no. 1–2 (January–April 1937): 28.

9. Brad Weiss, *The Making and Unmaking of the Haya Lived World: Consumption, Commoditization, and Everyday Practice* (Durham, NC: Duke University Press, 1996), 18; Priscilla Copeland Reining, "Haya Land Tenure: Landholding and Tenancy," *Anthropological Quarterly* 35, no. 2 (April 1962): 62.

10. Reining, "The Haya," 54–56; Weiss, *Making and Unmaking*, 86; interview with Odilia Kokujwara, Bugombe, Tanzania, 21 Aug. 2008.

11. See Cesard, "Le Muhaya," 27–28; Reining, "The Haya," chap. 3; Weiss, *Making and Unmaking*, 17–21, 57, 97. Alongside the distinction between grasslands (*orweya*) and villages (*ebyaro*) were further distinctions between land capable of sustaining perennial crops (*kisi*) and land planted with perennial crops (*mwate*), as well as "public" lands (*irungu*) available for grazing or harvesting wood, per Hans Cory and M. M. Hartnoll, *Customary Law of the Haya Tribe: Tanganyika Territory* (London: Percy Lund, Humphries, 1945), 116–43, 265.

12. Reining, "The Haya," 61.

13. Weiss, *Making and Unmaking*, 57.

14. Weiss, *Making and Unmaking*, 19; Reining, "Haya Land Tenure."

15. Weiss, *Making and Unmaking*, chaps. 3 and 4.

16. Weiss, 39–44.

17. Brad Weiss, *Sacred Trees, Bitter Harvests: Globalizing Coffee in Northwest Tanzania* (Portsmouth, NH: Heinemann, 2003), 37, 43–44.

18. Weiss, *Sacred Trees*, 15–18; David L. Schoenbrun, *A Green Place, a Good Place: Agrarian Change, Gender, and Social Identity in the Great Lakes Region to the 15th Century* (Portsmouth, NH: Heinemann, 1998), 181–95.

19. Rehse, *Kiziba*, 128, 130; Schmidt, *Historical Archaeology*, 62, 65, 81.

20. Rehse, *Kiziba*, 128–29; Schmidt, *Historical Archaeology*, 66. Schmidt argues tentatively that royal association with Mugasha may have cohered in part in opposition to the power of Bacwezi and cwezi mediums; Mugasha was not a Bacwezi in Haya myth.

21. Rehse, *Kiziba*, 136; Iris Berger, *Religion and Resistance: East African Kingdoms in the Precolonial Period* (Tervuren, Belgium: Musée royal de l'Afrique centrale, 1981), 23; Schoenbrun, *Green Place*, 108 10.

22. Schmidt, *Historical Archaeology*, 61–63. In Buhaya, Bacwezi had different meanings than elsewhere in the region.

23. Abel G. M. Ishumi, *Kiziba, the Cultural Heritage of an Old African Kingdom* (Syracuse, NY: Foreign and Comparative Studies Program, Syracuse University, 1980), 70–77.

24. M. Afr.–DSM, Kashozi Diary, 17 Mar. 1903, and Mugana Diary, 30 Oct. 1907; BArch R1001/5898, Robert Kudicke, Report from Kigarama, 31 July 1908; M. Afr.–Rome–EE, Marienberg (Kyamtwala), 1905, *CTSMA*, no. 125 (March 1906); Tanzania National Archives, Bukoba District book, "Witchcraft." Regarding Robert Kudicke and other medical officers or colonial staff: I identified the first names of German colonial officers and sleeping sickness campaign staff using personnel lists found in BArch R1001/9580, BArch R1001/9581, and the Bundesarchiv–Militärarchiv N14 Akte 15, as well as individual personnel files in BArch R1001/9585. Whenever full names were available, I have used them below; if unavailable, I have used last name only. See BArch R1001/9581, "Namentliches Verzeichnis" (undated) and "Anschriften-Verzeichnis der ehemaligen Schutztruppen-Offiziere und -Beamten" (undated).

25. Ishumi, *Kiziba*, 94n6.

26. Rehse, *Kiziba*, 137–38.

27. Rehse, 135.

28. Rehse, 138.

29. Rehse, 139.

30. Rehse, 138.

31. Lwamgira, *Amakuru ga Kiziba*, chap. 22, phase 3, 348–49; Richard J. Reid, *Political Power in Pre-colonial Buganda: Economy, Society, and Warfare in the Nineteenth Century* (Oxford: James Currey, 2002), 197–98; Richard J. Reid, "The Ganda on Lake Victoria: A Nineteenth-Century East African

Imperialism," *Journal of African History* 39, no. 3 (1998): 349, dates Suna's reign as c. 1830–56.

32. Lwamgira, *Amakuru ga Kiziba*, chap. 22, phase 4, 350–51; italics added.

33. Lwamgira, chap. 22, phase 4.

34. Lwamgira, chap. 22, phase 5, chap. 22, phase 6, 49; Ralph A. Austen, *Northwest Tanzania under German and British Rule: Colonial Policy and Tribal Politics, 1889–1939* (New Haven, CT: Yale University Press, 1968), 263.

35. Lwamgira, *Amakuru ga Kiziba*, chap. 22, phase 6.

36. *Rubunga* and *lubunga* are related to the Bantu root/verb *-bungo*, to be a vagabond, per André Coupez et al., *Inkoranya y íkinyarwaanda mu kinyarwaanda nó mu gifaraansá/Dictionnaire Rwanda–Rwanda et Rwanda–Français* (Butare, Rwanda: Institut de recherche scientifique et technologique; Tervuren, Belgium: Musée royal de l'Afrique Centrale, 2005).

37. Alois Meyer, *Kleines Ruhaya–Deutsches Wörterbuch* (Trier, Germany: Mosella-Verlag, 1914), 16, glosses *ku-bunga* as a verb meaning "to not let oneself be seen all day (as a child)," "to pay a visit," and, in the reflexive form, *ku-bungwa*, "to be eaten away by insects (as wood)."

38. I posit that Mugengere's destination Butumbi was possibly Butumbi/Rutshuru in southwestern Uganda, Lake Edward; Ishanje, on his return, is an obscure place-name and additionally not in modern use, but may refer to Ishanje Forest, sacred site of burial of Ankole's kings, near modern Mbarara, Lake Nakivale. Charles M. Good, "Salt, Trade, and Disease: Aspects of Development in Africa's Northern Great Lakes Region," *International Journal of African Historical Studies* 5, no. 4 (1972): 543–86.

39. Lwamgira, *Amakuru ga Kiziba*, chap. 22, phase 6.

40. BArch R1001/5936, Eggel, Report, 18 Apr. 1901.

41. Until recently, scholarly consensus generally supported this historical epidemiology of bubonic plague in eastern and central Africa focused on its late nineteenth-century arrival. Myron Echenberg, *Plague Ports: The Global Urban Impact of Bubonic Plague, 1894–1901* (New York: New York University Press, 2007), 4–6, offers the standard narrative of the third plague pandemic beginning in the 1890s and circulating out of Asia into Eastern Africa.

42. Médard's analysis of kaumpuli in Buganda and the narrative of rubunga here suggest that a chronology of *Y. pestis* in central and eastern Africa focused on arrival during the third pandemic is not accurate. Ongoing collaborations led by Monica Green, Gérard Chouin, and others strongly suggest that *Y. pestis* circulated in Africa prior to the third pandemic, particularly in eastern Africa. See Monica H. Green, "Taking 'Pandemic' Seriously: Making the Black Death Global," in *Pandemic Disease in the Medieval World: Rethinking the Black Death*, vol. 1., ed. Monica H. Green (Kalamazoo, MI: Arc Medieval Press, 2015), 35–45; Green, "Putting Africa on the Black Death Map: Narratives from Genetics and History," *Afriques: Débats, méthodes et terrains d'histoire* 9 (2018), https://doi.org/10.4000/afriques.2125; Gérard Chouin, "Reflections on Plague in African History (14th–19th c.)," *Afriques: Débats, méthodes et terrains d'histoire* 9 (2018), https://doi.org/10.4000/afriques.2228.

43. Lwamgira, *Amakuru ga Kiziba*, chap. 22, phase 6.

44. Andrew Cunningham, "Transforming Plague: The Laboratory and the Identity of Infectious Disease," in *The Laboratory Revolution in Medicine*, ed. Andrew Cunningham and Perry Williams (Cambridge: Cambridge University Press, 1992). White Fathers missionaries' use of *kaumpuli* likely relied on experiences in Buganda as a reference point. It is unclear whether the word circulated as an identifier of the illness/problem in Buhaya.

45. BArch R1001/1026, Carl Hermann, Report, June 1896; BArch R1001/5936, Eggel, Report from Bukoba, 25 Jan. 1900. BArch R1001/5936, Oskar Feldmann, "Pestbericht," 8 Aug. 1902, discusses Bishop Hirth of the White Fathers mission at Kashozi recounting that plague cases had a yearly uptick in June, when caravan traffic from Uganda increased as the rainy season ended.

46. M. Afr.–DSM, Kashozi Diary (1897), 5 Nov. 1897, 10 Nov. 1897.

47. M. Afr.–DSM, Kashozi Diary (1898), 19 July 1898.

48. Lwamgira, *Amakuru ga Kiziba*, chap. 22, phase 6.

49. BArch R1001/5936, Robert Ahlbory, Report from Bukoba, 1 Jan. 1901.

50. BArch R1001/5936, Gallus, Report, 8 June 1901.

51. BArch R1001/5936, Gallus, Report, 8 June 1901.

52. *Medizinal Berichte* über die *deutschen Schutzgebiete 1903/04* (Berlin: Mittler, 1905), 59.

53. Ziba movements to and from steppe homes appear to parallel the Ganda movements to and from homes and farms under the mediation of Kaumpuli's mediums I have discussed in chap. 1 above.

54. BArch R1001/5936, Gallus, Report from Bukoba, 1 June 1901.

55. M. Afr.–Rome–EE, Kiziba-Marienberg—Notre-Dame des Sept-Douleurs, 9 Oct. 1899, *CTSMA*, no. 85 (January 1900).

56. Megan Vaughan, *Curing Their Ills: Colonial Power and African Illness* (Cambridge: Polity Press, 1991), chap. 2.

57. BArch R1001/5936, Eggel, Report, 25 Jan. 1900.

58. Lwamgira, *Amakuru ga Kiziba*, chaps. 20–24.

59. Rehse (*Kiziba*, 119–20) notes that burial traditions included burying a person with his/her clothing, except a hide given to his eldest heir, but not otherwise destroying other possessions upon death. Lwamgira (*Amakuru ga Kiziba*, chap. 20, phase 1, 13) discusses changing burial practices in the mid-to-late nineteenth century, when interment, rather than placing bodies in forests, became conventional. Weiss (*Making and Unmaking*, 192–95) cites the modern practice of burial on family farms. Cory and Hartnoll (*Customary Law*, 267) discuss the ritual roles of a village *muharambwa*, a hereditarily invested position of high esteem associated with village first-comer clan affinity that appears to have been involved with ritually managing pollution and danger, as well as ensuring maintenance of land boundaries and fertility of land and women. Notably, the muharambwa bore responsibility for disposing of particular corpses: human bodies burned in fire, and dogs or snakes, all of which he would bury outside the village. He was also responsible, in cases of fire, for ritually protecting other homes so that fire did not spread.

60. BArch R1001/5936, Robert Ahlbory, Report, 12 Mar. 1902.

61. Vaughan, *Curing Their Ills*, 40–41; Echenberg, *Black Death, White Medicine: Bubonic Plague and the Politics of Public Health in Colonial Senegal, 1914–1945* (Portsmouth, NH: Heinemann, 2002), part 1.

62. BArch R1001/5936, Eggel, Report, 18 Apr. 1901.

63. BArch R1001/5936, Gallus, Report, 8 June 1901.

64. Alexandre Yersin's work connecting bubonic plague with a specific bacillus was published in 1894, but plague's transmission remained vague and it was presumed to be contagious, rather than infectious, at the time; the rat-flea-human transmission cycle was suggested in 1897 but not widely accepted until the early 1900s; Myron Echenberg, "Pestis Redux: The Initial Years of the Third Bubonic Plague Pandemic, 1894–1901," *Journal of World History* 13, no. 2 (Fall 2002): 437–38.

65. BArch R1001/5936, Eggel, Report, 18 Apr. 1901; Cory and Hartnoll, *Customary Law*, 267.

66. Cunningham, "Transforming Plague," 213–19, 235–36.

67. BArch R1001/5936, Eggel, Report, 18 Apr. 1901.

68. M. Afr.–DSM, Kashozi Diary, 9 Oct. 1897.

69. I follow Rehse, *Kiziba*, 266–67; Lwamgira, *Amakuru ga Kiziba*, 100–115; M. Afr., Kashozi Diary, 1897–1900, and Austen, *Northwest Tanzania*, 46–47. Patrick Malloy discusses this same episode in "Research Material and Necromancy: Imagining the Political-Economy of Biomedicine in Colonial Tanganyika," *International Journal of African Historical Studies* 47, no. 3 (2014): 425–43.

70. M. Afr.–DSM, Kashozi Diary, 5 Nov. 1897.

71. Echenberg, "Pestis Redux," 437–38; Cunningham, "Transforming Plague," 213–19.

72. M. Afr.–Rome–EE, Fr. Van Thiel, Marienberg, "Occasion et cause de la guerre," *CTSMA*, no. 79, 2nd Trimester 1898 (July 1898).

73. Weiss, *Making and Unmaking*, chaps. 3 and 4.

74. Rehse, *Kiziba*, 131.

75. M. Afr.–Rome–EE, Fr. Van Thiel, Marienberg, "Notice sur le pays," *CTSMA*, no. 79, 2nd Trimester 1898 (July 1898).

76. BArch R1001/5936, Eggel, "Pest in Bukoba," 18 Apr. 1901.

77. BArch R1001/5936, Eggel, "Pest in Bukoba," 18 Apr. 1901.

78. Steven Feierman, *The Shambaa Kingdom: A History* (Madison: University of Wisconsin Press, 1974), 31–32 and chap. 4.

79. Plague featured consistently as a concern in areas bordering Uganda through the outbreak of World War I. BArch R1001/5747, Hugo Meixner, Reports on Health in DOA, 1911–14.

80. BArch R1001/5936, Eggel and von Beringe, "Report about Quarantine Station," 12 Aug. 1899.

CHAPTER 4: GLAND-FEELERS, ELUSIVE PATIENTS, AND THE KIGARAMA CAMP

1. Kirk A. Hoppe, *Lords of the Fly: Sleeping Sickness Control in British East Africa, 1900–1960* (Westport, CT: Praeger, 2003); BArch R1001/5898,

Gudowius, "Bericht über die Ausbreitung der Schlafkrankheit im Bezirk Bukoba," 31 May 1908.

2. M. Afr.–DSM, Kashozi Diary, 24 Oct. 1903 and 26 Oct. 1903.

3. M. Afr.–Rome–EE, Vicariat apostolique du Nyanza méridional, Marienberg, *CTSMA*, no. 107 (June 1904).

4. BArch R1001/5833, Oskar Feldmann, undated Report (Oct. 1903), 1.

5. BArch R1001/5833, Oskar Feldmann, undated Report (Oct. 1903), 1.

6. BArch R1001/5833, Oskar Feldmann, undated Report (Oct. 1903), 1. On plague, BArch R86/2622, Oskar Feldmann, Report, 26 Nov. 1902.

7. BArch R86/2622, Robert Kudicke, "Bericht über das Schlafkrankenlager Kigarama für die Zeit vom 1.6. bis 1.10.1907 und Vorschläge zur Bekämpfung der Schlafkrankheit in den Sultanaten Kiziba und Bugabu," 1 Oct. 1907.

8. Hermann Rehse, *Kiziba: Land und Leute* (Stuttgart: Strecker and Schröder, 1910), 1–2.

9. BArch R86/2622, Robert Kudicke, Report, 1 Oct. 1907; see BArch R1001/5896, Koch, "Bericht über die Tätigkeit der Schlafkrankheits-Expedition bis zum 5. September 1907," 2.

10. BArch R86/2622, Robert Kudicke, Report, 1 Oct. 1907.

11. BArch R1001/5890, Robert Kudicke to administration, 8 Dec. 1907; BArch R1001/5890, R. Kudicke to Oberstabsarzt der Schutztruppe, Dar es Salaam, 31 Oct. 1907, including Anlage 1, "Verzeichnis der von der Schlafkrankheits-expedition des Herrn Geh. Rat Dr Koch nach hier überstandte Gegenstände und Medikamente," and Anlage 2, "Verzeichnis. Ausserdem sind von der Expedition übersandt und hier in Gebrauch genommen."

12. Francis X. Lwamgira, *Amakuru ga Kiziba: The History of Kiziba and Its Kings*, trans. E. R. Kamuhangire (1949; Kampala, Uganda: Makerere University College, Department of History, 1969), 137; Robert Koch's field notes from the Ssese Islands in 1906–7 (RKI AS/W6/006) indicate that Mutahangarwa was responsible for gathering sick people from the kingdom's villages at Kigarama.

13. Interview with Heslon Lutimba, 21 Aug. 2008, Kigarama, Tanzania.

14. Mari Webel, "Ziba Politics and the German Sleeping Sickness Camp at Kigarama, Tanzania, 1907–14," *International Journal of African Historical Studies* 47, no. 3 (2014): 411.

15. Rehse, *Kiziba*, 107; Priscilla Copeland Reining, "Haya Land Tenure: Landholding and Tenancy," *Anthropological Quarterly* 35, no. 2 (April 1962): 62.

16. Brad Weiss, *Sacred Trees, Bitter Harvests: Globalizing Coffee in Northwest Tanzania* (Portsmouth, NH: Heinemann, 2003), 110–12.

17. BArch R1001/5896, Koch, "Bericht über die Tätigkeit der Schlafkrankheits-Expedition bis zum 5. September 1907," 5, 7–8.

18. BArch R86/2622, Robert Kudicke, Report, 1 Oct. 1907, 2.

19. BArch R1001/5897, Robert Kudicke, Report, 19 Sept. 1907.

20. BArch R1001/5897, Robert Kudicke, Reports from Kigarama, 1 Oct 1907, 13 May 1908a, 13 May 1908b, 4 June 1908; BArch R1001/5897, Letter from R. Kudicke to Government, 19 Sept. 1907.

21. BArch R86/2622, Robert Kudicke, Report, 1 Oct. 1907.

22. BArch R1001/5898, Friedrich Kleine, Report, 30 Aug. 1908, 2.

23. Kudicke had advised putting people who fled from the camp into foot shackles as a deterrent to others, per BArch R1001/5899, Resident von Steumer to Friedrich Kleine, 26 Dec. 1908.

24. Hoppe, *Lords of the Fly*, 75–76. Kudicke noted numerous cases of *Irrsein* (madness) at Kigarama in 1908, per BArch R1001/5901.

25. BArch R1001/5876, "Aufzeichnung über die Sitzung des Reichs-Gesundheitsrats (Unterausschuß für Schlafkrankheit)," 5 Apr. 1909.

26. BArch R86/2622, Robert Kudicke, Report, 1 Oct. 1907.

27. Raised ground at the perimeter of the camp still remains at Kigarama today, though it is unclear from archival and oral sources whether it was enclosed by a permanent wall, barbed wire, or a fence. Interview with Heslon Lutimba at former camp location, Kigarama, Tanzania, 21 Aug. 2008.

28. BArch R1001/5897, Robert Kudicke, Report, 13 May 1908, 3.

29. BArch R1001/5903, Georg Ullrich, Report, 15 Oct. 1909.

30. BArch R1001/5892, Emil Steudel, "Auszug aus dem Bericht des Generaloberarztes Professor Dr. Steudel über seine von August 1911 bis Januar 1912 ausgeführte Dienstreise nach Deutsch-Ostafrika," Apr. 1912, 13.

31. *Medizinal Berichte über die Deutschen Schutzgebieten*, 1909/10 (Berlin: Mittler, 1911), 7.

32. BArch R1001/5892, Emil Steudel, "Auszug aus dem Bericht des Generaloberarztes Professor Dr. Steudel über seine von August 1911 bis Januar 1912 ausgeführte Dienstreise nach Deutsch-Ostafrika," Apr. 1912, 13.

33. Francis P. Nolan, *Mission to the Great Lakes: The White Fathers in Western Tanzania, 1878–1978* (Tabora, Tanzania: Tanganyika Mission Press, 1978), 24.

34. BArch R86/2622, Robert Kudicke, Report, 1 Oct. 1907. A colonial *Pflegeschwester* was seconded to Kigarama in 1908, though the duration of her tenure and participation in the campaign is unknown. See BArch R1001/5897, Winterfeld to Imperial Colonial Office, 9 May 1908.

35. Christian Burri and Reto Brun, "Human African Trypanosomiasis," in *Manson's Tropical Diseases*, 22nd ed., ed. Gordon C. Cook and Alimuddin I. Zumla (Philadelphia: W. B. Saunders, 2013), 1317.

36. Lumbar punctures were performed very seldom at Kigarama, and only with difficulty. BArch R1001/5905, Robert Kudicke, Report, 25 Oct. 1910; BArch R1001/5907, 20 Apr. 1911. Kudicke notes performing lumbar punctures on the dead in 1908 to identify trypanosomes, BArch R1001/5898, 31 Aug. 1908.

37. Interview with Heslon Lutimba, Kigarama, Tanzania, 21 Aug. 2008.

38. German doctors also used Swahili to name the illness; early twentieth-century dictionaries offered *botongo* and *isimagira* as names for the biomedical entity. My thanks to Professor Henry Muzale, a linguist at the University of Dar es Salaam, for his help in August 2008 with these and other disease names in Oluhaya. I have also consulted Fr. Paul Betbeder's *Dictionnaire Kihaya–Français* (undated), which defines *botongo* as sleeping sickness, and Alois Meyer, *Kleines Ruhaya–Deutsches Wörterbuch* (Trier, Ger.: Mosella-Verlag,

1914), both held at M. Afr.–Rome. I discussed disease names in interviews with Mzee Heslon Lutimba, Kigarama, Tanzania, 21 Aug. 2008, and Ma Odilia Kokujwara, Bugombe, Tanzania, 22 Aug. 2008. Lutimba also mentioned *mulalamo* as a Haya name for the Swahili *ugonjwa wa malale*. I do not presume that "isimagira" or "botongo" and "sleeping sickness" were equivalent, nor that names and definitions of disease were consistent among doctors, patients, and/or patients' caretakers in the early twentieth century. Persistent association of both botongo and isimagira with sleeping sickness over time, however, indicate that these words captured meaningful elements of observed changes to the body and temperament. See Julie Livingston, "Productive Misunderstandings and the Dynamism of Plural Medicine in Mid-century Bechuanaland," *Journal of Southern African Studies* 33, no. 4 (December 2007): 801–10. *Mulalamo* is currently translated as meningitis or cryptococcal meningitis; see Daniel P. Kisangau et al., "Use of Traditional Medicines in the Management of HIV/AIDS Opportunistic Infections in Tanzania: A Case in the Bukoba Rural District," *Journal of Ethnobiology and Ethnomedicine* 3, no. 1 (2007): 3–29. In the early twentieth century, European doctors associated meningitis with sleeping sickness (before 1902) and later with the pathology of trypanosomal infections (after 1902), but also differentiated between sleeping sickness and cerebrospinal meningitis in surveys of infectious disease in eastern Africa; see BArch R1001/5747, Exner, Report on General Health Situation in German East Africa for Third Quarter 1913, 16 Dec. 1913. It is unclear whether *mulalamo* as a disease name was in use contemporaneously to *sleeping sickness*, *botongo*, or *isimagira*.

39. Alois Meyer, *Wörterbuch*, 15.

40. Alois Meyer, 125.

41. See Michael W. Tuck, "Kabaka Mutesa and Venereal Disease: An Essay on Medical History and Sources in Precolonial Buganda," *History in Africa* 30 (2003): 314–18. See M. Afr.–Rome, Sese-Bumangi Diary, Feb. 1906, on social prohibitions around kabotongo.

42. I thank Neil Kodesh for his insights on this possible interpretation of linguistic evidence. See David Schoenbrun, *The Historical Reconstruction of Great Lakes Bantu Cultural Vocabulary: Etymologies and Distributions* (Cologne: Rüdiger Köppe, 1997), 15–16.

43. BArch R1001/5896, Winterfeld to Robert Kudicke, 25 Nov. 1907; BArch R1001/5897, R. Kudicke, Report, 13 May 1908.

44. BArch R1001/5897, Robert Kudicke, Report, 13 May 1908. People received ten heller, or 1/10 rupee per work day, in 1908.

45. BArch R1001/5901, Ruschhaupt, Report, 1 July 1909; interview with Bernard Mutekanga, Kashozi, Tanzania, 19 Aug. 2008.

46. John Iliffe, *Tanganyika under German Rule, 1905–1912* (Cambridge: Cambridge University Press, 1969), 160; BArch R1001/5898, Gudowius, Report, 31 May 1908.

47. BArch R86/2622, Robert Kudicke, Report, 1 Oct. 1907. Kudicke lists the camp as being staffed by "colored personnel" including two katikiros, two

fly catchers (with one on furlough), and four orderlies. The training of these orderlies is not detailed in colonial records or oral histories. Interview with Heslon Lutimba, Kigarama, Tanzania, 21 Aug. 2008.

48. BArch R1001/5897, Robert Kudicke, Report, 13 May 1908.

49. Iris Berger, *Religion and Resistance: East African Kingdoms in the Precolonial Period* (Tervuren, Belgium: Musée royal de l'Afrique centrale, 1981), 6.

50. Hans Cory and M. M. Hartnoll, *Customary Law of the Haya Tribe: Tanganyika Territory* (London: Percy Lund, Humphries, 1945), appendix V, "List of Clans," 2, 8, 15, 38, and appendix III, "Tribal Structure," 255, 258–60; Lwamgira, *Amakuru ga Kiziba*, chap. 4, phase 1, 22; Abel G. M. Ishumi, *Kiziba, the Cultural Heritage of an Old African Kingdom* (Syracuse, NY: Foreign and Comparative Studies Program, Syracuse University, 1980), 23–25, 37.

51. Ishumi, *Kiziba*, 21; Tanzanian National Archives, Bukoba District Book, vol. II, M. M. Hartnoll, "Heirlooms of the Chiefdom of Kiziba," 1941.

52. Weiss, *Sacred Trees*, 42–43.

53. Lwamgira, *Amakuru ga Kiziba*, 77, names Karutasigwa of Kanyigo.

54. Schoenbrun, *Historical Reconstruction*, 209.

55. Schoenbrun, 208.

56. BArch R1001/5898, Robert Kudicke, Report, 31 July 1908; Schoenbrun, *Historical Reconstruction*, 208.

57. Hermann Rehse, *Wörtersammlung des Ruziba* (Hamburg: Lucas Gräfe and Sillem, 1915), 107.

58. Peter R. Schmidt, *Historical Archaeology in Africa: Representation, Social Memory, and Oral Traditions* (Lanham, MD: AltaMira, 2006), chap. 5.

59. Reining, "Haya Land Tenure," 64–66.

60. Reining, 60.

61. Brad Weiss, *The Making and Unmaking of the Haya Lived World: Consumption, Commoditization, and Everyday Practice* (Durham, NC: Duke University Press, 1996), 21. Weiss notes that grassland plots adjacent to a woman's natal or marital were used under the rubric of *orweya*.

62. I. K. Katoke and P. Rwehumbiza, "The Administrator: Francis Lwamugira," in *Modern Tanzanians: A Volume of Biographies*, ed. John Iliffe (Nairobi: East African Publishing House, 1973); Weiss, *Sacred Trees*, 134, 143.

63. Ralph A. Austen, *Northwest Tanzania under German and British Rule: Colonial Policy and Tribal Politics, 1889–1939* (New Haven, CT: Yale University Press, 1968), 10. See Weiss, *Sacred Trees*; Brian K. Taylor, *The Western Lacustrine Bantu* (London: International African Institute, 1962), 134; Juhani Koponen, *Development for Exploitation: German Colonial Policies in Mainland Tanzania, 1884–1914* (Münster: Lit Verlag, 1994), 436–37. Cory and Hartnoll's study of Haya customary law, published in 1945, cites an explanation of squatting that overlaps with the transition from coffee as royal prerogative to coffee as cash crop, *Customary Law*, 149.

64. John Iliffe, *A Modern History of Tanganyika* (New York: Cambridge University Press, 1979), 122, 156; Iliffe, *Tanganyika under German Rule*, 171–74;

Koponen, *Development*, 285, 436–37; Austen, *Northwest Tanzania*, chaps. 5 and 6; Weiss, *Sacred Trees*, chaps. 1 and 2.

65. Lwamgira, *Amakuru ga Kiziba*, 135.

66. Weiss, *Sacred Trees*, 40. Mutahangarwa attempted an open partnership with planter-ethnographer Hermann Rehse and another man, Weber, in 1907, in a proposal to the colonial government that had each party sharing profits equally, per Iliffe, *Modern History of Tanganyika*, 156.

67. Austen, *Northwest Tanzania*, 170–71.

68. BArch R1001/5897, Friedrich Kleine, "Schlafkrankheit," 31 Mar. 1908.

69. Wolfgang Eckart, *Medizin und Kolonialimperialismus, Deutschland 1884–1945* (Paderborn, Ger.: Schöningh, 1997), 348; Hiroyuki Isobe, "Medizin und Kolonialgesellschaft: Die Bekämpfung der Schlafkrankheit in den deutschen 'Schutzgebieten' vor dem Ersten Weltkrieg" (PhD diss., University of Konstanz, Berlin, 2009), 115; John Iliffe, *East African Doctors: A History of the Modern Profession* (Cambridge: Cambridge University Press, 1998), 31–32.

70. Stephen J. Rockel, *Carriers of Culture: Labor on the Road in Nineteenth-Century East Africa* (Portsmouth, NH: Heinemann, 2006); Johannes Fabian, *Out of Our Minds: Reason and Madness in the Exploration of Central Africa* (Berkeley: University of California Press, 2000); Nancy Rose Hunt, *A Colonial Lexicon: Of Birth Ritual, Medicalization, and Mobility in the Congo* (Durham, NC: Duke University Press, 1999); Emily Lynn Osborn, "'Circle of Iron': African Colonial Employees and the Interpretation of Colonial Rule in French West Africa," *Journal of African History* 44, no. 1 (2003): 29–50; Benjamin N. Lawrance, Emily Lynn Osborn, and Richard L Roberts, eds., *Intermediaries, Interpreters, and Clerks: African Employees in the Making of Colonial Africa* (Madison: University of Wisconsin Press, 2006); Markku Hokkanen, "Towards a Cultural History of Medicine(s) in Colonial Central Africa," in *Crossing Colonial Historiographies: Histories of Colonial and Indigenous Medicines in Transnational Perspective*, ed. Anne Digby, Waltraud Ernst, and Projit B. Mukharji (Newcastle upon Tyne, UK: Cambridge Scholars, 2010).

71. Nancy J. Jacobs, "The Intimate Politics of Ornithology in Colonial Africa," *Comparative Studies in Society and History* 48, no. 3 (July 2006): 564–603; Jacobs, *Birders of Africa: History of a Network* (New Haven: Yale University Press, 2016); Lyn Schumaker, *Africanizing Anthropology: Fieldwork, Networks, and the Making of Cultural Knowledge in Central Africa* (Durham, NC: Duke University Press, 2001); Julia Cummiskey, "Placing Global Science in Africa: International Networks, Local Places, and Virus Research in Uganda, 1936–2000" (PhD diss., Johns Hopkins University, 2017).

72. BArch R1001/5876, Emil Steudel, Report, 4 Nov. 1908, 3–4.

73. David L. Schoenbrun, *A Green Place, a Good Place: Agrarian Change, Gender, and Social Identity in the Great Lakes Region to the 15th Century* (Portsmouth, NH: Heinemann, 1998), 166.

74. Rehse, *Kiziba*, 1–3.

75. BArch R86/2622, Robert Kudicke, Report, 1 Oct. 1907.

76. BArch R1001/5876, Imperial Health Office, Report, 28 Dec. 1908; BArch R1001/5876, Emil Steudel, Report, 4 Nov. 1908; BArch R1001/5898, Gudowius, Report, 31 May 1908; BArch R1001/5903, Georg Ullrich, Report, 31 Dec. 1909, 5.

77. BArch R1001/5903, Georg Ullrich, Report, 31 Dec. 1909, 5.

78. BArch R1001/5904, Georg Ullrich, Report, 16 Apr. 1910.

79. Iliffe, *Tanganyika under German Rule*, 160; Austen, *Northwest Tanzania*, 54, 91. The hut tax ranged from three to five rupees before World War I.

80. Interview with Bernard Mutekanga, Kashozi, Tanzania, 21 Aug. 2008.

81. BArch R1001/5901, Robert Kudicke, Chart, 9 Aug. 1908.

82. BArch R1001/5897, Robert Kudicke, Report, 31 May 1908, 1.

83. BArch R1001/5897, Robert Kudicke, Report, 31 May 1908, 1.

84. BArch R1001/5904, Scherschmidt, Report, 4 Jan. 1910; BArch R1001/5911, Scherschmidt, Report, 1 Jan. 1914; BArch R1001/5910, Lurz, Report, 1 July 1913; BArch R1001/5911, Lurz, Report, 1 Oct. 1913.

85. *Muteko* relates to the verb *ku-téka*, contemporarily meaning to join oneself in a group or to form a particular order. Alois Meyer, *Wörterbuch*, 134.

86. Alois Meyer, *Wörterbuch*, 124; Cory and Hartnoll, *Customary Law*, 125, 149. *Nsiku* correlated with "forced labor" by the time of their research in the interwar period. Schmidt discusses a similar institution known historically as *kikale* (a reference to the mukama's court) in *Historical Archaeology*, 29.

87. Rehse, *Kiziba*, 115; Ishumi, *Kiziba*, 40–43, 60.

88. Ishumi, *Kiziba*, 64–68, Schmidt, *Historical Archaeology*, 29.

89. Schmidt, *Historical Archaeology*, 29; Austen, *Northwest Tanzania*, 11, 144; Lesley Stevens, "Religious Change in a Haya Village, Tanzania," *Journal of Religion in Africa* 21, no. 1 (1991): 8; Nolan, *Mission to the Great Lakes*, 21; Goran Hyden, *Political Development in Rural Tanzania* (Nairobi: East African Publishing House, 1969), 79–82; Johnson M. Ishengoma, "African Oral Traditions: Riddles among the Haya of Northwestern Tanzania," *International Review of Education* 51, no. 2/3 (May 2005): 142.

90. Method M. P. Kilaini, *The Catholic Evangelization of Kagera in North-West Tanzania: The Pioneer Period, 1892–1912* (Rome: Pontificia Università Gregoriana, 1990), 12–13; Rehse, *Kiziba*, 110; Ishumi, *Kiziba*, 68.

91. Schmidt, *Historical Archaeology*, 29.

92. Alois Meyer, *Wörterbuch*, 124.

93. L. A. Fallers and S. B. K. Musoke, "Social Mobility, Traditional and Modern," in *The King's Men: Leadership and Status in Buganda on the Eve of Independence*, ed. L. A. Fallers (Oxford: Oxford University Press, 1964), 170–71; Richard J. Reid, "The Ganda on Lake Victoria: A Nineteenth-Century East African Imperialism," *Journal of African History* 39, no. 3 (1998): 349–50; Dorothy L. Hodgson, *Once Intrepid Warriors: Gender, Ethnicity, and the Cultural Politics of Maasai Development* (Bloomington: Indiana University Press, 2001); Mathieu Deflem, "Warfare, Political Leadership, and State Formation: The Case of the Zulu Kingdom, 1808–1879," *Ethnology* 38, no. 4 (Autumn 1999): 371–91.

94. The normal progress into royal service suggests that gland-feelers (whom Kudicke and Lwamgira referred to as "young men" and not "boys") would likely have been in their late teens or early twenties (following or in the latter years of their service in the court), younger than Kudicke's own age of thirty. See Cory and Hartnoll, *Customary Law*, 271.

95. BArch R1001/1029, Haber, Report, 30 June 1904, 6; Iliffe, *Tanganyika under German Rule*, 174; Iliffe, *Modern History of Tanganyika*, 122; Austen, *Northwest Tanzania*, 100–104; M. Afr.–DSM, Kashozi Diary, 1902, 27 Aug. 1902.

96. BArch R1001/5897, Oskar Feldmann, 1 Apr. 1908; BArch R1001/5910, Schnee to Government, Dar es Salaam, 24 Jan. 1913, with enclosed notes from Sacher, Rusizi Valley, 19 Oct. 1912 and Penschke, Usumbura, 6 Nov. 1912.

97. Brad Weiss, "Dressing at Death: Clothing, Time, and Memory in Buhaya, Tanzania," in *Clothing and Difference: Embodied Identities in Colonial and Post-colonial Africa*, ed. Hildi Hendrickson (Durham, NC, 1996).

98. Lwamgira, *Amakuru ga Kiziba*, chap. 25, phase 1, 137–38.

99. M. Afr.–Rome, Mugana Diary, April 1904, "Ibale"; Lwamgira, *Amakuru ga Kiziba*, chap. 25, phase 1.

100. Hunt, *Colonial Lexicon*, 2, 23.

101. BArch R1001/5898, Gudowius, Report, 31 May 1908, 59–60.

102. BArch R1001/5898, Robert Kudicke, "Bericht über die Bekämpfung der Schlafkrankheit im Bezirk Bukoba 1. Mai bis 31. Juli 1908," undated 1908, 3.

103. Olga Amsterdamska, "Demarcating Epidemiology," *Science, Technology, and Human Values* 30, no. 1 (2005): 17–51; Worboys, "The Emergence of Tropical Medicine: A Study in the Establishment of a Scientific Discipline," in *Perspectives on the Emergence of Scientific Disciplines*, ed. Gerard LeMaine et al. (Chicago: Aldine, 1976).

104. BArch R1001/5899, Robert Kudicke, Report, 18 Dec. 1908, 1; BArch R1001/5898, R. Kudicke, "Bericht über die Bekämpfung der Schlafkrankheit im Bezirk Bukoba 1. Mai bis 31. Juli 1908," undated 1908, 1.

105. Maryinez Lyons, *The Colonial Disease: A Social History of Sleeping Sickness in Northern Zaire, 1900–1940* (Cambridge: Cambridge University Press, 1992), 199–206; Rita Headrick and Daniel R. Headrick, *Colonialism, Health and Illness in French Equatorial Africa, 1885–1935* (Atlanta, GA: African Studies Association Press, 1994), 89–91.

106. BArch R1001/5899, Robert Kudicke, Report, 18 Dec. 1908, 4.

107. BArch R1001/5901, Georg Ullrich, Report, 31 Mar. 1909.

108. BArch R1001/5901, Ruschhaupt, Report, 1 July 1909.

109. Closser notes similar tensions in the modern polio eradication campaign; Svea Closser, *Chasing Polio in Pakistan: Why the World's Largest Public Health Initiative May Fail* (Nashville, TN: Vanderbilt University Press, 2010), chapters 3, 4.

110. BArch R1001/5901, Ruschhaupt, Report, 1 July 1909.

111. BArch R1001/5901, Ruschhaupt, Report, 1 July 1909.

112. BArch R1001/5903, Georg Ullrich, Report, 1 July 1909.

113. BArch R1001/5899, Robert Kudicke, Report, 18 Dec. 1908, 4.

114. BArch R1001/5901, Ruschhaupt, Report, 1 July 1909, 2; BArch R1001/5899, Marshall, 7 Jan. 1909.

115. Osborn, "Circle of Iron"; Lawrance, Osborn, and Roberts, *Intermediaries, Interpreters, and Clerks;* Nancy Rose Hunt, "Letter-Writing, Nursing Men and Bicycles in the Belgian Congo: Notes towards the Social Identity of a Colonial Category," in *Paths toward the Past: African Historical Essays in Honor of Jan Vansina,* ed. Robert W. Harms et al. (Atlanta, GA: African Studies Association Press, 1994).

116. Lwamgira, *Amakuru ga Kiziba,* 138.

117. Lyons, *Colonial Disease,* 184–85.

118. Weiss, *Making and Unmaking,* 31–35; Rehse, *Kiziba,* 92–98.

119. Rehse, *Kiziba,* 137–39.

120. Ziba healers would have marshaled a wide array of botanical and mineral extracts and mixtures for their curative power, and applied their knowledge to physical manifestations of social or personal imbalances. My research did not address how healers would have specifically addressed isimagira/botongo/ugonjwa wa malale. See Rehse, *Kiziba,* 137–41.

121. David L. Schoenbrun, "Conjuring the Modern in Africa: Durability and Rupture in Histories of Public Healing between the Great Lakes of East Africa," *American Historical Review* 111, no. 5 (December 2006): 1403–39.

122. Webel, "Ziba Politics," 221; BArch R1001/5897–5911 contain monthly and quarterly reports for Kigarama.

123. BArch R1001/5901, Georg Ullrich, Report, 31 Mar. 1909.

124. BArch R1001/5903, Georg Ullrich, Report, 1 July 1909.

125. Lwamgira, *Amakuru ga Kiziba,* 138.

126. Hyden, *Political Development,* 89; Cory and Hartnoll, *Customary Law,* 264; Berger, *Religion and Resistance,* 7; Holly Elisabeth Hanson, *Landed Obligation: The Practice of Power in Buganda* (Portsmouth, NH: Heinemann, 2003), 61–72.

127. James L. Giblin, *The Politics of Environmental Control in Northeastern Tanzania, 1840–1940* (Philadelphia: University of Pennsylvania Press, 1992), chapter 8; Hoppe, *Lords of the Fly,* 19.

128. BArch R1001/5898, Robert Kudicke, "Bericht über die Bekämpfung der Schlafkrankheit im Bezirk Bukoba 1. Mai bis 31. Juli 1908," undated 1908, 2.

129. "Paul Ehrlich's Colonial Connections: Scientific Networks and Sleeping Sickness Drug Therapy Research, 1900–1914," *Social History of Medicine* 22, no. 1 (April 2009): 61–77.

130. BArch R1001/5905, Robert Kudicke, Report, 25 Oct. 1910.

131. BArch R1001/5905, Robert Kudicke, Report, 25 Oct. 1910.

132. BArch R1001/5905, Robert Kudicke, Report, 25 Oct. 1910.

133. Deborah Neill, *Networks in Tropical Medicine: Internationalism, Colonialism, and the Rise of a Medical Specialty, 1890–1930* (Stanford, CA: Stanford University Press, 2012), 178–79.

134. Rehse, *Kiziba,* 131.

135. BArch R1001/5901, Georg Ullrich, Report, 31 Mar. 1909.

136. BArch R1001/5905, Robert Kudicke, Report, 25 Oct. 1910.

137. BArch R1001/5905, Robert Kudicke, Report, 25 Oct. 1910.

138. BArch R1001/5892, Emil Steudel, "Auszug aus dem Bericht," Apr. 1912, 13. Lwamgira writes (*Amakuru ga Kiziba*, chap. 25, phase 3) that Mutahangarwa himself was also treated at Kigarama by Dr. Otto Wittrock for a serious, unspecified illness for several months in 1913.

139. BArch R1001/5909, Bruno Eckard, Report, 1 Apr. 1912; BArch R1001/5909, Vorwerk, Report, 4 July 1912; BArch R1001/5908, d, Report, 1 Oct. 1911, 2–3.

140. Jeremy Rich, "Searching for Success: Boys, Family Aspirations, and Opportunities in Gabon, ca. 1900–1940," *Journal of Family History* 35, no. 1 (January 2010): 9.

PART III: THE SOUTHERN IMBO, C. 1890–1914

THE SOUTHERN IMBO, C. 1890: AN OVERVIEW

1. Michele D. Wagner, "Trade and Commercial Attitudes in Burundi before the Nineteenth Century," *International Journal of African Historical Studies* 26, no. 1 (1993): 150n4; Jean-Pierre Chrétien, *The Great Lakes of Africa: Two Thousand Years of History*, trans. Scott Strauss (New York: Zone Books, 2003); Marc Sommers, *Fear in Bongoland: Burundi Refugees in Urban Tanzania* (New York: Berghahn Books, 2001), 8; David Newbury, *The Land beyond the Mists: Essays on Identity and Authority in Precolonial Congo and Rwanda* (Athens: Ohio University Press, 2009), 287–88. A key exception is the area around Ras Magala (Cape Magala) between Rumonge and Bujumbura, with a far narrower strip of lowland and mountains rising much more sharply to the east. The coastal plain generally lay below a thousand meters in elevation, in contrast to heights above two thousand meters in the mountains nearby.

2. Newbury, *The Land beyond the Mists*, 287.

3. Raymonde Bonnefille and Françoise Chalié, "Pollen-Inferred Precipitation Time-Series from Equatorial Mountains, Africa, the Last 40 kyr BP," *Global and Planetary Change* 26, no. 1–3 (November 2000): 27; Jean-Pierre Chrétien, *Burundi: L'histoire retrouvée: 25 ans de métier d'historien en Afrique* (Paris: Karthala, 1993), 80.

4. Sommers, *Fear in Bongoland*, 54; Hans Meyer, *Die Barundi: Eine völkerkundliche Studie aus Deutsch-Ostafrika* (Leipzig: Otto Spamer, 1916), 53. Chrétien (*Burundi*, 82) argues that widespread banana cultivation did not take hold in Imbo until the later nineteenth century.

5. Chrétien, *Burundi*, 82–83

6. David Newbury, *The Land beyond the Mists*, 287.

7. Hubert Cochet, "Agrarian Dynamics, Population Growth and Resource Management: The Case of Burundi," *GeoJournal* 60, no. 2 (2004): 113–14; Cochet, "A Half Century of Agrarian Crisis in Burundi (1890–1945): The Incapacity of the Colonial Administration in Managing the Agrarian Crisis of the Late Eighteen-Hundreds," *African Economic History* 31 (2003): 20–21; David Newbury, *The Land beyond the Mists*, 330.

8. Chrétien, *Burundi*, 131.

9. Hans Meyer, *Die Barundi*, 88–90; Liisa Malkki, *Purity and Exile: Violence, Memory, and National Cosmology among Hutu Refugees in Tanzania* (Chicago: University of Chicago Press, 1995), 20–30, addresses the context of and ongoing debate over the characterization of Burundi as a feudal state.

10. René Lemarchand, *Burundi: Ethnocide as Discourse and Practice* (Cambridge: Cambridge University Press, 1994), 37.

11. David Newbury, *The Land beyond the Mists*, 308; David Newbury, *Kings and Clans: Ijwi Island and the Lake Kivu Rift, 1780–1840* (Madison: University of Wisconsin Press, 1991), 151–52.

12. Chrétien, *Great Lakes*, 249–51; René Lemarchand, *Rwanda and Burundi* (New York: Praeger, 1970), 53–56; William Roger Louis, *Ruanda-Urundi, 1884-1919* (Oxford: Clarendon, 1963), chaps. 12 and 13, which provide a detailed history of the colonial politics of early twentieth-century Burundi.

13. Bishikwabo Chubaka, "Aux origines de la ville d'Uvira selon les explorateurs et les pionniers de la colonisation Belge au Zaire (1840–1914)," *Civilisations* 37, no. 1 (1987): 95.

14. Jacques Depelchin, "From Pre-Capitalism to Imperialism: A History of Social and Economic Formations in Eastern Zaïre (Uvira Zone, c. 1800–1965)" (PhD diss., Stanford University, 1974), 41; Chubaka, "Aux Origines," 94–95.

15. Daniel P. Biebuyck, "Bembe Art," *African Arts* 5, no. 3 (Spring 1972): 15–17. Scholarly attention to the Bwari and Sanse group is scanty. Biebuyck groups the Sanse and Bwari populations along the lake, along with others, into the Zoba cultural group, linked to the heritage of the better-known Bembe cultural group that dominated near Baraka and environs reaching inland to the northwest. Linguistically, Kabwari speakers had deep historical associations with Kirundi speakers on the opposite lakeshore, per David Schoenbrun, *The Historical Reconstruction of Great Lakes Bantu Cultural Vocabulary: Etymologies and Distributions* (Cologne: Rüdiger Köppe, 1997), map 1, 12–13; Jacques Depelchin, "A Contribution to the Study of Pre-capitalist Modes of Production: Uvira Zone (Eastern Zaïre) c. 1800–1937," *African Economic History Review* 2, no. 1 (Spring 1975): 3; Depelchin, "From Pre-Capitalism to Imperialism," chap. II.

16. J. M. M. van der Burgt, *Dictionnaire Français–Kirundi* (Bois-le-Duc, Holland: Société "l'Illustration Catholique," 1903), 124.

17. Hans Meyer, *Die Barundi*, 38–39, 73–74.

18. Richard Burton, *The Lake Regions of Central Africa: A Picture of Exploration* (New York: Harper and Brothers, 1860), 322; van der Burgt, *Dictionnaire*, 126, 446; M. Afr.–Rome–EE, Révérend Père Coulbois, Journal, *CTSMA*, no. 26 (April 1885); AA RA/RU (0)7, "Resume des Rapports Economiques," 4 Dec. 1917, 5; M. Afr.–Rome–EE, Journal de Kibanga, 19 Oct. 1886 and 23 Oct. 1886, *CTSMA*, no. 36 (October 1887).

19. Burton, *Lake Regions*, 347; Henry M. Stanley, *Through the Dark Continent* (London: Sampson Low, 1890), 328, on cassava; Verney Lovett Cameron, *Across Africa* (London : George Phillip, 1885), 176, on hemp; AA, GG

7072, Lieutenant Drôs, "Rapport sur la reconnaisance qui a été effectuée du 16 au 27 November 1908," 30 Nov. 1908.

20. Chrétien, *Great Lakes*, 256; Thaddeus Sunseri, "The Hide Trade from German East Africa, ca. 1840–1916: Violence, Disease, and Social Change," *International Journal of African Historical Studies* 51, no. 3 (2018): 373–402.

21. Wagner, "Trade and Commercial Attitudes," 155–59; Philip Gooding, "Lake Tanganyika: Commercial Frontier in the Era of Long-Distance Commerce, East and Central Africa, c. 1830–1890" (PhD diss., SOAS University of London, 2017), chap. 4.

22. Chubaka, "Aux Origines," 98–99, 102–3.

23. S. Bimangu and Tshishiku Tshibangu, "Contribution à l'histoire de l'implantation de l'islam au Zaïre," *Paideuma* 24 (1978): 226; Beverly Brown, "Muslim Influence on Trade and Politics in the Lake Tanganyika Region," *African Historical Studies* 4, no. 3 (1971): 617.

24. Sheryl McCurdy, "Fashioning Sexuality: Desire, Manyema Ethnicity, and the Creation of the *Kanga*, ca. 1880–1900," *International Journal of African Historical Studies* 39, no. 3 (2006): 461; Thomas F. McDow, *Buying Time: Debt and Mobility in the Western Indian Ocean* (Athens: Ohio University Press, 2018), 139–40; Chubaka, "Aux Origines," 101–3.

25. Chubaka, "Aux Origines," 98; McDow, *Buying Time*, chap. 5.

26. Chrétien, *Great Lakes*, 217

27. Chrétien, *Great Lakes*, 218; Louis, *Ruanda-Urundi*.

28. Chrétien, *Great Lakes*, 218.

29. David Newbury, *The Land beyond the Mists*, 313–14.

30. Chrétien, *Great Lakes*, 252–53; Louis, *Ruanda-Urundi*, 128–30.

31. Lemarchand, *Burundi: Ethnocide*, 42.

32. For Burundi and particularly Imbo, the seminal work is Chrétien's "La crise écologique du lac Tanganyika entre 1890 et 1916," in *Burundi*.

33. Van der Burgt, *Dictionnaire*, 287.

CHAPTER 5: MOBILITY, ILLNESS, AND COLONIAL PUBLIC HEALTH ON THE TANGANYIKA LITTORAL

1. Michele D. Wagner, "Trade and Commercial Attitudes in Burundi before the Nineteenth Century," *International Journal of African Historical Studies* 26, no. 1 (1993): 149–66.

2. Bimangu and Tshibangu suggest the relative isolation of Ubwari and Ubembe with regard to areas *other* than Uvira and the Imbo shore on the opposite side of the lake; S. Bimangu and Tshishiku Tshibangu, "Contribution à l'histoire de l'implantation de l'islam au Zaïre," *Paideuma* 24 (1978): 226.

3. Beverly Brown, "Muslim Influence on Trade and Politics in the Lake Tanganyika Region," *African Historical Studies* 4, no. 3 (1971): 619.

4. Bishikwabo Chubaka, "Aux origines de la ville d'Uvira selon les explorateurs et les pionniers de la colonisation Belge au Zaire (1840–1914)," *Civilisations* 37, no. 1 (1987): 94-97.

5. M. Afr.–Rome–EE, Diaire de Kibanga, 26 Nov. 1884, *CTSMA*, no. 27 (July 1885); M. Afr.–Rome–EE, Letter from Père Coulbois (Kibanga) to Supérieur général, 1 Feb. 1885, *CTSMA*, no. 28 (Oct. 1885).

6. M. Afr.–Rome–EE, Letter from Père Coulbois (Kibanga) to Supérieur général, 1 Feb. 1885, *CTSMA*, no. 28, (Oct. 1885).

7. M. Afr.–Rome–EE, Journal de Kibanga, 7 Jan. and 16 Jan. 1887, *CTSMA*, no. 36 (October 1887).

8. Philip Gooding, "Lake Tanganyika: Commercial Frontier in the Era of Long-Distance Commerce, East and Central Africa, c. 1830–1890" (PhD diss., SOAS University of London, 2017).

9. BArch R1001/5911, Vorwerk, Quarterly Report for Rumonge and Romangu, 14 Oct. 1913.

10. Archives de l'État en Belgique, Archives générales du Royaume 2 (dépôt Joseph Cuvelier), Archives africaines, Administration coloniale, Cartothèque, 67/12, Ruanda-Urundi, "Annexe no. 4 relative à la maladie du sommeil," (August 1925).

11. M. Afr.–Rome–EE, Journal du Révérend Père Coulbois, 1884, *CTSMA*, no. 26 (April 1885); see François Coulbois, *Dix Années au Tanganyka* (Limoges: Dumont, 1901), 74-96.

12. Edward C. Hore, *Tanganyika: Eleven Years in Central Africa* (London: Edward Stanford, 1892), 152.

13. BArch R1001/5907, "Bekanntmachung" from von Stegmann, Usumbura, 2 May 1911; Jean-Pierre Chrétien, *The Great Lakes of Africa: Two Thousand Years of History*, trans. Scott Strauss (New York: Zone Books, 2003); Chrétien, *Burundi: L'histoire retrouvée: 25 ans de métier d'historien en Afrique* (Paris: Karthala, 1993), 131.

14. Bimangu and Tshibangu, "Contribution à l'histoire," 225.

15. BArch R155F/81454, Otto Wittrock to Resident, 25 Apr. 1911; BArch R155F/81454, Otto Wittrock to Resident, 9 May 1911.

16. BArch R1001/5895, Karl Neubert, "Bericht über die zwecks Erforschung des Vorkommens der Schlafkrankheit unternommene Bereisung des Tanganyikaküstengebiets des Bezirks Ujiji, I. Teil: Nördliches Küstengebiet," 1 May 1906.

17. BArch R86/2630, Oskar Feldmann, Report from Ujiji, 28 Nov. 1906.

18. BArch R155F/81454, Correspondence between Dr. Otto Wittrock and Residentur Stegmann, 25 Apr. 1911, 2 May 1911, and 9 May 1911.

19. Hubert Cochet, "A Half Century of Agrarian Crisis in Burundi (1890–1945): The Incapacity of the Colonial Administration in Managing the Agrarian Crisis of the Late Eighteen-Hundreds," *African Economic History* 31 (2003): 20–21; Thomas Laely, "Peasants, Local Communities, and Central Power in Burundi," *Journal of Modern African Studies* 35, no. 4 (December 1997): 695–716; David Newbury, *The Land beyond the Mists: Essays on Identity and Authority in Precolonial Congo and Rwanda* (Athens: Ohio University Press, 2009), 329–30.

20. Chubaka, "Aux Origines," 95.

21. BArch R86/2630, Oskar Feldmann, Report from Ujiji, 28 Nov. 1906.

22. BArch R155F/81454, Otto Wittrock to Res. Stegmann, 9 May 1911.

23. *Waungwana* also connoted "general qualities of urbane gentility" attached with urban Swahili culture on the coast in other contexts, as Jonathon Glassman argues in *Feasts and Riot: Revelry, Rebellion, and Popular Consciousness on the Swahili Coast, 1856–1888* (Portsmouth, NH: Heinemann, 1995), 62. Bimangu and Tshibangu, writing about the penetration of Islam into the Congo, use "*Wangwana*" to refer to "la classe des autochtones qui participaient au commerce des Arabes" in the towns of Manyema in the mid-to-late nineteenth century, Bimangu and Tshibangu, "Contribution à l'histoire," 228. It is possible that by using *waungwana*, German doctors intended to convey that a person was both not considered culturally/ethnically local, as well as connected in some fashion to Indian Ocean coastal trading networks.

24. BArch R86/2630, Oskar Feldmann, Report from Ujiji, 28 Nov. 1906.

25. Jean-Pierre Chrétien, "The Slave Trade in Burundi and Rwanda at the Beginning of German Colonisation 1890–1906," in *Slavery in the Great Lakes Region of East Africa*, ed. Henri Médard and Shane Doyle (Athens: Ohio University Press, 2007), 225.

26. Wagner, "Trade and Commercial Attitudes."

27. Philip Gooding, "Slavery, 'Respectability,' and Being 'Freeborn' on the Shores of Nineteenth-Century Lake Tanganyika," *Slavery and Abolition* 40, no. 1 (2019): 147–67.

28. Roger Botte, "Rwanda and Burundi, 1889-1930: Chronicle of a Slow Assassination, Part I," *International Journal of African Historical Studies* 18, no. 1 (1985): 59–60.

29. Extractive economies in the region were different than in Uele, but the impact of disruptions of war and slave raiding echo those described in Maryinez Lyons, *The Colonial Disease: A Social History of Sleeping Sickness in Northern Zaire, 1900–1940* (Cambridge: Cambridge University Press, 1992).

30. AA, Hyg 849, Report (copy) from Dr. Trolli, 24 May 1910.

31. Feldmann's May 1905 letter and August 1905 report note the disease as established on the Tanganyika coast; BArch R86/2622, Oskar Feldmann, Report/Letter to Dar es Salaam, 2 May 1905; BArch R1001/5895, Feldmann, Report, 21 Aug. 1905.

32. BArch R1001/5895, Leupolt, Report, 24 Jan. 1906.

33. BArch R1001/5895, Leupolt, Report, 24 Jan. 1906. Leupolt offered "Malale" and "Marare" as alternate spellings of *malali*.

34. Scholarly debate on the origins of "malale" is ongoing; it seems most plausible that it is related to *kulala*, "to sleep," in Kiswahili and some western Great Lakes Bantu languages. David Schoenbrun, *The Historical Reconstruction of Great Lakes Bantu Cultural Vocabulary: Etymologies and Distributions* (Cologne: Rüdiger Köppe, 1997), s.v. "-lala".

35. BArch R1001/5895, Leupolt, Report, 24 Jan. 1906, for instance.

36. BArch R1001/5895, Leupolt, Report, 24 Jan. 1906.

37. See BArch R1001/5895, Leupolt, Report, 24 Jan. 1906, and AA, Hyg/841, "Rapport sanitaire du 2e semestre 1909."

38. M. Afr.–Rome–EE, Nouvelles générales des Missions, Marienheim (Buhonga), July–Aug. 1905, *CTSMA*, no. 125 (March 1906).

39. In 1905, only two doctors were stationed at either end of Imbo, Dr. Leupolt in Ujiji and Dr. Karl Neubert in Usumbura.

40. BArch R86/2622, Oskar Feldmann, Report/Letter to Government, 2 May 1905.

41. BArch R1001/5895, Karl Neubert, "Bericht über die zwecks Erforschung des Vorkommens der Schlafkrankheit unternommene Bereisung des Tanganyikaküstengebiets des Bezirks Ujiji, I. Teil: Nördliches Küstengebiet," 1 May 1906.

42. BArch R1001/5895, Leupolt, Report, Jan. 1906.

43. AA, HYG/846, Excerpt from Report of Sous-Officier Stevens, 4 Sept. 1904. "Poret" is an alternate spelling of Pori.

44. AA, HYG/846, Commissaire du District, Tombeur, regarding 4 Sep. 1904 Report of Sous-Officier Stevens.

45. AA, HYG 845 (171), Derche to Governeur-General, 26 June 1907.

46. M. Afr.–Rome–EE, Mission du lac Tanganyka, Lettre adressée au R. P. Supérieur général, 2° Oujiji sur Tanganyka, Fr. Deniaud, 24 Nov. 1879, *CTSMA*, no. 6 (Apr. 1880); AA, GG 7072, Lieutenant Drôs, "Rapport sur la reconnaisance qui a été effectuée du 16 au 27 November 1908," 30 Nov. 1908; Richard Burton, *The Lake Regions of Central Africa: A Picture of Exploration* (New York: Harper and Brothers, 1860), 325; Hore, *Eleven Years*, 152-61.

47. AA, GG 7072, Lieutenant Drôs, "Rapport sur la reconnaisance qui a été effectuée du 16 au 27 November 1908," 30 Nov. 1908; Thaddeus Sunseri, "Famine and Wild Pigs: Gender Struggles and the Outbreak of the Majimaji War in Uzaramo (Tanzania)," *Journal of African History* 38, no. 2 (1997): 250–52.

48. Liisa Malkki, *Purity and Exile: Violence, Memory, and National Cosmology among Hutu Refugees in Tanzania* (Chicago: University of Chicago Press, 1995), 27–31.

49. BArch R1001/5898, Oskar Feldmann, Report, 20 May 1908.

50. BArch R1001/5898, Oskar Feldmann, "Bericht über die erste Bereisung der inficierten Urundiküste von Kap Magalla bis Niansa," 12 July 1908.

51. Chrétien, *Burundi*, 143–50.

52. BArch R1001/5898, Friedrich Kleine, Report to Government, Dar es Salaam, 30 Aug. 1908.

53. BArch R1001/5900, Oskar Feldmann, "Glossina palpalis und ihre Ausrottung zur Bekämpfung der Schlafkrankheit," 15 Apr. 1909.

54. BArch R86/2622, Karl Neubert, Report, 1 May 1906; BArch R86/2622, K. Neubert, Report, 22 Aug. 1906; BArch R1001/5904, Letter from Friedrich Kleine to Government, Dar es Salaam, 12 May 1910.

55. BArch R1001/5876, "Aufzeichnung über die Sitzung des Reichs-Gesundheitsrats (Unterausschuß für Schlafkrankheit)," Section 3, Lake Tanganyika, 5

Apr. 1909; Friedrich Kleine, "Bericht über die Schlafkrankheit," *Medizinal Berichte über die Deutschen Schutzgebiete* 1911/12 (Berlin: Mittler, 1915), 95–96.

56. See BArch R1001/5936, Eggel and von Beringe, 12 Aug. 1899; Juhani Koponen, *Development for Exploitation: German Colonial Policies in Mainland Tanzania, 1884–1914* (Münster: Lit Verlag, 1994), 492; Wolfgang Eckart, *Medizin und Kolonialimperialismus, Deutschland* 1884–1945 (Paderborn, Germany: Schöningh, 1997), 301.

57. BArch R155F/81454, Max Taute, "Bestimmungen über die Verkehrsüberwachung am Tanganyika," Aug. 1911.

58. BArch R155F/81454, Max Taute, "Bestimmungen über die Verkehrsüberwachung am Tanganyika," Aug. 1911; BArch R1001/5908, Penschke, Quarterly Report for Usumbura, 8 Jan. 1912.

59. BArch R155F/81454, Hauptmann Panken (Ujiji) to Leader of Sleeping Sickness Campaign Max Taute (Niansa), 15 Oct. 1911.

60. AA, GG 3598, "Rumonge," reproduced 1918. A letter from Kigoma dated 24 June 1918 that accompanied the set of maps indicates that they were reproductions based on maps and a 1913–14 hydrographic survey found in German archives in Kigoma.

61. BArch R1001/5904, Otto Wittrock, Quarterly Report for Urambi and Rumonge, 3 Apr. 1910.

62. BArch R1001/5906, Otto Wittrock, Quarterly Report from North and South Urundi, 3 Jan. 1911.

63. BArch R1001/5911, Otto Fehlandt, Quarterly Report for Rugufu/Ujiji, 25 Oct. 1913.

64. See BArch R1001/5884, Letter from Bumm (Imperial Health Office) to Secretary of Interior Ministry, 6 Apr. 1909; BArch R1001/5903, Friedrich Kleine and Hugo Meixner, Report, 22 Oct. 1909; BArch R1001/5898, F. Kleine, Report, 30 Aug. 1908.

65. Kirk A. Hoppe, *Lords of the Fly: Sleeping Sickness Control in British East Africa, 1900–1960* (Westport, CT: Praeger, 2003), 64–66.

66. BArch R1001/5898, Friedrich Kleine, Report to Government, Dar es Salaam, 30 Aug. 1908.

67. BArch R1001/5900, Oskar Feldmann, "Glossina palpalis und ihre Ausrottung zur Bekämpfung der Schlafkrankheit," 15 Apr. 1909.

68. Friedrich Kleine, "Bericht über die Schlafkrankheitsbekämpfung im Jahre 1911/12," *Medizinal Berichte* über die Deutschen Schutzgebiete 1911–12, 97–98.

69. BArch R155F/81454, Draft Letter from von Langenn-Steinkeller to Belgian Administration, 6 May 1912.

70. BArch R1001/5910, Otto Fehlandt, Quarterly Report for Ujiji, 5 Jan. 1913.

71. Friedrich Kleine, "Bericht über die Schlafkrankheitsbekämpfung im Jahre 1911/12," *Medizinal Berichte* über die Deutschen Schutzgebiete 1911–12, 97–98; BArch R1001/5911, Otto Fehlandt, Quarterly Report for Rugufu/Ujiji, 25 Oct. 1913; BArch R1001/5911, Scherschmidt, Quarterly Report for Urambi/North Urundi, 1 Jan. 1914.

72. See BArch R1001/5906, Otto Wittrock, Report on Meeting with Belgian Authorities, 18 Mar. 1911; BArch R1001/5908, Penschke, Quarterly Report for Usumbura, 8 Jan. 1912; BArch R1001/5910, Penschke, Quarterly Report for Usumbura, 1 Jan. 1913.

73. BArch R1001/5884, Letter from Bumm (Imperial Health Office) to Secretary of Interior Ministry, 6 Apr. 1909.

74. BArch R1001/5906, Otto Wittrock, Report on Meeting with Belgian Authorities, 18 Mar. 1911.

75. BArch R1001/5898, Von Grawert, "Bericht über Massnahmen und Erfahrungen einer Reise im Interesse der Schlafkrankheitsbekämpfung von Kap Magalla bis Njansa," 12 July 1908.

76. See Michelle R. Moyd, "'All People Were Barbarians to the *Askari* . . .': *Askari* Identity and Honor in the Maji Maji War, 1905–1907," in *Maji Maji: Lifting the Fog of War*, ed. James Giblin and Jamie Monson (Leiden: Brill, 2014).

77. BArch R1001/5904, Otto Wittrock, Quarterly Report for Urambi and Rumonge, 3 Apr. 1910.

78. BArch R1001/5904, Friedrich Breuer, Quarterly Report for Urambi and Rumonge, 1 Jan. 1910; BArch R1001/5904, Otto Wittrock, Quarterly Report for Urambi and Rumonge, 3 Apr. 1910.

79. Marcia Wright, "Swahili Language Policy, 1890–1940," *Swahili: Journal of the Institute of Swahili Research* 35, no. 1 (March 1965): 40–48; John M. Mugane, *The Story of Swahili* (Athens: Ohio University Press, 2015), chap. 9.

80. BArch R1001/5901, Oskar Feldmann, 1 July 1909; BArch R1001/5903, Rechenberg to Imperial Colonial Office, 4 Dec. 1909.

81. BArch R1001/5905, Otto Wittrock, Report, 4 July 1910.

82. BArch R1001/5898, Friedrich Kleine, Report to Government, Dar es Salaam, 30 Aug. 1908.

83. BArch R1001/5897, Oskar Feldmann, Report, 4 Apr. 1908.

84. BArch R1001/5902, Friedrich Breuer, Quarterly Report for Niansa and Kiguena, 10 July 1909.

85. BArch R1001/5906, Otto Wittrock, Quarterly Report for Kiguena and Rumongu, 4 Apr. 1911.

86. BArch R155F/81454, Memo, von Langenn-Steinkeller, 13 Dec. 1911.

87. Chrétien discusses this episode briefly in *Burundi*, 146. BArch R155F/81454, Vorwerk, 28 May 1912, indicates that all people involved in the controversy were not the same as the group who refused to work.

88. BArch R155F/81454, Vorwerk, Urgent Memo to Resident, 21 May 1912.

89. BArch R155F/81454, Vorwerk, Urgent Memo to Resident, 21 May 1912.

90. BArch R155F/81454, Vorwerk, Urgent Memo to Resident, 21 May 1912. Chrétien notes that the group's leader was a Bwari man named Hamiri in *Burundi*, 146.

91. See Thomas Laely, *Autorität und Staat in Burundi* (Berlin: D. Reimer, 1995), 186–95; Laely, "Peasants," 695–716.

92. BArch R155F/81454, Vorwerk, Memo to Resident, 25 May 1912.

93. BArch R1001/5904, Otto Wittrock, Quarterly Report for Urambi and Rumonge, 3 Apr. 1910.

94. BArch R155F/81454, Vorwerk, Memo to Resident, 25 May 1912; emphasis in original.

95. Michael Pesek, "Cued Speeches: The Emergence of *Shauri* as Colonial Praxis in German East Africa, 1850–1903," *History in Africa* 33 (2006): 395–412.

96. BArch R1001/81454, Vorwerk, Memorandum, 28 May 1912.

97. BArch R1001/81454, Vorwerk, Memorandum, 28 May 1912.

98. BArch R155F/81454, von Langenn-Steinkeller, Draft Act, 4 June 1912.

99. BArch R155F/81454, von Langenn-Steinkeller, Draft Act, 4 June 1912.

100. BArch R86/2632, Max Taute and Hugo Meixner, Report from Niansa, 12 May 1914.

CONCLUSION

1. BArch R155F/81454, von Langenn-Steinkeller to Government, 10 Jan. 1913.

2. BArch R1001/5747, Meixner, Report, 19 Aug. 1912.

3. Interview with Heslon Lutimba, Kigarama, Tanzania, 21 Aug. 2008.

4. Friedrich Kleine, *Ein deutscher Tropenarzt* (Hannover: Schmorl and Von Seefeld Nachf., 1949), 59.

5. AA RA/RU (o)a, Doss (o)10, Lejeune, "Note sur l'activité du service médical des territoire occupes [sic]," 3 Oct. 1918.

6. Nancy Rose Hunt, *A Nervous State: Violence, Remedies, and Reverie in Colonial Congo* (Durham: Duke University Press, 2016), 5–8.

7. Randall M. Packard, *A History of Global Health: Interventions into the Lives of Other Peoples* (Baltimore: Johns Hopkins University Press, 2016).

8. Vanja Kovacic et al., "We Remember . . . Elders' Memories and Perceptions of Sleeping Sickness Control Interventions in West Nile, Uganda," *PLoS Neglected Tropical Diseases* 10, no. 6 (2016): e0004745.

9. Binyavanga Wainaina, *Beyond River Yei: A Journey into Sudan* (Nairobi: Kwani Trust, 2006), 17.

10. Maryinez Lyons, *The Colonial Disease: A Social History of Sleeping Sickness in Northern Zaire, 1900–1940* (Cambridge: Cambridge University Press, 1992), 47.

11. Dietmar Steverding, "The Development of Drugs for Treatment of Sleeping Sickness: A Historical Review," *Parasites and Vectors* 3, no. 15 (2010): 3; Noémi Tousignant, "Trypanosomes, Toxicity and Resistance: The Politics of Mass Therapy in French Colonial Africa," *Social History of Medicine* 25, no. 3 (2012): 625–43.

12. Kimberly Layne Collins, "Profitable Gifts: A History of the Merck Mectizan Donation Program and Its Implications for International Health," *Perspectives in Biology and Medicine* 47, no. 1 (Winter 2004): 100–109.

13. The example of eflornithine is the most glaring of recent years. Eflornithine is a compound that kills trypanosomes, and was temporarily taken

out of production as an anti–sleeping sickness drug in the 1990s while still being manufactured as a component of a treatment for excessive facial hair growth. See Albert Sjoerdsma and Paul J. Schechter, "Eflornithine for African Sleeping Sickness," *Lancet* 354, no. 9174 (July 17, 1999): 254; Tobias Luppe, "Pharmaceutical Research and Development for Neglected Diseases: Market Failures and the Case of Sleeping Sickness," in *Limits to Privatization: How to Avoid Too Much of a Good Thing*, ed. Ernst Ulrich von Wiezsäcker, Oran Young, and Matthias Finger (London: Earthscan, 2005), 141–44; Peter J. Hotez, *Forgotten People, Forgotten Diseases: The Neglected Tropical Diseases and their Impact on Global Health and Development* (Washington, DC: ASM Press, 2008), 87–88; Charles Ebikeme, "The Death and Life of the Resurrection Drug," *PloS Neglected Tropical Diseases* 8, no. 7 (2014): e2910.

14. Michael P. Pollastri, "Fexinidazole: A New Drug for African Sleeping Sickness on the Horizon," *Trends in Parasitology* 34, no.3 (2018): 178–79.

15. See, for instance, Paul Farmer, "Rethinking Emerging Infectious Diseases," in *Partner to the Poor: A Paul Farmer Reader*, ed. Haun Saussy (Berkeley: University of California Press, 2010), 155–74. The 2014–15 Ebola virus disease epidemic has reinvigorated critical conversations about the limitations of "emerging" diseases as a useful paradigm, particularly in resource-poor contexts. See Annie Wilkinson and Melissa Leach, "Briefing: Ebola—Myths, Realities, and Structural Violence," *African Affairs* 114, no. 454 (2015): 136–48.

Bibliography

ARCHIVES

Ministère des Affaires Etrangères et de Commerce Extérieur–Archives Africaines, Brussels (AA)
Archives de l'État en Belgique, Brussels
Archives of the London School of Hygiene and Tropical Medicine, London (LSHTM)
Bethel Mission Archives, Wuppertal
Bundesarchiv, Berlin-Lichterfelde (BArch)
Church Missionary Society Archive, Cadbury Research Library, Birmingham, UK (CMS)
Geheimes Staatsarchiv-Preußisches Kulturbesitz, Berlin-Dahlem
Leibniz-Institut für Länderkunde, Leipzig
National Archives of Britain, Kew (NAB)
Robert Koch Institute Archives, Berlin (RKI)
Royal Commonwealth Society Library, Cambridge (RCS)
Society of the Missionaries of Africa (White Fathers) Archives, Generalate, Rome (M. Afr.–Rome)
Society of the Missionaries of Africa (White Fathers) Archives, Provincial, Dar es Salaam, Tanzania (M. Afr.–DSM)
Society of the Missionaries of Africa (White Fathers) Archives, Generalate, Rome, Electronic Edition, http://www.mafrome-archivio.org/ (M. Afr.–Rome–EE)
Tanzania National Archives, Dar es Salaam
Wellcome Library, London (WL)

Note on Digitized and Manuscript Archival Materials

My research draws upon diverse archival source materials, most of which I accessed in traditional libraries and archives, and some of which I accessed as digitized versions of original texts available through online databases. In the course of my research, the Society of the Missionaries of Africa (White Fathers) Generalate in Rome made the complete text of the *Chronique Trimestrielle de la Société des Missionnaires d'Afrique* and *Société des Missionnaires d'Afrique—Rapports Annuels* available through an online database, the *Éditions électronique*, accessible as of June 2019 at http://www.mafrome-archivio.org/. Where possible, I have indicated when

I cite a source accessed in manuscript form in the White Fathers Generalate Archives in Rome (noted as M. Afr.–Rome) or in the White Fathers Provincial Archives in Dar es Salaam (noted as M. Afr.–DSM), or in digitized form via the *Éditions électronique* database of materials held in the Generalate in Rome (noted as M. Afr.–Rome–EE).

PERIODICALS

Chronique Trimestrielle de la Société des Missionaires d'Afrique (CTSMA)
Church Missionary Intelligencer and Record
Medizinal Berichte über die Deutschen Schutzgebiete
Mission d'Alger
Reports of Sleeping Sickness Commission of the Royal Society
Société des Missionnaires d'Afrique—Rapports Annuels (RA)

BOOKS AND ARTICLES

Abramowitz, Sharon, and Catherine Panter-Brick, eds. *Medical Humanitarianism: Ethnographies of Practice.* Philadelphia: University of Pennsylvania Press, 2015.

Adams, Vincanne, ed. *Metrics: What Counts in Global Health.* Durham, NC: Duke University Press, 2016.

Alpers, Edward A. *Ivory and Slaves: Changing Pattern of International Trade in East Central Africa to the Later Nineteenth Century.* Berkeley: University of California Press, 1975.

Amsterdamska, Olga. "Demarcating Epidemiology." *Science, Technology, and Human Values* 30, no. 1 (2005):17–51.

Anderson, David M. "Cow Power: Livestock and the Pastoralist in Africa." Review article. *African Affairs* 92, no. 366 (1993): 121–33.

Austen, Ralph A. *Northwest Tanzania under German and British Rule: Colonial Policy and Tribal Politics, 1889–1939.* New Haven, CT: Yale University Press, 1968.

Bashford, Alison, and Carolyn Strange, eds. *Isolation: Places and Practices of Exclusion.* London: Routledge, 2003.

Batchelor, Nicola A., Peter M. Atkinson, Peter W. Gething, Kim Picozzi, Eric M. Fèvre, Abbas S. L. Kakembo, and Susan C. Welburn. "Spatial Predictions of Rhodesian Human African Trypanosomiasis (Sleeping Sickness) Prevalence in Kaberamaido and Dokolo, Two Newly Affected Districts of Uganda." *PLoS Neglected Tropical Diseases* 3, no. 12 (2009): e563.

Bauche, Manuela. "Medizin und staatliche Herrschaft im deutschen Kaiserreich: Die Bekämpfung von Malaria in Kamerun, Ostafrika und Ost-Friesland, 1890–1919." PhD diss., Universität Leipzig, 2015.

Bell, Heather. *Frontiers of Medicine in the Anglo-Egyptian Sudan, 1899–1940.* Oxford: Clarendon Press, 1999.

Benton, Adia. *HIV Exceptionalism: Development through Disease in Sierra Leone.* Minneapolis: University of Minnesota Press, 2015.

Berger, Iris. *Religion and Resistance: East African Kingdoms in the Precolonial Period.* Tervuren, Belgium: Musée royal de l'Afrique centrale, 1981.

Berrang-Ford, Lea, David Waltner-Toews, Dominique Charron, Martin Odiit, John McDermott, and Barry Smit. "Sleeping Sickness in Southeastern Uganda: A Systems Approach." *Ecohealth* 2, no. 3 (2005): 183–94.

Biebuyck, Daniel P. "Bembe Art." *African Arts* 5, no. 3 (Spring 1972): 12–19 and 75–84.

Biehl, João, and Adriana Petryna, eds. *When People Come First: Critical Studies in Global Health.* Princeton, NJ: Princeton University Press, 2013.

Bimangu, S., and Tshishiku Tshibangu. "Contribution à l'histoire de l'implantation de l'Islam au Zaïre." *Paideuma* 24 (1978): 225–30.

Bonnefille, Raymonde, and Françoise Chalié. "Pollen-Inferred Precipitation Time-Series from Equatorial Mountains, Africa, the Last 40 kyr BP." *Global and Planetary Change* 26, no. 1–3 (November 2000): 25–50.

Botte, Roger. "Rwanda and Burundi, 1889–1930: Chronicle of a Slow Assassination, Part I." *International Journal of African Historical Studies* 18, no. 1 (1985): 53–91.

———. "Rwanda and Burundi, 1889–1930: Chronology of a Slow Assassination, Part II." *International Journal of African Historical Studies* 18, no. 2 (1985): 289–314.

Boyd, John. "Sleeping Sickness: The Castellani-Bruce Controversy." *Notes and Records of the Royal Society of London* 28, no. 1 (June 1973): 93–110.

Brock, Thomas D. *Robert Koch: A Life in Medicine and Bacteriology.* Washington, DC: ASM Press, 1999.

Brown, Beverly. "Muslim Influence on Trade and Politics in the Lake Tanganyika Region." *African Historical Studies* 4, no. 3 (1971): 617–29.

Burri, Christian, and Reto Brun. "Human African Trypanosomiasis." In *Manson's Tropical Diseases*, 22nd ed., edited by Gordon C. Cook and Alimuddin I. Zumla, 1307–25. Philadelphia: W. B. Saunders, 2013.

Burton, Antoinette, Madhavi Kale, Isabel Hofmeyr, Clare Anderson, Christopher J. Lee, and Nile Green. "Sea Tracks and Trails: Indian Ocean Worlds as Method." In "Tracks and Trails: Indian Ocean Worlds as Method." Special issue, *History Compass* 11, no. 7 (2013): 497–535.

Burton, Richard. *The Lake Regions of Central Africa: A Picture of Exploration.* New York: Harper and Brothers, 1860.

Büscher, Philippe, and Veerle Lejon. "Diagnosis of Human African Trypanosomiasis." In *The Trypanosomiases*, edited by Ian Maudlin, Peter H. Holmes, and Michael A. Miles, 203–18. Cambridge, MA: CABI Publishing, 2004.

Cesard, Edmond. "Le Muhaya (l'Afrique orientale) (fin)." *Anthropos* 32, no. 1–2 (January–April 1937): 15–60.

Chouin, Gérard. "Reflections on Plague in African History (14th–19th c.)." *Afriques: Débats, méthodes et terrains d'histoire* 9 (2018). https://doi.org/10.4000/afriques.2228.

Chrétien, Jean-Pierre. *Burundi. L'histoire retrouvée: 25 ans de métier d'historien en Afrique.* Paris: Karthala, 1993.

——. *The Great Lakes of Africa: Two Thousand Years of History*. Translated by Scott Strauss. New York: Zone Books, 2003.

Christy, Cuthbert. "Bubonic Plague ('Kaumpuli') in Central East Africa." *British Medical Journal* 2, no. 2237 (November 14, 1903): 1265–67.

Chubaka, Bishikwabo. "Aux origines de la ville d'Uvira selon les explorateurs et les pionniers de la colonisation Belge au Zaire (1840–1914)." *Civilisations* 37, no. 1 (1987): 83–126.

Closser, Svea. *Chasing Polio in Pakistan: Why the World's Largest Public Health Initiative May Fail*. Nashville, TN: Vanderbilt University Press, 2010.

Cochet, Hubert. "Agrarian Dynamics, Population Growth and Resource Management: The Case of Burundi." *GeoJournal* 60, no. 2 (2004): 111–22.

——. "A Half Century of Agrarian Crisis in Burundi (1890–1945): The Incapacity of the Colonial Administration in Managing the Agrarian Crisis of the Late Eighteen-Hundreds." *African Economic History* 31 (2003): 19–42.

Cochi, Stephen L., and Walter R. Dowdle. *Disease Eradication in the 21st Century: Implications for Global Health*. Cambridge, MA: MIT Press, 2013.

Cohen, David W. "Peoples and States of the Great Lakes Region." In *Africa in the Nineteenth Century until the 1880s*, edited by J. F. Ade Ajayi, 270–93. UNESCO General History of Africa 6. Paris: UNESCO, 1989.

——. "Review: The Cwezi Cult." *Journal of African History* 9, no. 4 (October 1968): 651–57.

——. *Womunafu's Bunafu: A Study of Authority in a Nineteenth-Century African Community*. Princeton, NJ: Princeton University Press, 1977.

Collins, Kimberly Layne. "Profitable Gifts: A History of the Merck Mectizan Donation Program and Its Implications for International Health." *Perspectives in Biology and Medicine* 47, no. 1 (Winter 2004): 100–109.

Cook, Albert R. *Uganda Memories, 1897–1940*. Kampala: Uganda Society, 1945.

Cook, J. Howard. "Notes on Cases of 'Sleeping Sickness' Occurring in the Uganda Protectorate." *Journal of Tropical Medicine* 4 (July 15, 1901): 236–39.

Cooper, Frederick. *Colonialism in Question: Theory, Knowledge, History*. Berkeley: University of California Press, 2005.

——. "Conflict and Connection: Rethinking African Colonial History." *American Historical Review* 99, no. 5 (December 1994): 1516–45.

——. *Plantation Slavery on the East Coast of Africa*. Portsmouth, NH: Heinemann, 1997.

Cooper, Frederick, and Ann Laura Stoler, eds. *Tensions of Empire: Colonial Cultures in a Bourgeois World*. Berkeley: University of California Press, 1997.

Cory, Hans. *Historia ya Wilaya Bukoba*. Mwanza, Tanzania: Lake Printing Works, 1956.

Cory, Hans, and M. M. Hartnoll. *Customary Law of the Haya Tribe: Tanganyika Territory*. London: Percy Lund, Humphries, 1945.

Coulbois, François. *Dix années au Tanganyka*. Limoges: Dumont, 1901.

Coupez, André, Thomas Kamanzi, Simon Bizimana, et al. *Inkoranya y'ikinyarwaanda mu kinyarwaanda nó mu gifaraansá: Dictionnaire Rwanda–Rwanda et Rwanda–Français*. 3 vols. Butare, Rwanda: Institut de recherche scientifique et technologique; Tervuren, Belgium: Musée royal de l'Afrique centrale, 2005.

Courtin, F., V. Jamonneau, G. Duvallet, A. Garcia, B. Coulibaly, J. P. Doumenge, G. Cuny, and P. Solano. "Sleeping Sickness in West Africa (1906–2006): Changes in Spatial Repartition and Lessons from the Past." *Tropical Medicine and International Health* 13, no. 3 (2008): 334–44.

Crabtree, W. A. *Elements of Luganda Grammar*. London: Society for Promoting Christian Knowledge, 1902.

Crane, Johanna Tayloe. *Scrambling for Africa: AIDS, Expertise, and the Rise of American Global Health Science*. Ithaca, NY: Cornell University Press, 2013.

Crombé, Xavier, and Jean-Hervé Jézéquel, eds. *A Not-So-Natural Disaster: Niger 2005*. New York: Columbia University Press, 2009.

Crosby, Alfred W. *The Columbian Exchange: Biological and Cultural Consequences of 1492*. Westport, CT: Greenwood, 1972.

Cummiskey, Julia. "Placing Global Science in Africa: International Networks, Local Places, and Virus Research in Uganda, 1936–2000." PhD diss., Johns Hopkins University, 2017.

Cunningham, Andrew. "Transforming Plague: The Laboratory and the Identity of Infectious Disease." In *The Laboratory Revolution in Medicine*, edited by Andrew Cunningham and Perry Williams, 209–44. Cambridge: Cambridge University Press, 1992.

Cunningham, James Francis. *Uganda and Its Peoples: Notes on the Protectorate of Uganda, Especially the Anthropology and Ethnology of Its Indigenous Races*. London: Hutchinson, 1905.

Curtin, Philip D. *Death by Migration: Europe's Encounter with the Tropical World in the Nineteenth Century*. Cambridge: Cambridge University Press, 1989.

———. "The White Man's Grave: Image and Reality, 1780–1850." *Journal of British Studies* 1, no. 1 (November 1961): 94–110.

Curtis, Kenneth. "Capitalism Fettered: State, Merchant, and Peasant in Northwestern Tanzania." PhD diss., University of Wisconsin–Madison, 1989.

Dawson, Marc H. "Socioeconomic Change and Disease: Smallpox in Colonial Kenya, 1880–1920." In *The Social Basis of Health and Healing in Africa*, edited by Steven Feierman and John M. Janzen, 90–103. Berkeley: University of California Press, 1992.

Deflem, Mathieu. "Warfare, Political Leadership, and State Formation: The Case of the Zulu Kingdom, 1808–1879." *Ethnology* 38, no. 4 (Autumn 1999): 371–91.

de Luna, Kathryn M. *Collecting Food, Cultivating People: Subsistence and Society in Central Africa*. New Haven, CT: Yale University Press, 2016.

Depelchin, Jacques. "A Contribution to the Study of Pre-capitalist Modes of Production: Uvira Zone (Eastern Zaïre) c. 1800–1937." *African Economic History Review* 2, no. 1 (Spring 1975): 1–6.

———. "From Pre-Capitalism to Imperialism: A History of Social and Economic Formations in Eastern Zaïre (Uvira Zone, c. 1800–1965)." PhD diss., Stanford University, 1974.

Des Forges, Alison L. *Defeat Is the Only Bad News: Rwanda under Musinga, 1896–1931.* Edited by David Newbury. Madison: University of Wisconsin Press, 2011.

Digby, Anne, Waltraud Ernst, and Projit B. Muhkarji, eds. *Crossing Colonial Historiographies: Histories of Colonial and Indigenous Medicines in Transnational Perspective.* Newcastle upon Tyne, UK: Cambridge Scholars, 2010.

Dilger, Hansjörg, Abdoulaye Kane, and Stacey A. Langwick, eds. *Medicine, Mobility, and Power in Global Africa: Transnational Health and Healing.* Bloomington: Indiana University Press, 2012.

Doyle, Shane. *Before HIV: Sexuality, Fertility and Mortality in East Africa, 1900–1980.* Oxford: Oxford University Press, 2013.

———. *Crisis and Decline in Bunyoro: Population and Environment in Western Uganda 1860–1955.* Athens: Ohio University Press, 2006.

———. "The Cwezi-Kubandwa Debate: Gender, Hegemony and Pre-colonial Religion in Bunyoro, Western Uganda." *Africa* 77, no. 4 (2007): 559–81.

Duke, H. Lyndhurst. "On the Zoological Status of the Polymorphic Mammalian Trypanosomes of Africa and Their Relation to Man." *Parasitology* 13, no. 4 (November 1921): 352–97.

———. "Trypanosomiasis in Northern Uganda." *Journal of Hygiene* 15, no. 3 (September 1916): 372–87.

Dunbar, A. R. *A History of Bunyoro-Kitara.* Nairobi: Oxford University Press, 1965.

Ebikeme, Charles. "The Death and Life of the Resurrection Drug." *PloS Neglected Tropical Diseases* 8, no. 7 (2014): e2910.

Echenberg, Myron J. *Africa in the Time of Cholera: A History of Pandemics from 1817 to the Present.* Cambridge: Cambridge University Press, 2011.

———. *Black Death, White Medicine: Bubonic Plague and the Politics of Public Health in Colonial Senegal, 1914–1945.* Portsmouth, NH: Heinemann, 2002.

———. "Pestis Redux: The Initial Years of the Third Bubonic Plague Pandemic, 1894–1901." *Journal of World History* 13, no. 2 (Fall 2002): 429–49.

———. *Plague Ports: The Global Urban Impact of Bubonic Plague, 1894–1901.* New York: New York University Press, 2007.

Eckart, Wolfgang U. "The Colony as Laboratory: German Sleeping Sickness Campaigns in German East Africa and in Togo, 1900–1914." *History and Philosophy of the Life Sciences* 24, no. 1 (February 2002): 69–89.

———. *Medizin und Kolonialimperialismus, Deutschland 1884–1945.* Paderborn, Ger.: Schöningh, 1997.

Endfield, Georgina H., David B. Ryves, Keely Mills, and Lea Berrang-Ford. "'The Gloomy Forebodings of This Dread Disease': Climate, Famine and Sleeping Sickness in East Africa." *Geographical Journal* 175, no. 3 (September 2009): 181–95.

Engels, Dagmar, and Shula Marks, eds. *Contesting Colonial Hegemony: State and Society in Africa and India*. London: British Academic Press, 1994.

Fabian, Johannes. *Out of Our Minds: Reason and Madness in the Exploration of Central Africa*. Berkeley: University of California Press, 2000.

Fallers, L. A., and S. B. K. Musoke. "Social Mobility, Traditional and Modern." In *The King's Men: Leadership and Status in Buganda on the Eve of Independence*, edited by L. A. Fallers, 158–210. Oxford: Oxford University Press, 1964.

Farmer, Paul. *Partner to the Poor: A Paul Farmer Reader*. Edited by Haun Saussy. Berkeley: University of California Press, 2010.

Feierman, Steven. "A Century of Ironies in East Africa (ca. 1780–1890)." In *African History from Earliest Times to Independence*, edited by Philip Curtin, Steven Feierman, Leonard Thompson, and Jan Vansina, 352–76. London: Longman, 1995.

———. *Peasant Intellectuals: Anthropology and History in Tanzania*. Madison: University of Wisconsin Press, 1990.

———. *The Shambaa Kingdom: A History*. Madison: University of Wisconsin Press, 1974.

———. "Struggles for Control: The Social Roots of Health and Healing in Modern Africa." *African Studies Review* 28, no. 2–3 (June–September 1985): 73–147.

Feierman, Steven, and John M. Janzen, eds. *The Social Basis of Health and Healing in Africa*. Berkeley: University of California Press, 1992.

Fèvre, Eric M., Paul G. Coleman, Susan C. Welburn, and Ian Maudlin. "Reanalyzing the 1900–1920 Sleeping Sickness Epidemic in Uganda." *Emerging Infectious Diseases* 10, no. 4 (April 2004): 567–73.

Fischer, Walter. *Unter der Geissel der Schlafkrankheit*. Berlin: Weller Verlag, 1938.

Foley, Ellen E. *Your Pocket Is What Cures You: The Politics of Health in Senegal*. New Brunswick, NJ: Rutgers University Press, 2010.

Ford, John. *The Role of the Trypanosomiases in African Ecology: A Study of the Tsetse-Fly Problem*. Oxford: Clarendon, 1971.

Gaffky, Georg, and Eduard Pfuhl, eds. *Gesammelte Werke von Robert Koch*. Leipzig: Thieme, 1912.

Geissler, P. Wenzel, ed. *Para-States and Medical Science: Making African Global Health*. Durham, NC: Duke University Press, 2015.

Geissler, P. Wenzel, and Catherine Molyneux, eds. *Evidence, Ethos, and Experiment: The Anthropology and History of Medical Research in Africa*. New York: Berghahn Books, 2011.

Giblin, James. *The Politics of Environmental Control in Northeastern Tanzania, 1840–1940*. Philadelphia: University of Pennsylvania Press, 1992.

———. "Trypanosomiasis Control in African History: An Evaded Issue?" *Journal of African History* 31, no. 1 (1990): 59–80.

Gilbert, Erik. *Dhows and the Colonial Economy of Zanzibar, 1860–1970.* Athens: Ohio University Press, 2004.

Giles-Vernick, Tamara, and Stephanie Rupp. "People, Great Apes, Disease, and Global Health in the Northern Forests of Equatorial Africa." In *Global Health in Africa: Historical Perspectives on Disease Control*, edited by Tamara Giles-Vernick and James L. A. Webb, Jr., 117–37. Athens: Ohio University Press, 2013.

Giles-Vernick, Tamara, and James L. A. Webb, Jr., eds. *Global Health in Africa: Historical Perspectives on Disease Control.* Athens: Ohio University Press, 2013.

Glassman, Jonathon. *Feasts and Riot: Revelry, Rebellion, and Popular Consciousness on the Swahili Coast, 1856–1888.* Portsmouth, NH: Heinemann, 1995.

Good, Charles M. "Salt, Trade, and Disease: Aspects of Development in Africa's Northern Great Lakes Region." *International Journal of African Historical Studies* 5, no. 4 (1972): 543–86.

Gooding, Philip. "Lake Tanganyika: Commercial Frontier in the Era of Long-Distance Commerce, East and Central Africa, c. 1830–1890." PhD diss., SOAS University of London, 2017.

———. "Slavery, 'Respectability,' and Being 'Freeborn' on the Shores of Nineteenth-Century Lake Tanganyika." *Slavery and Abolition* 40, no. 1 (2019): 147–67.

Graboyes, Melissa. *The Experiment Must Continue: Medical Research and Ethics in East Africa, 1940–2014.* Athens: Ohio University Press, 2015.

Gradmann, Christoph. "'It Seemed About Time to Try One of Those Modern Medicines': Animal and Human Experimentation in the Chemotherapy of Sleeping Sickness 1905–1908." In *Twentieth Century Ethics of Human Subjects Research: Historical Perspectives on Values, Practices, and Regulations*, edited by Volker Roelcke and Giovanni Maio, 83–97. Stuttgart: Fritz Steiner, 2004.

———. *Laboratory Disease: Robert Koch's Medical Bacteriology.* Baltimore: Johns Hopkins University Press, 2009.

Green, Monica H., ed. *Pandemic Disease in the Medieval World: Rethinking the Black Death.* Kalamazoo, MI: Arc Medieval, 2015.

———. "Putting Africa on the Black Death Map: Narratives from Genetics and History." *Afriques: Débats, méthodes et terrains d'histoire* 9 (2018). https://doi.org/10.4000/afriques.2125.

———. "Taking 'Pandemic' Seriously: Making the Black Death Global." In *Pandemic Disease in the Medieval World: Rethinking the Black Death.* Vol. 1. Edited by Monica H. Green. Kalamazoo, MI: Arc Medieval Press, 2015.

Hall, Martin J. *Through My Spectacles in Uganda; Or, the Story of a Fruitful Field.* London: Church Missionary Society, 1898.

Hanson, Holly Elisabeth. *Landed Obligation: The Practice of Power in Buganda.* Portsmouth, NH: Heinemann, 2003.

Haraway, Donna. "Situated Knowledges: The Science Question in Feminism and the Privilege of Partial Perspective." *Feminist Studies* 14, no. 3 (Fall 1988): 575–99.

Hartwig, Gerald W. "The Victoria Nyanza as a Trade Route in the Nineteenth Century." *Journal of African History* 11, no. 4 (1970): 535–52.

Headrick, Daniel R. "Sleeping Sickness Epidemics and Colonial Responses in East and Central Africa, 1900–1940: E2772." *PLoS Neglected Tropical Diseases* 8, no. 4 (2014): e2772.

Headrick, Rita, and Daniel R. Headrick. *Colonialism, Health and Illness in French Equatorial Africa, 1885–1935*. Atlanta, GA: African Studies Association, 1994.

Herbert, Eugenia W. "Smallpox Inoculation in Africa." *Journal of African History* 16, no. 4 (1975): 539–59.

Hermann, Carl. "Die Wasiba und Ihr Land." *Mitteilungen von Forschungsreisenden und Gelehrten aus den deutschen Schutzgebieten* 7 (1894): 43–59.

Hide, Geoff. "History of Sleeping Sickness in East Africa." *Clinical Microbiology Reviews* 12, no. 1 (1999): 112–25.

Hodgson, Dorothy L. *Once Intrepid Warriors: Gender, Ethnicity, and the Cultural Politics of Maasai Development*. Bloomington: Indiana University Press, 2001.

———, ed. *Rethinking Pastoralism in Africa: Gender, Culture, and the Myth of the Patriarchal Pastoralist*. Athens: Ohio University Press, 2000.

Hoesing, Peter. "Kusamira Ritual Music and the Social Reproduction of Wellness in Uganda." PhD diss., Florida State University, 2011.

Hokkanen, Markku. "Towards a Cultural History of Medicine(s) in Colonial Central Africa." In *Crossing Colonial Historiographies: Histories of Colonial and Indigenous Medicines in Transnational Perspective*, edited by Anne Digby, Waltraud Ernst, and Projit B. Mukharji, 143–64. Newcastle upon Tyne, UK: Cambridge Scholars, 2010.

Holmes, C. F. "Zanzibari Influence at the Southern End of Lake Victoria: The Lake Route." *African Historical Studies* 4, no. 3 (1971): 477–503.

Hoppe, Kirk A. *Lords of the Fly: Sleeping Sickness Control in British East Africa, 1900–1960*. Westport, CT: Praeger, 2003.

Hore, Edward C. *Tanganyika: Eleven Years in Central Africa*. London: Edward Stanford, 1892.

Hotez, Peter J. *Forgotten People, Forgotten Diseases: The Neglected Tropical Diseases and Their Impact on Global Health and Development*. Washington, DC: ASM Press, 2008.

Hunt, Nancy Rose. "'Le bébé en brousse': European Women, African Birth Spacing and Colonial Intervention in Breast Feeding in the Belgian Congo." *International Journal of African Historical Studies* 21, no. 3 (1988): 401–32.

———. *A Colonial Lexicon: Of Birth Ritual, Medicalization, and Mobility in the Congo*. Durham, NC: Duke University Press, 1999.

———. "Letter-Writing, Nursing Men and Bicycles in the Belgian Congo: Notes towards the Social Identity of a Colonial Category." In *Paths toward the Past: African Historical Essays in Honor of Jan Vansina*, edited by Robert W. Harms, Joseph C. Miller, David S. Newbury, and Michele D. Wagner, 187–210. Atlanta, GA: African Studies Association Press, 1994.

———. *A Nervous State: Violence, Remedies, and Reverie in Colonial Congo*. Durham, NC: Duke University Press, 2016.

Hyden, Goran. *Political Development in Rural Tanzania*. Nairobi: East African Publishing House, 1969.

Iliffe, John. *East African Doctors: A History of the Modern Profession*. Cambridge: Cambridge University Press, 1998.

———. *A Modern History of Tanganyika*. New York: Cambridge University Press, 1979.

———. *Tanganyika under German Rule, 1905–1912*. Cambridge: Cambridge University Press, 1969.

Ishengoma, Johnson M. "African Oral Traditions: Riddles among the Haya of Northwestern Tanzania." *International Review of Education* 51, no. 2/3 (May 2005): 139–53.

Ishumi, Abel G. M. *Kiziba, the Cultural Heritage of an Old African Kingdom*. Syracuse, NY: Foreign and Comparative Studies Program, Syracuse University, 1980.

Isobe, Hiroyuki. "Medizin und Kolonialgesellschaft: die Bekämpfung der Schlafkrankheit in den deutschen 'Schutzgebieten' vor dem Ersten Weltkrieg." PhD diss., University of Konstanz, Berlin, 2009.

Jacobs, Nancy J. *Birders of Africa: History of a Network*. New Haven: Yale University Press, 2016.

———. "The Intimate Politics of Ornithology in Colonial Africa." *Comparative Studies in Society and History* 48, no. 3 (July 2006): 564–603.

Janzen, John M. *Lemba, 1650–1930: A Drum of Affliction in Africa and the New World*. New York: Garland, 1982.

Johnson, Jennifer Lee. "Fish, Family, and the Gendered Politics of Descent along Uganda's Southern Littorals." *History in Africa* 45 (June 2018): 445–71.

———. "Fishwork in Uganda: A Multispecies Ethnohistory about Fish, People, and Ideas about Fish and People." PhD diss., University of Michigan, 2014.

Jones, Adam, and Isabel Voigt. "'Just a First Sketchy Makeshift': German Travellers and Their Cartographic Encounters in Africa, 1850–1914." *History in Africa* 39 (2012): 9–39.

Kagwa, Apolo. *The Customs of the Baganda*. Translated by Ernest B. Kalibala. Edited by May M. Edel. New York: Columbia University Press, 1934.

Kaijage, Frederick J. "AIDS Control and the Burden of History in Northwestern Tanzania." *Population and Environment* 14, no. 3 (January 1993): 279–300.

Kasozi, A. B. K. "The Impact of Islam on Ganda Culture, 1844–1894." *Journal of Religion in Africa* 12, no. 2 (1981): 127–35.

Katoke, I. K., and P. Rwehumbiza. "The Administrator: Francis Lwamugira." In *Modern Tanzanians: A Volume of Biographies*, edited by John Iliffe, 43–65. Nairobi: East African Publishing House, 1973.

Kenny, Michael G. "Mutesa's Crime: Hubris and the Control of African Kings." *Comparative Studies in Society and History* 30, no. 4 (October 1988): 595–612.

———. "The Powers of Lake Victoria." *Anthropos* 72, no. 5/6 (1977): 717–33.

Kilaini, Method M. P. *The Catholic Evangelization of Kagera in North-West Tanzania: The Pioneer Period, 1892–1912*. Rome: Pontificia Università Gregoriana, 1990.

Kisangau, Daniel P., Herbert V. M. Lyaruu, Ken M. Hosea, and Cosam C. Joseph. "Use of Traditional Medicines in the Management of HIV/AIDS Opportunistic Infections in Tanzania: A Case in the Bukoba Rural District." *Journal of Ethnobiology and Ethnomedicine* 3, no. 1 (2007): 3–29.

Kleine, Friedrich. *Ein deutscher Tropenarzt*. Hannover: Schmorl and Von Seefeld Nachf., 1949.

Kodesh, Neil. *Beyond the Royal Gaze: Clanship and Public Healing in Buganda*. Charlottesville: University of Virginia Press, 2010.

Koerner, Thorsten, Peter de Raadt, and Ian Maudlin. "The 1901 Uganda Sleeping Sickness Epidemic Revisited: A Case of Mistaken Identity?" *Parasitology Today* 11, no. 8 (August 1995): 303–6.

Koponen, Juhani. *Development for Exploitation: German Colonial Policies in Mainland Tanzania, 1884–1914*. Münster: Lit Verlag, 1994.

Kovacic, Vanja, Inaki Tirados, Johan Esterhuizen, Clement T. N. Mangwiro, Michael J. Lehane, Stephen J. Torr, and Helen Smith. "We Remember . . . Elders' Memories and Perceptions of Sleeping Sickness Control Interventions in West Nile, Uganda." *PLoS Neglected Tropical Diseases* 10, no. 6 (2016): e0004745.

Kuhanen, Jan. *Poverty, Health, and Reproduction in Early Colonial Uganda*. Joensuu, Finland: Joensuu University Publications, 2005.

Lachenal, Guillaume. "A Genealogy of Treatment as Prevention (TasP): Prevention, Therapy, and the Tensions of Public Health in African History." In *Global Health in Africa: Historical Perspectives on Disease Control*, edited by Tamara Giles-Vernick and James L. A. Webb, Jr., 70–91. Athens: Ohio University Press, 2013.

———. *The Lomidine Files: The Untold Story of a Medical Disaster in Colonial Africa*. Baltimore: Johns Hopkins University Press, 2017.

Laely, Thomas. *Autorität und Staat in Burundi*. Berlin: D. Reimer, 1995.

———. "Peasants, Local Communities, and Central Power in Burundi." *Journal of Modern African Studies* 35, no. 4 (December 1997): 695–716.

Langwick, Stacey A. *Bodies, Politics, and African Healing: The Matter of Maladies in Tanzania*. Bloomington: Indiana University Press, 2011.

Latour, Bruno. "Give Me a Laboratory and I Will Raise the World." In *Science Observed: Perspectives on the Social Study of Science*, edited by Karin Knorr-Cetina and Michael Mulkay, 141–70. London: Sage, 1983.

———. *The Pasteurization of France*. Cambridge, MA: Harvard University Press, 1988.

Lawrance, Benjamin N., Emily Lynn Osborn, and Richard L Roberts, eds. *Intermediaries, Interpreters, and Clerks: African Employees in the Making of Colonial Africa*. Madison: University of Wisconsin Press, 2006.

Lemarchand, René. *Burundi: Ethnocide as Discourse and Practice.* Cambridge: Cambridge University Press, 1994.

———. *Rwanda and Burundi.* New York: Praeger, 1970.

Livingston, Julie. *Debility and the Moral Imagination in Botswana.* Bloomington: Indiana University Press, 2005.

———. "Productive Misunderstandings and the Dynamism of Plural Medicine in Mid-century Bechuanaland." *Journal of Southern African Studies* 33, no. 4 (December 2007): 801–10.

Louis, William Roger. *Ruanda-Urundi, 1884–1919.* Oxford: Clarendon, 1963.

Low, D. A. *Fabrication of Empire: The British and the Uganda Kingdoms, 1890–1902.* Cambridge, MA: Cambridge University Press, 2009.

Luppe, Tobias. "Pharmaceutical Research and Development for Neglected Diseases: Market Failures and the Case of Sleeping Sickness." In *Limits to Privatization: How to Avoid Too Much of a Good Thing,* edited by Ernst Ulrich von Wiezsäcker, Oran Young, and Matthias Finger, 140–44. London: Earthscan, 2005.

Lwamgira, Francis X. *Amakuru ga Kiziba: The History of Kiziba and Its Kings.* Translated by E. R. Kamuhangire. Kampala, Uganda: Makerere University College, Department of History, 1969. First appeared as "Amakuru ga Kiziba na abakama bamu," PhD diss., Makerere University College, 1949.

Lyons, Maryinez. "African Trypanosomiasis." In *Cambridge World History of Human Disease,* edited by Kenneth Kiple et al., 552–61. Cambridge: Cambridge University Press, 1993.

———. *The Colonial Disease: A Social History of Sleeping Sickness in Northern Zaire, 1900–1940.* Cambridge: Cambridge University Press, 1992.

———. "The Power to Heal: African Auxiliaries in Colonial Belgian Congo and Uganda." In *Contesting Colonial Hegemony: State and Society in Africa and India,* edited by Dagmar Engels and Shula Marks, 202–23. London: British Academic Press, 1994.

Malkki, Liisa H. *Purity and Exile: Violence, Memory, and National Cosmology among Hutu Refugees in Tanzania.* Chicago: University of Chicago Press, 1995.

Malloy, Patrick Thomas. "Holding [Tanganyika] by the Sindano: Networks of Medicine in Colonial Tanganyika." PhD diss., University of California–Los Angeles, 2003.

———. "Research Material and Necromancy: Imagining the Political-Economy of Biomedicine in Colonial Tanganyika." *International Journal of African Historical Studies* 47, no. 3 (2014): 425–43.

Manderson, Lenore. *Sickness and the State: Health and Illness in Colonial Malaya, 1870–1940.* Cambridge: Cambridge University Press, 1996.

Martin, Gustave, Alexis Leboeuf, and Émile Roubaud. *Rapport de la Mission d'Études de la Maladie du Sommeil au Congo Français, 1906–08.* Paris: Masson, 1909.

Matete, Ndyabahika, and Bakama B. BakamaNume. "Climate of Uganda." In *Contemporary Geography of Uganda,* edited by Bakama B. BakamaNume, 1–33. Dar es Salaam, Tanzania: Mkuki na Nyota, 2010.

McCurdy, Sheryl. "Fashioning Sexuality: Desire, Manyema Ethnicity, and the Creation of the *Kanga*, ca. 1880–1900." *International Journal of African Historical Studies* 39, no. 3 (2006): 441–69.

——. "Transforming Associations: Fertility, Therapy, and the Manyema Diaspora in Urban Kigoma, Tanzania, c. 1850–1993." PhD diss., Columbia University, 2000.

McDow, Thomas F. *Buying Time: Debt and Mobility in the Western Indian Ocean.* Athens: Ohio University Press, 2018.

McNeill, William H. *Plagues and Peoples.* New York: Anchor Books, 1989.

Médard, Henri. "La peste et les missionnaires: Maladies et syncrétisme médical au royaume du Buganda à la fin du XIXe siècle." *Outre-Mers Revue d'Histoire*, no. 346–47 (First Semester, 2005): 79–102.

——. *Le royaume du Buganda au XIXe siècle: Mutations politiques et religieuses d'un ancien état d'Afrique de l'Est.* Paris: Karthala, 2007.

Médard, Henri, and Shane Doyle, eds. *Slavery in the Great Lakes Region of East Africa.* Athens: Ohio University Press, 2007.

Meyer, Alois. *Kleines Ruhaya–Deutsches Wörterbuch.* Trier, Ger.: Mosella-Verlag, 1914.

Meyer, Hans. *Die Barundi: Eine völkerkundliche Studie aus Deutsch-Ostafrika.* Leipzig: Otto Spamer, 1916.

Mertens, Myriam. "Chemical Compounds in the Congo: Pharmaceuticals and the 'Crossed History' of Public Health in Belgian Africa (ca. 1905–1939)." PhD diss., University of Ghent, 2014.

Mika, Marissa. "Fifty Years of Creativity, Crisis, and Cancer in Uganda." *Canadian Journal of African Studies/Revue canadienne des études africaines* 50, no. 3 (September 2016): 395–413.

Moed, Lisa, Tor A. Shwayder, and Mary Wu Chang, "Cantharidin Revisited: A Blistering Defense of an Ancient Medicine." *Archives of Dermatolology* 137, no. 10 (October 2001): 1357–60.

Mölders, Monika. "Formeln gegen Flagellaten: Der pharmakologische Kampf gegen die afrikanische Trypanosomiasis, 1900 bis 1950." PhD diss., Ruprecht-Karls-Universität Heidelberg, 2003.

Möllers, Bernhard. *Robert Koch: Persönlichkeit und Lebenswerk.* Hannover: Schmorl and Von Seefeld, 1951.

Monnais, Laurence, and Noémi Tousignant. "The Values of Versatility: Pharmacists, Plants, and Place in the French (Post)Colonial World." *Comparative Studies in Society and History* 58, no. 2 (April 2016): 432–62.

Moore-Sheeley, Kirsten. "'Nothing but Nets': The History of Insecticide-Treated Nets in Africa, 1980s–Present." PhD diss., Johns Hopkins University, 2017.

Moran-Thomas, Amy. "A Salvage Ethnography of the Guinea Worm: Witchcraft, Oracles and Magic in a Disease Eradication Program." In *When People Come First: Critical Studies in Global Health*, edited by João Biehl and Adriana Petryna, 207–39. Princeton, NJ: Princeton University Press, 2013.

Moyd, Michelle R. "'All People Were Barbarians to the *Askari* . . .': *Askari* Identity and Honor in the Maji Maji War, 1905–1907." In *Maji Maji:*

Lifting the Fog of War, edited by James Giblin and Jamie Monson, 149–79. Leiden: Brill, 2014.

———. *Violent Intermediaries: African Soldiers, Conquest, and Everyday Colonialism in German East Africa*. Athens: Ohio University Press, 2014.

Mpesha, Nyambura. *Mugasha: Epic of the Bahaya*. Nairobi: East African Educational Publishers, 2000.

Mugane, John M. *The Story of Swahili*. Athens, Ohio: Ohio University Press, in association with the Ohio University Center for International Studies, 2015.

Musambachime, Mwelwa C. "The Social and Economic Effects of Sleeping Sickness in Mweru-Luapula 1906–1922." *African Economic History* 10 (1981): 151–73.

Neill, Deborah. *Networks in Tropical Medicine: Internationalism, Colonialism, and the Rise of a Medical Specialty, 1890–1930*. Stanford, CA: Stanford University Press, 2012.

———. "Paul Ehrlich's Colonial Connections: Scientific Networks and Sleeping Sickness Drug Therapy Research, 1900–1914." *Social History of Medicine* 22, no. 1 (April 2009): 61–77.

Newbury, Catharine. *The Cohesion of Oppression: Clientship and Ethnicity in Rwanda, 1860–1960*. New York: Columbia University Press, 1988.

Newbury, David. *Kings and Clans: Ijwi Island and the Lake Kivu Rift, 1780–1840*. Madison: University of Wisconsin Press, 1991.

———. *The Land beyond the Mists: Essays on Identity and Authority in Precolonial Congo and Rwanda*. Athens: Ohio University Press, 2009.

Nguyen, Vinh-Kim. *The Republic of Therapy: Triage and Sovereignty in West Africa's Time of AIDS*. Durham, NC: Duke University Press, 2010.

Nolan, Francis P. *Mission to the Great Lakes: The White Fathers in Western Tanzania, 1878–1978*. Tabora, Tanzania: Tanganyika Mission Press, 1978.

Osborn, Emily Lynn. "'Circle of Iron': African Colonial Employees and the Interpretation of Colonial Rule in French West Africa." *Journal of African History* 44, no. 1 (2003): 29–50.

Osseo-Asare, Abena Dove. *Bitter Roots: The Search for Healing Plants in Africa*. Chicago: University of Chicago Press, 2014.

Packard, Randall M. *Chiefship and Cosmology: An Historical Study of Political Competition*. Bloomington: Indiana University Press, 1981.

———. *A History of Global Health: Interventions into the Lives of Other Peoples*. Baltimore: Johns Hopkins University Press, 2016.

———. *White Plague, Black Labor: Tuberculosis and the Political Economy of Health and Disease in South Africa*. Berkeley: University of California Press, 1989.

Parker, Melissa, and Tim Allen. "De-Politicizing Parasites: Reflections on Attempts to Control the Control of Neglected Tropical Diseases." *Medical Anthropology* 33, no. 3 (2014): 223–39.

Patterson, K. David. "Disease and Medicine in African History: A Bibliographical Essay." *History in Africa* 1 (1974): 141–48.

———. *Pandemic Influenza, 1700–1900: A Study in Historical Epidemiology.* Totowa, NJ: Rowman and Littlefield, 1986.

Pesek, Michael. "Cued Speeches: The Emergence of *Shauri* as Colonial Praxis in German East Africa, 1850–1903." *History in Africa* 33 (2006): 395–412.

———. "The Boma and the Peripatetic Ruler: Mapping Colonial Rule in German East Africa, 1889–1903." *Western Folklore* 66, no. 3/4 (2007): 233–57.

Pentreath, Victor W., and Peter G. E. Kennedy. "Pathogenesis of Human African Trypanosomiasis." In *The Trypanosomiases*, edited by Ian Maudlin, Peter H. Holmes, and Michael A. Miles, 283–301. Cambridge, MA: CABI Publishing, 2004.

Petryna, Adriana, Andrew Lakoff, and Arthur Kleinman, eds. *Global Pharmaceuticals: Ethics, Markets, Practices.* Durham, NC: Duke University Press, 2006.

Pham, Quy, and Sharon E. Jacob, "Cantharidin: Relevance of an Ancient Remedy." *Journal of the Dermatology Nurses' Association* 8, no. 2 (March 2016): 141–44.

Picozzi, Kim, Eric M. Fèvre, Martin Odiit, Mark Carrington, Mark C. Eisler, Ian Maudlin, and Susan C. Welburn. "Sleeping Sickness in Uganda: A Thin Line between Two Fatal Diseases." *British Medical Journal* 331, no. 7527 (November 26, 2005): 1238–41.

Pilkington, George L. *Luganda-English and English-Luganda Vocabulary.* London: Society for Promoting Christian Knowledge, 1892.

Pollastri, Michael P. "Fexinidazole: A New Drug for African Sleeping Sickness on the Horizon." *Trends in Parasitology* 34, no. 3 (2018): 178–79.

Prevost, Elizabeth. "Married to the Mission Field: Gender, Christianity, and Professionalization in Britain and Colonial Africa, 1865–1914." *Journal of British Studies* 47, no. 4 (October 2008): 796–826.

Prince, Ruth J., and Rebecca Marsland, eds. *Making and Unmaking Public Health in Africa: Ethnographic and Historical Perspectives.* Athens: Ohio University Press, 2014.

Pringle, Yolanda. "Crossing the Divide: Medical Missionaries and Government Service in Uganda, 1897–1940." In *Beyond the State: The Colonial Medical Service in British Africa*, edited by Anna Greenwood, 19–38. Manchester, UK: Manchester University Press, 2015.

Raviña, Enrique. *The Evolution of Drug Discovery: From Traditional Medicines to Modern Drugs.* Weinheim, Germany: Wiley VCH Verlag, 2011.

Redfield, Peter. *Life in Crisis: The Ethical Journey of Doctors without Borders.* Berkeley: University of California Press, 2013.

Rehse, Hermann. *Kiziba: Land und Leute.* Stuttgart: Strecker and Schröder, 1910.

———. *Wörtersammlung des Ruziba.* Hamburg: Lucas Gräfe and Sillem, 1915.

Reid, Andrew. "The Lake, Bananas and Ritual Power in Buganda." In *A History of Water: Series 3, vol. 3, Water and Food*, edited by Terje Tvedt, Terje Oestigaard, and R. Coopey, 277–98. London: I. B. Tauris, 2010.

Reid, Richard J. "The Ganda on Lake Victoria: A Nineteenth-Century East African Imperialism." *Journal of African History* 39, no. 3 (1998): 349–63.

———. *A History of Modern Uganda.* Cambridge: Cambridge University Press, 2017.

———. *Political Power in Pre-colonial Buganda: Economy, Society, and Warfare in the Nineteenth Century.* Athens: Ohio University Press, 2002.

Reining, Priscilla Copeland. "Haya Land Tenure: Landholding and Tenancy." *Anthropological Quarterly* 35, no. 2 (April 1962): 58–73.

———. "The Haya: The Agrarian System of a Sedentary People." PhD diss., University of Chicago, 1967.

Renne, Elisha P. *The Politics of Polio in Northern Nigeria.* Bloomington: Indiana University Press, 2010.

Reynolds-Whyte, Susan, ed. *Second Chances: Surviving AIDS in Uganda.* Durham, NC: Duke University Press, 2014.

Rich, Jeremy. "Searching for Success: Boys, Family Aspirations, and Opportunities in Gabon, ca. 1900–1940." *Journal of Family History* 35, no. 1 (January 2010): 7–24.

Riethmiller, Steven. "From Atoxyl to Salvarsan: Searching for the Magic Bullet." *Chemotherapy* 51, no. 5 (August 2005): 234–42.

Robertshaw, Peter, and David Taylor. "Climate Change and the Rise of Political Complexity in Western Uganda." *Journal of African History* 41, no. 1 (2000): 1–28.

Rockel, Stephen J. *Carriers of Culture: Labor on the Road in Nineteenth-Century East Africa.* Portsmouth, NH: Heinemann, 2006.

———. "Enterprising Partners: Caravan Women in Nineteenth Century Tanzania." *Canadian Journal of African Studies/Revue canadienne des études africaines* 34, no. 3 (2000): 748–78.

Roscoe, John. *The Baganda: An Account of Their Native Customs and Beliefs.* London: Macmillan, 1911.

———. "Further Notes on the Manners and Customs of the Baganda." *Journal of the Anthropological Institute of Great Britain and Ireland* 32 (January–June 1902): 25–80.

Schmidt, Peter R. *Historical Archaeology in Africa: Representation, Social Memory, and Oral Traditions.* Lanham, MD: AltaMira, 2006.

Schneider, William H. "Smallpox in Africa during Colonial Rule." *Medical History* 53, no. 2 (April 2009): 193–227.

Schoenbrun, David L. "Cattle Herds and Banana Gardens: The Historical Geography of the Western Great Lakes Region, ca. AD 800–1500." *African Archaeological Review* 11 (December 1993): 39–72.

———. "Conjuring the Modern in Africa: Durability and Rupture in Histories of Public Healing between the Great Lakes of East Africa." *American Historical Review* 111, no. 5 (December 2006): 1403–39.

———. "Ethnic Formation with Other-Than-Human Beings: Island Shrine Practice in Uganda's Long Eighteenth Century." *History in Africa* 45 (June 2018): 397–443.

———. *A Green Place, a Good Place: Agrarian Change, Gender, and Social Identity in the Great Lakes Region to the 15th Century.* Portsmouth, NH: Heinemann, 1998.

———. *The Historical Reconstruction of Great Lakes Bantu Cultural Vocabulary: Etymologies and Distributions.* Cologne: Rüdiger Köppe, 1997.

———. "A Mask of Calm: Emotion and Founding the Kingdom of Bunyoro in the 16th Century." *Comparative Studies in Society and History* 55, no. 3 (July 2013): 634–64.

———. "Pythons Worked: Constellating Communities of Practice with Conceptual Metaphor in Northern Lake Victoria, ca. A.D. 800 to 1200." In *Knowledge in Motion: Constellations of Learning across Time and Place,* edited by Andrew P. Roddick and Ann B. Stahl, 216–46. Tucson: University of Arizona Press, 2016.

———. "We Are What We Eat: Ancient Agriculture between the Great Lakes." *Journal of African History* 34, no. 1 (1993): 1–31.

Schumaker, Lyn. *Africanizing Anthropology: Fieldwork, Networks, and the Making of Cultural Knowledge in Central Africa.* Durham, NC: Duke University Press, 2001.

Sheriff, Abdul. *Slaves, Spices, and Ivory in Zanzibar: Integration of an East African Commercial Empire into the World Economy, 1770–1873.* Athens: Ohio University Press, 1987.

Shorter, Aylward. *Cross and Flag in Africa: The "White Fathers" during the Colonial Scramble (1892–1914).* Maryknoll, NY: Orbis Books, 2006.

Simarro, Pere P., Giuliano Cecchi, José R. Franco, Massimo Paone, Abdoulaye Diarra, José Antonio Ruiz-Postigo, Eric M. Fèvre, Raffaele C. Mattioli, and Jean G. Jannin. "Estimating and Mapping the Population at Risk of Sleeping Sickness." *PLoS Neglected Tropical Diseases* 6, no. 10 (October 2012): e1859.

Simpson, Donald H. *Dark Companions: The African Contribution to the European Exploration of East Africa.* New York: Barnes and Noble, 1976.

Sjoerdsma, Albert, and Paul J. Schechter. "Eflornithine for African Sleeping Sickness." *Lancet* 354, no. 9174 (July 17, 1999): 254.

Smail, Daniel Lord, and Andrew Shryock. "History and the 'Pre.'" *American Historical Review* 118, no. 3 (June 2013): 709–37.

Smythe, Kathleen R. "African Women and White Sisters at the Karema Mission Station, 1894–1920." *Journal of Women's History* 19, no. 2 (June 2007): 59–84.

Snoxall, Ronald A. *Luganda–English Dictionary.* London: Oxford University Press, 1967.

Sommers, Marc. *Fear in Bongoland: Burundi Refugees in Urban Tanzania.* New York: Berghahn Books, 2001.

Spinage, Clive A. *African Ecology: Benchmarks and Historical Perspectives.* New York: Springer, 2012.

Ssekamwa, John C. "Witchcraft in Buganda Today." *Transition* 30 (April–May 1967): 30–39.

Sseguya, L. "The Concept of Death and Future Life among the Ganda." *Occasional Research Papers in African Religions and Philosophies* [Makerere University, Department of Religious Studies and Philosophy] 30 (September 1974): 1–30.

Stepan, Nancy Leys. *Eradication: Ridding the World of Diseases Forever?* Ithaca, NY: Cornell University Press, 2011.

Stephens, Rhiannon. *A History of African Motherhood: The Case of Uganda, 700–1900*. Cambridge: Cambridge University Press, 2013.

Stevens, Lesley. "Religious Change in a Haya Village, Tanzania." *Journal of Religion in Africa* 21, no. 1 (1991): 2–25.

Steverding, Dietmar. "The Development of Drugs for Treatment of Sleeping Sickness: A Historical Review." *Parasites and Vectors* 3, no. 15 (2010): 1–9.

Summers, Carol. "Intimate Colonialism: The Imperial Production of Reproduction in Uganda, 1907–1925." *Signs* 16, no. 4 (Summer 1991): 787–807.

Sunseri, Thaddeus. "Famine and Wild Pigs: Gender Struggles and the Outbreak of the Majimaji War in Uzaramo (Tanzania)." *Journal of African History* 38, no. 2 (1997): 235–59.

———. "The Hide Trade from German East Africa, ca. 1840–1916: Violence, Disease, and Social Change." *International Journal of African Historical Studies* 51, no. 3 (2018): 373–402.

Tappan, Jennifer. *The Riddle of Malnutrition: The Long Arc of Biomedical and Public Health Interventions in Uganda*. Athens: Ohio University Press, 2017.

Taylor, Brian K. *The Western Lacustrine Bantu*. London: International African Institute, 1962.

Tilley, Helen. *Africa as a Living Laboratory: Empire, Development, and the Problem of Scientific Knowledge: 1870–1950*. Chicago: University of Chicago Press, 2011.

———. "Ecologies of Complexity: Tropical Environments, African Trypanosomiasis, and the Science of Disease Control in British Colonial Africa, 1900–1940." *Osiris*, 2nd ser., 19 (2004): 21–38.

Todd, John L. "The Treatment of Human Trypanosomiasis by Atoxyl." *British Medical Journal* 1, no. 2366 (1906): 1037.

Tousignant, Noémi. "Trypanosomes, Toxicity and Resistance: The Politics of Mass Therapy in French Colonial Africa." *Social History of Medicine* 25, no. 3 (2012): 625–43.

Travis, Anthony S. "Anilines: Historical Background." In *The Chemistry of Anilines*, part 1, edited by Zvi Rappoport, 1–73. Chichester, UK: Wiley, 2007.

Tschakert, Petra, Vincent Ricciardi, Erica Smithwick, Mario Machado, David Ferring, Heidi Hausermann, and Leah Bug. "Situated Knowledge of Pathogenic Landscapes in Ghana: Understanding the Emergence of Buruli Ulcer through Qualitative Analysis." *Social Science & Medicine* 150 (2016): 160–71.

Tuck, Michael W. "Kabaka Mutesa and Venereal Disease: An Essay on Medical History and Sources in Precolonial Buganda." *History in Africa* 30 (2003): 309–25.

Twaddle, Michael. "The Emergence of Politico-Religious Groupings in Late Nineteenth-Century Buganda." *Journal of African History* 29, no. 1 (1988): 81–92

———. *Kakungulu and the Creation of Uganda, 1868–1928*. Athens: Ohio University Press, 1993.

Ureña Valerio, Lenny A. "The Stakes of Empire: Colonial Fantasies, Civilizing Agendas, and Biopolitics in the Prussian-Polish Provinces (1840–1914)." PhD diss., University of Michigan, 2010.

van der Burgt, J. M. M. *Dictionnaire Français–Kirundi*. Bois-le-Duc, Holland: Société "l'Illustration Catholique," 1903.

Vansina, Jan. *Antecedents to Modern Rwanda: The Nyiginya Kingdom*. Madison: University of Wisconsin Press, 2004.

Vaughan, Megan. *Curing Their Ills: Colonial Power and African Illness*. Cambridge: Polity Press, 1991.

Vongsathorn, Kathleen. "'First and Foremost the Evangelist'? Mission and Government Priorities for the Treatment of Leprosy in Uganda, 1927–1948." *Journal of Eastern African Studies* 6, no. 3 (August 2012): 544–60.

Wagner, Michele D. "Environment, Community, and History: 'Nature in the Mind' in Nineteenth- and Early Twentieth-Century Buha, Tanzania." In *Custodians of the Land: Ecology and Culture in the History of Tanzania*, edited by Gregory Maddox, James Giblin, and Isaria N. Kimambo, 175–99. Athens: Ohio University Press, 1996.

———. "Trade and Commercial Attitudes in Burundi before the Nineteenth Century." *International Journal of African Historical Studies* 26, no. 1 (1993): 149–66.

Wainaina, Binyavanga. *Beyond River Yei: A Journey into Sudan*. Nairobi: Kwani Trust, 2006.

Webb, James L. A., Jr. "Historical Epidemiology and Infectious Disease Processes in Africa." *Journal of African History* 54, no. 1 (March 2013): 3–10.

———. "The Historical Epidemiology of Global Disease Challenges." *Lancet* 385, no. 9965 (January 24, 2015): 322–23.

———. *Humanity's Burden: A Global History of Malaria*. Cambridge: Cambridge University Press, 2009.

Webel, Mari K. "Borderlands of Research: Medicine, Empire, and Sleeping Sickness in East Africa, 1902–1914." PhD diss., Columbia University, 2012.

———. "Mapping the Infected Landscape: Sleeping Sickness Prevention and the African Production of Colonial Knowledge in the Early Twentieth Century." *Environmental History* 20, no. 4 (October 2015): 722–35.

———. "Ziba Politics and the German Sleeping Sickness Camp at Kigarama, Tanzania, 1907–14." *International Journal of African Historical Studies* 47, no. 3 (2014): 399–423.

Weiss, Brad. "Dressing at Death: Clothing, Time, and Memory in Buhaya, Tanzania." In *Clothing and Difference: Embodied Identities in Colonial and Post-colonial Africa*, edited by Hildi Hendrickson, 133–54. Durham, NC: Duke University Press, 1996.

———. *The Making and Unmaking of the Haya Lived World: Consumption, Commoditization, and Everyday Practice.* Durham, NC: Duke University Press, 1996.

———. *Sacred Trees, Bitter Harvests: Globalizing Coffee in Northwest Tanzania.* Portsmouth, NH: Heinemann, 2003.

Welburn, Susan C., Eric Fèvre, and Paul G. Coleman, "Sleeping Sickness Rediscovered." Review article. *Parasitology Today* 15, no. 8 (August 1999): 303–5.

Welburn, Susan C., Eric Fèvre, Paul G. Coleman, Martin Odiit, and Ian Maudlin. "Sleeping Sickness: A Tale of Two Diseases." *Trends in Parasitology* 17, no. 1 (January 2001): 19–24.

Weiskopf, Julie. "Living in 'Cold Storage': An Interior History of Tanzania's Sleeping Sickness Concentrations, 1933–1946." *International Journal of African Historical Studies* 49, no. 1 (2016): 1–22.

Wigen, Kären, Peregrine Horden, Nicholas Purcell, Matt K. Matsuda, and Alison Games. "AHR Forum: Oceans of History." *American Historical Review* 111, no. 3 (June 2006): 717–57.

White, Luise. *Speaking with Vampires: Rumor and History in Colonial Africa.* Berkeley: University of California Press, 2000.

———. "'They Could Make Their Victims Dull': Genders and Genres, Fantasies and Cures in Colonial Southern Uganda." *American Historical Review* 100, no. 5 (December 1995): 1379–1402.

———. "Tsetse Visions: Narratives of Blood and Bugs in Colonial Northern Rhodesia, 1931–9." *Journal of African History* 36, no. 2 (1995): 219–45.

Wilkinson, Annie, and Melissa Leach. "Briefing: Ebola—Myths, Realities, and Structural Violence." *African Affairs* 114, no. 454 (2015): 136–48.

Williams, B. I. "African Trypanosomiasis." In *The Wellcome Trust Illustrated History of Tropical Medicine*, edited by F. E. G. Cox, 184–87. London: Wellcome Trust, 1996.

Worboys, Michael. "The Comparative History of Sleeping Sickness in East and Central Africa." *History of Science* 32, no. 1 (1994): 89–102.

———. "The Emergence of Tropical Medicine: A Study in the Establishment of a Scientific Discipline." In *Perspectives on the Emergence of Scientific Disciplines*, edited by Gerard Lemaine, Roy Macleod, Michael Mulkay, and Peter Weingart , 75–98. Chicago: Aldine, 1976.

———. *Spreading Germs: Disease Theories and Medical Practice in Britain, 1865–1900.* Cambridge: Cambridge University Press, 2000.

Wright, Marcia. *Strategies of Slaves and Women: Life-Stories from East/Central Africa.* London: James Currey, 1993.

———. "Swahili Language Policy, 1890–1940." *Swahili: Journal of the Institute of Swahili Research* 35, no. 1 (March 1965): 40–48.

Wrigley, Christopher. *Kingship and State: The Buganda Dynasty.* Cambridge: Cambridge University Press, 1996.

———. "The Story of Rukidi." *Africa: Journal of the International African Institute* 43, no. 3 (July 1973): 219–35.

Zeller, Diane Leinwand. "The Establishment of Western Medicine in Buganda." PhD diss., Columbia University, 1974.

Index

The letter f following a page number denotes a figure or map on that page.